Site Engineers Manual
second edition

Edited by
David Doran
Consultant

Formerly Chief Structural
Engineer, Wimpey plc

Whittles Publishing

CRC Press
Taylor & Francis Group

Published by
Whittles Publishing,
Dunbeath,
Caithness KW6 6EY,
Scotland, UK

www.whittlespublishing.com

Distributed in North America by
CRC Press LLC,
Taylor and Francis Group,
6000 Broken Sound Parkway NW, Suite 300,
Boca Raton, FL 33487, USA

© 2009 Whittles Publishing

ISBN 978-1904445-70-8
USA ISBN 978-1-4398-0869-6

Permission to reproduce extracts from BS 8666:205 is granted by BSI. British Standards can be obtained in PDF of hard copy formats from the BSI online shop: www.bsigroup.com/Shop or by contacting BSI Customer Services for hardcopies only: Tel: +44 (0)20 8996 9001, Email: cservices@bsigroup.com

Printed and bound in the United Kingdom by Bell & Bain Ltd., Glasgow

PREFACE TO SECOND EDITION

The first edition of this book was well received by many within the industry, including contractors and also proved very useful to students and newly-qualified graduates. However the construction industry has moved on and this second edition will help to keep pace with that advance.

Two new chapters dealing with the topics of glass and occupational health have been added. The health and safety chapter has been completely recast to recognise changes in legislation; in particular the updated Construction [Design & Management] Regulation Rev 2007 which now places more responsibility with the client. In addition a section has been added covering the management of demolition projects. All other chapters have been reviewed and revised to differing extents, as appropriate.

I am grateful to former and new authors for their work and also to Dr Keith Whittles and Linsey Gullon at Whittles Publishing for their encouragement in bringing this book to press.

Our industry is at the forefront of redevelopment and can boast of some spectacular achievements. However a recent report that 30% of contracts require remedial wok before contract completion suggests there is room for improvement. I therefore commend this book to practitioners and others as an aid in their struggle to achieve better standards.

David Doran,
London

FOREWORD

David Doran and I were longstanding colleagues, enjoying extensive and successful careers within the International Construction Company, George Wimpey plc.

I was, therefore, familiar with the early prototype of the *Site Engineers Manual* introduced into GW plc and witnessed the positive reaction from engineers at all levels within the company to what became an essential aide-memoire and reference book for a typical site engineer. Therefore, when David asked me if I would write the Foreword for his updated and modern version, I readily agreed.

The role and responsibilities of the site engineer are, as in all professions, ever changing: today's engineer needs to be IT literate, to understand the interdependency of other professions within the built environment, and to keep his or her competencies up to date and relevant. The concept of *lifelong learning* is real and I believe essential to today's successful engineer.

The production of this manual draws on the essential procedures and advice necessary for the successful execution of projects, both large and small, and is wide ranging in its reference. Its contributors are expert in their respective fields, their knowledge honed by both academic and practical experience, is written in plain English and is pragmatic in its advice.

As a clear book of reference, an aide-memoire to necessary procedure, it is, in my opinion, a *must have* for every engineer's desk or pocket and I am delighted to commend it to you.

Sir Joseph A. Dwyer, FREng FICE FCIOB
former Chairman of Wimpey plc
and past President of the Institution of Civil Engineers
and the Chartered Institute of Building.

INTRODUCTION

From those who know to those who need to know

In 1980 I was, as Chief Structural Engineer to George Wimpey, responsible for Quality Assurance over a large section of the Group. As part of a Group initiative to improve quality I pioneered the introduction of a **Company Construction Manual**. This was an A4 loose leaf volume and was distributed to all construction sites. Site Agents found it helpful and young trainees were known to take a copy home with them in the evening for further study. In 1987 the style of the publication was changed to a bound book that could slip easily into the pocket of an engineer's donkey jacket. Four editions were produced in this format, the last in 1995. This format has been adopted and expanded in what is essentially a 5th edition.

This new edition makes the work available to the whole of the construction industry.

Site engineers operate in an increasingly technological and legalistic environment. It is hoped that this book will provide them with a road map to guide them through that complex journey. Although this book is essentially for site engineers, some basic design information has been included where appropriate.

I would like to take this opportunity to thank all my erstwhile Wimpey colleagues for their help in launching this project and also to express my appreciation to those whose expertise has enhanced the quality of this latest edition. In particular I would like to thank my publisher Dr Keith Whittles and Bill Black (Drivers Jonas) for their helpful comments which have added value to this book.

David Doran, Editor.

Note

1. In this manual the male sex is used throughout. The term is intended to indicate both sexes and should be construed as such.
2. Cover illustrations by courtesy of BAM Nuttall.
3. **Disclaimer: The recommendations contained in this book are of a general nature and should be customised to suit the project under consideration. Although every care has been taken in assembling this document no liability for negligence or otherwise can be accepted by the editor, authors or the publisher.**
4. In the text, considerable use has been made of acronyms (e.g. ICE Institution of Civil Engineers). These are explained in full in Appendix B3 with, where appropriate, the addresses of useful websites.
5. In general, references are quoted briefly in the text and in full in the 'Further reading' sections at the end of each chapter.

6. British Standards are now giving way to European Standards and, where possible, this has been reflected in this book. The Institution of Structural Engineers has produced a report entitled *Implementation of the Structural Eurocodes* which is recommended as an initial point of reference.
7. The text contains many definitions: for further information, the reader should consult the latest editions of available dictionaries of building and civil engineering.

CONTENTS

Biographical notes of authors

John Armstrong DLC DMS MPhil CEng MIMechE MCIBSE MBIFM
John is a Property Consultant specialising in Facilities Management. He has spent his professional life in the management, care and maintenance of buildings and their engineering services and is a member of CIBSE and BSI Committees. His previous experience was with Ove Arup, Barclays Bank, BSRIA and the Health Service. He has written numerous technical books and publications.

Brian Barnes MCIOB
Brian is a chartered builder. He has worked in construction for all his professional career, initially with Higgs & Hill and latterly as an independent consultant specialising in providing training courses for the industry. His background is in estimating and site management in the UK and abroad. He has actively researched construction defects, particularly those related to the building envelope. In the course of that research he has investigated the cause and frequency of several thousand problems and written a thesis in which he analysed 208 flat roof defects. He believes passionately that today's site managers need a knowledge of basic building construction to enable them to manage specialist subcontractors effectively.

David Doran BSc (Eng) DIC FCGI CEng FICE FIStructE (Editor)
David is a chartered civil and structural engineering consultant who, for twenty years, was Chief Structural Engineer to George Wimpey plc. He also held directorships in Wimpey Laboratories and other subsidiaries where he was responsible for Quality Assurance. He is the editor of several technical books and was, for several years, on the Council of the IStructE becoming Honorary Secretary. He has served as Task Group Chairman for five Institution reports including those entitled: *Aspects of Cladding, Structural effects of alkali-silica reaction* and *The operation and maintenance of bridge access gantries.* He is currently retained by GB GEOTECHNICS Ltd.

John Fullwood BSc (Phys)
John, in his early youth, kept weather records near his home in the Pennines, the environment of which led him to a career in the study of the weather. He graduated at Bangor University then joined the Met Office to be trained in applied meteorology. He enjoys working with the public and specialises in hydrology and construction climatology.

Neil Henderson BSc PhD CEng MIMMM MCS AMICT
Neil is a Principal Engineer with Mott MacDonald with over 10 years experience in concrete technology and the durability of concrete structures. He coordinates a variety of projects within Mott MacDonald using his specialist knowledge to provide construction advice, specifications and durability assessments. He has authored many technical reports, publications and articles covering reinforced concrete durability, durability modelling, self-compacting concrete and sustainable construction. He sits on various industry committees and is a member of the Board for SPECC Ltd., a registration scheme for specialist concrete contractors.

Robert Langridge BSc MIAT
Robert is a Consultant to the construction industry. He has over 45 years experience with asphalts. His experience includes making a wide range of tests on binders, aggregates and bituminous mixtures, the development of laboratory equipment and test methods, site inspections, preparation of specifications, and membership of BSI Technical Committees. From 1981 to 1996 he provided an in-house asphalt advisory service to the divisions of the Wimpey Group. In 1996 he left Wimpey and set up the Asphalt Quality Advisory Service. Robert has been a visiting lecturer at the University of Westminster, Southall College and Lewisham College. At Lewisham he pioneered the work on the Asphalt Laboratory Technicians' Certificate Course. For his paper on *The sampling and testing of bituminous mixtures* he was awarded the Argent 1st Prize by the Institute of Asphalt Technology.

Francesca Machen BSc (Hons) RNOH
Francesca is a specialist occupational health practitioner who has worked in the construction industry for ten years. As a director of Sarsen Health Ltd which specialises in delivering occupational health services to the construction sector, Francesca currently works with a number of organisations designing, implementing and managing health assessment programmes to help companies with health and safety compliance.

Revd Malcolm James CEng MICE
Malcolm is a consultant providing advice on health and safety matters, particularly that dealing with working at height. He appears as an expert witness, carries out research contracts, writes technical articles and drafts guidance documents. He is Chairman of the BSI Committee dealing with safety nets and associated equipment. His previous experience was with the Health & Safety Executive as a principal specialist inspector where he also wrote guidance documents for BSI and CEN committees of which he was member. In earlier times he worked for Laings as a site and temporary works engineer.

Paul Newman BSc MSc PhD
Paul is manager of the Timber Technology Group at TRADA where he is responsible for technology for internal and external clients. His scope ranges from furniture to structural work on bridges and includes expert witness work in litigation. He was previously a lecturer in Wood Science at Guildhall University, London where he did research into the chemical and mechanical changes in ancient and historic dry wood.

Anthony Oakhill BSc(Hons) CEng MICE
Tony has over 40 years' experience in the design and construction of steel structures in the UK and oveseas. After training with Freeman Fox and Husbands he became Chief Designer at Redpath Dorman Long Contracting, involved with the engineering of complex construction operations. Later, research and consultancy work at BCSA and Warwick University included the writing of technical guidance notes and design codes, particularly Eurocode 4 Part 2, *Composite Bridges.* Following steel box girder bridge construction supervision work with Mott MacDonald in Seoul, and employment with WSP Civils, Tony now works with Gifford & Partners.

Eur. Ing. Peter Pallett BSc CEng FICE FCS
Peter is an engineering consultant specialising in temporary works and is regularly commissioned as an expert witness. He previously worked with Arup, A. E. Farr and for many years as Technical Manager with Rapid Metal Developments (RMD). He is extensively involved with BSI and CEN on Falsework committees, is Chairman of the Concrete Society's Construction Group and was responsible as Chairman of the joint CS/IStructE committee for the publication of the award winning report *Formwork—A guide to good practice.* Peter currently lectures to professional organisations and universities.

Les Richardson CEng MICE FIWEM
Les was for eight years with the drainage department of a London Borough from where he moved to the Clay Pipe Development Association (CPDA) where he has worked as an engineer for twenty eight years. Additionally, he currently works as an indepenent consultant undertaking drainage design, analysis and training commissions. He is active on a number of British and European Standards dealing with drainage, sewerage and pipe technology.

Chris Shaw CEng FICE FIET MIStructE MCMI
Chris is a Chartered Civil and Structural Engineer practising as a consultant. He had more than 25 years' experience in achieving the specified cover to the reinforcement in reinforced concrete structures, and devised the system for achieving this first time, every time, which was subsequently published as British Standard 7973 in 2001. He is now Chairman of the committee that prepared the Standard and has authored Papers and contributed to many publications on cover. Chris gives advice, CPD training, lectures and talks on the subject, as well as being an expert

witness. He is a visiting lecturer at the University of Dundee and continues to carry out research and development on the products, their applications and innovative uses worldwide.

Martin Sheward MInst CES
Martin studied civil engineering at Colchester Technical College before going to Trent Polytechnic to study Engineering Surveying. His work covers a wide range of building and civil engineering where he has managed and carried out the survey and setting out on major projects. This work has ranged from primary survey in the Sultanate of Oman, new motorway construction and, more recently, major projects in Central London including office blocks and complex tunnelling work under the congested capital. He is currently a Project Manager with Sir Robert McAlpine for whom he has worked since 1986.

Michael Tomlinson CEng FICE FIStructE
Michael was Technical Director of Wimpey Laboratories where he worked for 29 years after which he set up his own practice. He is a specialist in foundations and earthworks and has written several books and many reports and papers for learned societies. He has been involved as an advisor in a number of major projects including construction facilities for North Sea oil platforms; the River Lagan weir in Belfast; the Jeddah Urban ring road and the Jamuna river bridge in Bangladesh.

In addition to the chapter authors listed, the editor and publishers gratefully acknowledge the assistance of the following in updating chapters for this new edition: Joe McQuade (Mainland Construction), Michael Driver (Brick Development Association), Terry Quarmby (Institute of Demolition Engineers) and Dr A. Pitman of TRADA.

Chapter 1 QUALITY MANAGEMENT

David Doran

Basic requirements for best practice

- Company policy supported by Chief Executive
- Accurate and timely information from design team
- Strict control of variations
- Good communications
- Experienced and well-briefed staff
- Rigorous and traceable checking procedures

1.1 Company structure

For the information and education of staff there should be available, to all, a simple statement outlining the aims and objectives of the company and an organisation chart explaining the interrelationships between different parts of the company. For example, a project manager will need to know if a central engineering department exists which is capable of carrying out temporary works design. It will also be important for him to know who to contact should he need advice and how the cost of that advice will be reimbursed.

1.2 Company policy

Quality, safety, completion on time and profitability are essential ingredients of a successful contract. Provided these targets are met then both client and contractor will be satisfied.

Much has been written on the need for quality assurance (QA). For many, the means to achieve satisfactory quality has become unnecessarily synonymous with excessive bureaucracy typified by overburdening paperwork. It is unfortunate that little emphasis has been placed on the need for well-educated, trained and dedicated people to achieve these objectives.

The term quality assurance has been used in recent times to describe the management process by which appropriate quality can be achieved. Definitions proliferate.

1

The process has not been helped by the fact that much of the early work on the topic was based in the manufacturing industry. Two definitions of QA will suffice:

- Quality will not be indicative of special merit, excellence or high status. It will be used in its engineering sense in which it conveys the concepts of compliance with a defined requirement of value for money, of fitness for purpose or customer satisfaction. With this definition, a palace or a bicycle shed may be of equal quality if both function as they should and both give their owners an equal feeling of having received their money's worth (Ashford, 1989).
- All those planned actions and systematic actions necessary to provide adequate confidence that a product or service will satisfy requirements for quality (BSI, 1987).

It is essential that any QA system has the support of the Chief Executive of a company. This support should be stated in a signed policy statement. An example of this is shown below:

The corporate purpose of Construction Incorporated Ltd is to provide service of outstanding quality and efficiency in the construction industry.

Quality, profitability, programme and safety are essential to our business. All four are interrelated and the achievement of quality is, together with the others, the responsibility of line management.

The Managing Director is responsible to me for implementation of the quality assurance policy. He will monitor and report to me on its implementation within the company.

This should be accompanied by and related to a company QA Manual which should address the following:

- Chairman's statement
- policy
- organisation
- review and audit
- company instructions.

This list implies that an independent (but usually in-house) review and audit of site quality arrangements will take place. The extent to which this happens and method of doing this will depend on specific company arrangements. In a large company, dedicated staff will be recruited to carry out this function. In small companies senior personnel will perform this function within a portfolio of other duties. As the final responsibility rests with the Chief Executive, he must set in place an arrangement to satisfy his requirements.

At site level, a further document (see Fig. 1.1) should be prepared which reflects company QA policy but spells out in detail how the appropriate quality will be achieved on site.

Contractors should be conversant with existing industry-wide QA schemes covering a number of products. Examples of these include the British Board of Agrément (BBA) and the UK Certification Authority for Reinforcing Steels (CARES). A full list of these may be obtained from the Construction Industry Research & Information Association (CIRIA).

1.3 Health and safety

The following is a typical statement of the general health and safety policy for a company in accordance with the Health & Safety at Work etc. Act (1974). It can be modified to suit the requirements of a specific company:

> *It is the Policy of the company to ensure, so far as is reasonably practicable, the health, safety and welfare of their employees while they are at work and of others who may be affected by their undertakings.*

> *We believe that safety is synonymous with quality. Since we are committed to excellence, it follows that minimising risk of injury to people and damage to equipment and products is inseparable from our overall objective.*

Chapter 19 deals further with health and safety.

1.4 Documentation

Typical of the documentation that might circulate within a company are the following. Please note that these are illustrative and may vary from one company to another. Some sections have been expanded.

Group management circulars (dependent on company structure)

Group managers' instructions (dependent on company structure)

Health and safety instructions (see also Chapter 19)

Group quality manual (dependent on company structure)

Company quality manual (see Fig. 1.1)

This would normally be produced by head office and authorised by the Chief Exccutivc. It should sct out thc policy, organisation and rules to implement company quality policy.

Company standing instructions or procedures (see Fig. 1.1)

Issued by the appropriate company director or manager, these would be authorised documents to promulgate company policy in a series of instructions. They would specify the limits and extent of individual authority. All issues and amendments of this document must be logged in a central register; distribution arrangements must ensure that site personnel can be certain of working to the latest version.

Project quality plan (see Fig. 1.1)

This is a document usually produced by the site manager describing how the quality management system is applied to a particular project and indicating responsibilities of key site staff. It takes into account the contract documents and any statutory requirements. It will also indicate what method statements, work instructions, inspection and test plans, records and sub-contractors' quality plans are required with details of who is to prepare them and by when.

Work instructions (See Fig. 1.1)

Issued by the company director, manager or site manager, these may either be incorporated as standard documents in the Company Standing Instructions or Procedures Manual or may be written to suit particular or unusual requirements for specific contracts. With certain types of contract (e.g. for the Ministry of Defence) it is likely that the client will have an influence on these instructions. In all cases they document in greater and more precise detail the operations that are required. Method statements can be used as work instructions in most cases.

Inspection and test plans (see Fig. 1.1)

These are prepared from the technical specifications, working drawings and method statements and list all the check points for each stage in the particular work for which each one is prepared. In its simplest form, it is nothing more than a list of things to check but should also indicate hold points and specification references.

Working drawings specifications and schedules

These represent the output or 'product' from the designer. For constructing the permanent works, only use documents marked **For Construction.**

Manufacturers' and suppliers' literature and instructions

Specifications often contain the requirement '... *to be fixed strictly in accordance with the manufacturer's instructions*'. It follows, therefore, that these instructions become part of the contractual requirements, so when starting a project make sure that all such manufacturer or supplier information is available on site.

Even if not contractual, make sure that literature is obtained for all the significant materials and components with which you will be working. The standard of such information is usually very high and can play an important part in understanding the work. If site personnel are unhappy with any of the instructions contained in this literature then it is essential that doubts are raised with the design team before proceeding. Any new instruction must be confirmed in writing.

Planning and information control manual

This will usually be issued by Head Office. It will cover methods of successful planning, progress monitoring and information control. It will deal with pre- and post-tender experiences and the role of planning in the formulation of claims. It may also contain help in the use of computers and other equipment.

Site manager's technical file

A digest of technical information, available on site, relating to certain products and the use of different types of building materials. The book, of which this chapter is part, should be a key item in that file.

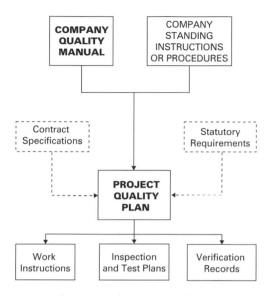

Figure 1.1 Sequence of company quality documentation.

Checklists

Throughout this book examples have been shown of checklists that can be used by engineers as an aide-memoire and also a record that essential items have been checked. However these must be used conscientiously otherwise they are of very limited value. It is impossible to illustrate every type of checklist so it is recommended that a review occurs before the commencement of each major phase of work to determine what is available and to draw up bespoke lists to cover any shortcomings. Good examples appear from time to time in technical journals such as *The Building Engineer* published by the ABE.

1.5 Duties and responsibilities

1.5.1 Employer

The employer is the organisation or person who commissions the works and pays the bills. The employer usually appoints a representative to organise the works on his behalf. This person may be from the employer's own staff or may be a professional consultant. According to the contract conditions used, the employer's representative may have one of the following titles:

- Architect
- Engineer
- Project Manager
- Surveyor
- Contract Administrator.

The design of the permanent works will normally be the responsibility of the employer's representative. However, in certain instances, dependent on the terms of contract, this may be undertaken in whole or part by the contractor and his subcontractors. The employer's representative usually has a site supervisory role, which may be delegated to a site representative. The site representative may have one of the following titles:

- Engineer's Representative
- Resident Engineer
- Architect's Representative
- Clerk of Works
- Inspector.

1.5.2 Contractor

The role and responsibilities of the **Contracts Manager** are:

- the quality, profitability, programme and safety on a number of contracts or projects
- forward planning and detection and avoidance of potential problems
- ensuring that each agent or project manager employs economic use of resources to fulfil the requirements of quality plans and contractual obligations
- selection, motivation, training and development of site staff
- job reviews of subordinate staff
- negotiating with clients
- negotiating with subcontractors and is ultimately responsible for their selection.

Role and responsibilities of the **Agent, Site Manager** and **Project Manager**

- Under contract law a contractor is required to maintain a full time Agent at the place of the works.
- The Agent is responsible to the contracts manager for quality, programme, profitability and safety and may sometimes be referred to as a Site Manager.
- The Agent must understand his Company Standing Instructions (CSIs/procedures or their equivalents) to be able to make maximum use of resources available.
- Job reviews of site-based subordinate staff.
- Ensuring that the work is carried out in accordance with the contract documents and any subsequent Architect's or similar instructions.
- The Agent is responsible for ensuring that his team meet at an early stage to establish the project quality plan (Fig. 1.1). This should preferably have been drafted as part of the project assessment at tender stage. This draft can then be developed for the working project.
- The title of Project Manager is often given to an Agent on a single multi-million pound contract involving a wide variety of disciplines and specialists. Responsibilities are as those for an Agent, but because of the complexity of the work more time needs to be spent on administration, information co-ordination and overall organisation and control.
- So far as is possible, most detailed responsibility should be delegated to create conditions whereby detailed day-to-day problems may be dealt with by others, giving the Agent or Project Manager freedom to liaise with other parties such as the client, statutory authorities, Resident Engineer and other members of the design and construction team.

Sub-agent

A sub-agent is responsible to his Agent or Project Manager for all work, including that by subcontractors, in a specific area or subsection of the construction work. Responsibilities include:

- ensuring that the requirements of quality plans, inspection and test plans and specifications are met
- control of quality, methods, safety, progress and workmanship to achieve the specified requirements
- requisition, control and distribution of plant, materials and consumable stores
- completion of paperwork in accordance with CSIs/procedures
- checking of drawings
- maintenance of site diary
- supervision, training and development of junior staff
- job review of subordinate staff.

General Foreman

He is responsible to the Agent or Project Manager for:

- organising the workforce
- making sure that work complies with the specification
- keeping work to programme
- making sure that plant, operatives and materials are available where and when required and arc not wasted
- making sure that work is done safely
- monitoring performance of subcontractors
- appreciating conditions of subcontractors
- maintaining written record of works.

Engineer

On any one contract there may be a number of engineers with varying amounts and types of experience. The duties and responsibilities of a senior engineer are typically as follows, many of these will be delegated to other engineers on the site according to their experience and ability:

- setting out the works in accordance with the drawings and specification
- liaising with the project planning engineer regarding construction programmes
- checking materials and work in progress for compliance with the specified
- requirements • observance of safety requirements
- resolving technical issues with employer's representatives, suppliers, subcontractors and statutory authorities
- quality control in accordance with CSIs/procedures method statements, quality plans and inspection and test plans, all prepared by the project management team and by subcontractors
- liaising with company or project purchasing department to ensure that purchase orders adequately define the specified requirements
- supervising and counselling junior or trainee engineers

- measurement and valuation (in collaboration with the project quantity surveyor where appropriate)
- providing data in respect of variation orders and site instructions
- preparing record drawings, technical reports, site diary
- job review of subordinate staff.

Site Office Manager

The Site Office Manager may be resident on larger sites, or may travel between a number of smaller sites. He is typically responsible for day to day control of site administration including:

- recruitment and dismissal, catering and statutory returns
- completion of accident and damage reports
- first aid
- completion of plant and equipment returns as required by the CSIs/Procedures
- maintenance and control of site stores
- accommodation
- control of petty cash
- time and pay administration for hourly paid operatives
- checking and issuing payment of material invoices
- maintenance of site security.

Quantity Surveyor

Normally the quantity surveyor is only visiting or covering several jobs and resident only on the larger, more complicated jobs. Typical tasks are:

- commercial oversight of site
- preparation of financial reports, budgets and forecasts
- supervision, training and development of some junior staff
- job review of subordinate staff
- application of health and safety requirements
- completion of paperwork as required by the CSIs/procedures
- co-ordination of enquiries for and assessment of subcontractors (in collaboration with designers, buyers and other disciplines as necessary)
- ensuring that all subcontract documentation adequately defines the specified requirements
- insurance cover (quotations and implications)
- measurement and presentation of monthly valuation for certificate
- formation of claims, variation orders and daywork
- fixing new rates for additional works
- measurement of work done and verification of compliance with specified requirements for subcontractor payment.

Quality Manager

If appointed, the Quality Manager will usually cover several sites. His tasks include:

- providing technical advice on construction materials
- advising line management on quality management matters

- assisting agents or project managers with the preparation of their quality plans for individual projects
- verifying the implementation of quality systems in accordance with the CSIs/ procedures by audit and surveillance, and informing appropriate management of any corrective actions required and monitoring their implementation
- conducting or arranging audits on vendors, suppliers and subcontractors in collaboration with purchasing managers
- educating all staff on their quality management responsibilities.

Planning Engineer

The planning engineer is resident on large contracts, but visits small- to medium-sized jobs. Typical tasks are:

- pre-tender and project planning, method statements
- advising on construction techniques and optimisation of plant
- leading progress meetings with subcontractors and others to compare performance with that laid down in the programme
- co-ordinating design of temporary works
- completing paperwork in accordance with the CSIs/procedures
- implementing systems for distribution of design information and drawings
- recognition of construction/design problems in advance
- application of health and safety requirements
- supervision, training and development of junior staff
- carrying out job reviews of subordinate staff.

Storeman/Materials Controller

The Storeman ensures that the applicable CSIs/procedures for receipt, quality checking, storage and issue of materials are complied with. Responsibilities include:

- agreeing with site management which items require physical checks on delivery and before acceptance on site; making arrangements for appropriate individuals to be notified of deliveries
- assisting with and facilitating checking procedures
- the storage and disposition of materials, tools and tackle, small plant, scaffolding, formwork etc. delivered to site and completion of related paperwork
- storing and issuing components from store and ensuring that likely material short-ages are notified to the site management as early as possible, so that materials may be called forward to meet site requirements
- compiling materials reconciliation schedules and other relevant forms before submission to the Regional or Head Office
- stock-taking.

1.6 Familiarisation

Upon joining a site team, an employee should familiarise himself with the overall scope and concept of the works, meet and understand the role of fellow members, and other external parties such as the employer's representatives and subcontractors.

Certain documentation is essential to the setting up and running of a contract, and should be available on the site at the start of work. These include:

- contract conditions
- the specification
- a bill of quantities
- drawings and allied schedules (including setting out drawings and the original survey to establish the site condition at the start of work)
- site investigation reports and any other reports concerning the environment
- the method statement upon which the cost of the works has been estimated
- project quality plan comprising, among other things, method statements, work instructions, inspection and test plans, check sheets
- construction programmes
- significant codes of practice, standards and manufacturers instructions referred to in the contract documents
- schedule of subcontracts
- material schedule (where appropriate)
- documents relating to industrial relations, safety, health and welfare—in particular you will find the BS 8000 series *Workmanship on Building Sites* an invaluable reference
- copy of building regulations and other appropriate statutory instruments.

An understanding of the above documents is essential if you and your colleagues are to carry out your tasks in a professional and workmanlike manner.

Reference to such documents is second nature to those experienced in site work. If you are inexperienced or under training, make a point of requesting to see them, not only as part of your training, but also to appreciate and understand methods adopted and to foresee problems which may occur in respect of your duties. From the specification and working drawings, compile a list of British Standards, Codes of Practice etc. to which the work is to comply. Select the main standards (e.g. BS 8110 if the work is mainly reinforced concrete) and ensure that copies are avail-able on site for the duration of the contract.

A complete set of British Standards is held at over 150 locations such as Public Reference Libraries and Technical Colleges in the UK and in many overseas locations. (Call BSI on 0908 221166 for the location nearest to you.) There is no excuse for not being aware of specified requirements. Membership of the BSI entitles members to hard copy and CD format yearbooks giving details of all British, European and international standards and codes of practice.

A large project may have its own technical information microfile and reader on site. The range of information stored is selected by the user and can include the complete set of British Standards (contact for example: Technical Indexes Ltd, Willoughby Road, Bracknell, Berkshire RG12 4DW; Tel: 01344 426311).

On working drawings, check the following points in addition to the details shown:

- Does the updated revision suffix to the drawing number match the entry in the revision column?
- Has the designer cancelled all 'standard notes' that do not apply?
- Has the designer added sufficient particular notes to supplement or replace standard notes to explain the requirements fully?

- Are plans of the same area but drawn by various disciplines to the same orientation?
- Is each document marked with its status e.g. **For Comment, For Information, For Construction?** The status, with date of issue and authority, should be logged in the Revision column to provide a record of revision and issues in sequence. This information is often of contractual significance in cases of dispute.
- Has the designer clearly indicated in the body of the drawing what details have been revised? Much time is wasted on site hunting for items of revision.

Queries should be raised with the appropriate design organisation, getting them to establish working procedures to address these matters at least for the remainder of their work on your project.

It is essential that a site log of all incoming drawings and schedules is maintained.

1.7 Communications

Communication on a construction site must take place at all levels for a variety of reasons, and may involve persons not resident on the site or even directly involved with the contract. All meetings involving the contractor should be accurately minuted for future reference. Meetings, discussions, correspondence can occur for the following purposes:

- site meetings to discuss financial and contractual matters, progress of the work, variations, staffing and procurement
- technical discussions between senior experienced members of the team and persons responsible to them, including subcontractors
- to ensure the attainment of the specified standard of quality, performance and safety
- to maintain good industrial relations
- to deal with disputes and grievances
- to deal with day to day technical problems.

Whenever a technical query or problem arises on site the person concerned should pose the question **Where do I go, and who do I ask?** and follow it through. Historically, either it has been too much trouble to climb down scaffolding to find someone, or the other party is difficult to approach and even then doesn't want to know. In the end it is all too much trouble and work is substandard: it is 'bodged'. Everyone on site must remember that bodging today may mean millions of pounds of cost in remedial work to your company in ten, fifteen or even twenty years' time if negligence is proved.

Errors on site or in documentation should be openly reported. Although temporarily embarrassing, immediate correction will often avoid costly remedial work. Likewise, queries should be encouraged so that all members of the team are working to **Get it right—first time.** The first recourse is to try to resolve the problem at the work place by approaching the person to whom you are directly responsible. If that person cannot provide a solution, he should take up the problem on your behalf with other members of the site team. When a solution cannot be found on site, the Agent should consult further company resources.

If you are in any doubt or do not understand, Ask.

1.8 Contractual obligations

In addition to complying with the specified requirements, work must comply with established current good practice. Provided these conditions are met and any deviations are within permissible limits (contract, specification, BS 5606, or other standard), the contractor has fulfilled his obligations under the terms of the contract.

Nevertheless, if faults are subsequently found and a contractor claims they have been caused by design faults outside his control, the response is usually that *the situation should never have been allowed to develop*. This brings forward once again the importance of scrutinising and appraising designers' details at as early a stage as possible.

In the absence of an express requirement in the contract for the contractor to advise the Architect of design defects or failure of the detail to comply with statutory requirements (building regulations etc.), there is no obligation for the contractor to warn the Architect of such defects. There is no such duty to be found in the JCT (Joint Contracts Tribunal) standard forms. Nevertheless, it is still obviously sensible to warn the Architect of the consequences of proceeding with work in a particular way, where it is clear that it will cause problems in the future or is not in accordance with accepted standards of good practice. It is also possible that non-action may result in latent defects and company liability in later years.

However, it is important that while the warnings should be explicit and in writing, means of overcoming the problems should be offered only as suggestions. Any changes arising from these warnings must be adopted by the Architect. Should a contractor become involved in giving detailed advice on these matters, that may invoke liability that might not otherwise have arisen.

1.9 Subcontractors

On most construction sites, some of the work is sublet to subcontractors. The degree to which this is carried out varies enormously, the higher proportion of work being sublet on building as opposed to civil engineering sites.

Potential subcontractors (vendors) must be thoroughly assessed, and contentious items such as access, construction methods, scaffolding, programme etc. agreed before a subcontract is let. Arrangements and organisation for quality management must also be established.

Remember, even when operating as a main contractor, your company is *ultimately* responsible for the quality of goods, materials, services and workmanship provided by subcontractors and suppliers. Members of the construction team must ensure that all aspects of subcontracted work comply with the specified requirements, both during construction and on completion.

However, the main contractor's team on site should avoid taking actions that may relieve the subcontractor of his contractual responsibilities.

1.10 Accuracy in construction

Modern methods of construction make increasing use of preformed components either specially designed or selected from a standard range. They may be manufactured on, adjacent to, or remote from the site, and provision to accommodate such units is made within the structure during design and construction.

Some building systems consist entirely of preformed components columns, beams, floor slabs and wall panels, having their own specification for materials used, their function and methods of manufacture. Where possible, contractors should insist on trial assemblies to reduce the incidence of on-site problems. Ideally these facilities should be negotiated at the tender or immediate post-tender stage.

If the designer has merely specified all tolerances to be as in BS 5606, it may mean he does not understand the interrelationship between manufacturing and construction and allowances required for adjustment and movements due to inherent and applied conditions. Working strictly to BS 5606 may lead to difficulty by possibly not achieving the designer's expectations.

It is essential that designers and constructors have an appreciation of matching manufacturing and construction requirements if assembly on site is to be straightforward. Queries should be resolved *before work starts*. The following points are fundamental:

- Study the specified requirements to find out what standards of accuracy are demanded.
- If BS 5606 is referred to, see that a copy is available on your site.
- Exact dimensions cannot be achieved on site. They may appear to be exact if the degree of accuracy of measurement is poor.
- Unless noted otherwise, sizes and positions marked on drawings represent the *specified dimensions*.
- The difference between a specified dimension and that actually achieved is the *deviation*.
- If the limit of deviation is specified (e.g. ±10 mm) then that is the *permissible deviation* (PD).
- The *tolerance* is the full extent of the PD e.g. as in the above example ±10mm =20 mm tolerance.

Further reading

Standards and Codes of Practice

BS 5606: 1990 *Guide to Accuracy in Building*.
BS ISO 9000: 1994 to 1997 *Quality Management and Quality Assurance Standards*.
Note: BS ISO 9000: 1994 to 1997 and its related standards 9001: 1994, 9002: 1994, 9003:1994 and 9004: 1994 deal in a general way with topics such as generic guidelines for selection and use; quality assurance in design, development, production, installation and servicing; quality assurance in final inspection and test; and quality management and quality assurance standards.
BS 8000: 1989 to 1997 *Workmanship on Building Sites*.
Note: This BS is in several parts dealing with such topics as excavation and filling; concrete mixing and placing; *in-situ* and precast concrete; masonry; waterproofing; carpentry, joinery and general fixings; slating and tiling; glazing; plasterboard partitions and dry linings; cementitious leveling screeds and wearing screeds; plastering and rendering; wall and floor tiling, ceramic tiles, terrazzo tiles and mosaics; wall and floor tiling natural stone tiles; decorative wall coverings and painting; above ground drainage and sanitary appliances; below ground drainage, hot and cold water services (domestic scale); and sealing joints in buildings using sealants.

Other related texts

The Management of Quality Assurance: J. L. Ashford, London, E&FN. Spon, 1989.

Quality Management in Construction in Construction: State of the Art Reports from 13 Countries (Australia; Denmark; Finland; France; Germany; Hungary; Netherlands; Norway; Portugal; Spain; Sweden; GBr; USA): P. Barrett, Salford UK, Salford University, 1994.

Offsite Fabrication: G. F. Gibb, Latheronwheel UK, Whittles Publishing, 1999.

The Impact of European Policies on Quality Management in Construction: R. Grover, London, CIRIA, 1993.

Quality Assurance in Construction (2nd ed.) B. Thorpe *et al.*, Aldershot, Gower Publishing, 1996.

Quality Management in Construction—Survey of Experiences with BS5750: Report of Key Findings: M. Ward, London, CIRIA, 1996.

An Update on Quality Assurance in Construction: Dr. P. Ward *et al.*, Northampton, Building Engineer, 2000.

Quality in the Constructed Project, Proceedings of the Workshop. Chicago 1984: Reston VA USA, ASCE, 1985.

Construction Inspection Handbook: QA/QC. (3rd ed.) J. J. O'Brien, New York, van Nostrand Reinhold, 1989.

Quality Control of Concrete Structures: Rilem, London, Chapman Hall Rilem/CEB, 1991.

Quality Management in Construction; Contractual Aspects, Proceedings of 2nd International Symposium Ghent 1992: CIRIA, London, CIRIA, 1992.

The Role of Safety in Total Quality Management: A. Deacon, Safety & Health Practitioner, 1994.

Total Quality in Construction: Stage 2 Report of the Total Quality Management Task Force: ECI, 1993.

Quality Management in Construction—Survey of Experiences with BS5750, Report of Key Findings: CIRIA, London, CIRIA, 1996.

The Control of Quality on Construction Sites: D. W. Churcher *et al.*, London, CIRIA, 1995.

Modern Construction Management (4th ed.) G. Harris *et al.*, Oxford, Blackwell Science, 1995.

Total Quality Management in Civil Engineering: ASCE, Reston VA USA, ASCE Journal of Management in Engineering, 1993.

Implementing Total Quality Management in the Construction Industry: a Practical Guide: ECI, London, Telford, 1996.

Right First Time: Quality Management for the Small Builder: BEC, London, BEC, 1991.

Improving Construction Methods: a Story about Quality: G. W. Chase, ASCE, Reston VA USA, ASCE Journal of Management in Engineering Vol. 14, 1998.

Quality Management Systems for Post-tensioned Concrete Structures According to ISO 9001: Guide to Good Practice: FIP, London, SETO, Vol. 15, 1999.

ISO 9000 Quality Standards in Construction: ASCE, Reston VA USA, ASCE Journal of Management of Engineering Vol. 15, 1999.

Integrating Safety, Quality and Environmental Management: CIRIA, London, CIRIA, 2000.

Survey of Total Quality Management in the Construction Industry in Upper Midwest (United States): C. McCintyre *et al.*, ASCE, Reston VA USA, ASCE Journal of Management in Engineering Vol. 6 No. 5, 2000.

ISO 9000 within the Swedish Construction Sector: A Landin, Construction Management & Economics Vol. 18 No. 5, 2000.

Understanding Quality Assurance in Construction: a Practical Guide to ISO 9000: H. W. Chung, London, Spon, 1990.

Quantifying the Causes and Costs of Rework in Construction: P. E. D. Love *et al.*, Construction Management & Economics Vol. 18, 2000.

The Contractor's Guide to Quality Concrete Construction: ACI, Detroit Michigan USA, ACI, 1992.

Constructing the Team: Sir M. Latham, London, HMSO, 1994.

Re-thinking Construction: Sir J. Egan, London, DETR, 1998.

Defects in Buildings: Symptoms, Investigation, Diagnosis and Cure: Carillion Services, London, HMSO, 2001.

Chapter 2 WORKING WITH THE WEATHER

John Fullwood

Basic requirement for best practice

- Prudent contractors will make best use of Met Office data to predict weather patterns and to establish accurate information in support of claims

---

Note: This chapter has been compiled from an original draft supplied by John Fullwood of the Met Office and augmented with information from other Met Office staff. Met Office services are continually improving, so it is recommended that readers acquaint themselves with the website (*www.metoffice.com*) to obtain up-to-date information of available services.

2.1 Introduction

The weather is an important consideration throughout most construction projects (see Fig. 2.1). The Met Office can assist at all stages.

- In the initial planning, at the design stage and during the preparation of a tender. For these stages, the Met Office can provide expertise on the range of weather conditions likely to be encountered during construction and during the intended life span.
- During the construction stage, when weather prospects for 1–5 days ahead are a useful tool in site management. All outdoor work is a gamble on the weather, but knowing the odds beforehand can reduce losses in the long run.
- When monitoring progress on site. Local weather records help put any weather-related delays into perspective and can settle disputes over downtime.

These services include past weather information and weather forecasts for world-wide land-based and marine construction, both coastal and offshore.

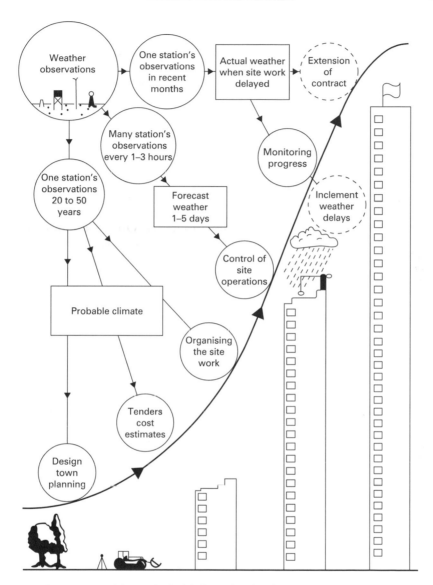

Figure 2.1 Meteorological information for the construction industry.

2.2 Initial design tendering and planning

2.2.1 Design

The weather patterns, particularly the extremes, expected during the life span of a proposed structure need to be considered in order to ensure that the type of construction, the chosen materials and the services made available are suitable. With the tools of contemporary climatological analysis at their disposal, the designer of the ill-fated Tay Bridge that opened in June 1878 would doubtless not have based their calculations on wind speeds likely to be exceeded, on average, once every two years!

This bridge collapsed in a gale in December 1879 with considerable loss of life. Matching design to the local climate can be done using information from the network of UK weather stations, many of which have continuous records extending back over 30 or more years. The network is made up of different types of station. For example, there are several thousand rainfall stations, several hundred stations recording a range of elements (e.g. temperatures and weather as well as rainfall) once a day, about 150 hourly wind stations and some 60 places recording a very wide range of elements each hour. A computer archive of data has been built up, available for analysis by flexible programs. Furthermore, it has become increasingly possible to estimate climatological statistics for sites remote from a weather observing station.

The challenges that can arise in weather-related design are diverse; therefore the majority of weather analyses for design are tailor-made 'one-offs'. For example, by taking into account the local topography and land usage around a site of construction, it is often possible to refine the wind climatology initially derived from a weather station some distance away. If you have a weather-related design problem it is very likely that the Met Office will be able to help. To make your initial contact with the Met Office see Section 2.6.

2.2.2 Tendering and planning

The archive of climatological data can also be used to assess the likely duration of the construction period, the methods to be used and, hence, the tender value. Records of past weather can provide hard information about the incidence of weather adverse to building work; remember that, when tendering for a contract, the weather to be expected is part of the risk accepted—only *exceptionally* adverse weather being admissible in claims for extra time. Our subconscious notion of what is normal depends a lot on chance! Using good information can reduce the odds of being taken by surprise (see Fig. 2.2 for location of UK weather stations).

This information can range from the incidence of, perhaps, damaging frosts, strong winds or heavy rain to the numbers of working hours with adverse weather. The threshold of 'adverse weather' can be chosen to suit the operation—for example, temperatures less than 2 °C for concreting or wind gusts over 17 m/s for using tower cranes. The way that the information is presented can also be chosen—for example, long period averages of interrupted 'working hours', spells of consecutive hours or days without adverse weather or information about year-to-year variability.

For some sites, particularly those in the more remote parts of the UK, no wholly representative weather station will be available and help in interpreting the available data may be required.

Information is also held concerning weather conditions overseas. For these locations a more restricted range of information is on offer, but useful comparisons with UK conditions can usually be made. Information for coastal and marine projects is dealt with separately in Section 2.5.

Two products specifically geared to this stage in the proceedings are *Building Downtime Averages* and *NEC Planning Averages*, described below.

Figure 2.2   Map of climatological stations (please contact the MET Office for a full list of site availability).

2.2.3 Building Downtime Averages for tendering and planning

These are sets of monthly averages of the numbers of working hours (taken as 0700–1700 GMT each day) when the weather may affect operations, using various temperature, wind speed, rain, snow and humidity thresholds. They are available for each of 50 UK weather stations. Illustrations of a complementary monthly service, the *Building Downtime Summary (MetBuild)* are shown at Fig. 2.3 (1 to 4).

2.2.4 New Engineering Contract planning averages

These are statistics designed specifically for users of the Institution of Civil Engineers (ICE) *New Engineering Contract* (NEC) and now available for about 150 locations. This contract is designed in a very specific way. For the weather to be considered exceptionally adverse during a contract it must have been worse than what could have been expected one year out of ten in terms of specified monthly parameters. This product gives a set of 1-in-10 year thresholds for each of the four weather variables most commonly used in these contracts—total rainfall, number of days exceeding 5 mm rainfall, number of air frosts and number of days with snow lying at 0900 GMT. Details of a complementary service the *NEC Monthly Summary* are given in Section 2.4.2.

2.3 The construction phase

While construction work is in progress, weather forecasting services can help reduce time losses, prevent damage to materials and assess the risk of dangerous working conditions. Project Managers may have on their minds such questions as, "How windy will it be for the tower crane?" or "When will snowfall begin and how heavy will it be here?" To help cope with changing local conditions, a variety of forecast services are available, providing varying degrees of specialisation. Contact the MET Office to discuss the possibilities.

2.3.1 Tower crane forecast services

Reliable crane deployment can make a significant improvement to project execution.

The Met Office tower crane forecast service offers a range of forecasts from 24 hours out to five days ahead giving mean wind speeds, gust and direction at the required heights specific to the project. The service helps project managers to identify weather windows and plan critical lifts with confidence whilst maximising crew safety and reducing equipment hire time.

2.3.2 Specialist consultancy

The Met Office offers a range of consultany services both during the project planning phase to highlight areas of climate change and meteorology that should be considered and during the execution of the project for updates and more detailed information to support standard forecasting services. Telephone consultancy is available 24 hours a day, seven days per week offering direct access to experienced forecasters.

Building Downtime Summary (MetBUILD)

Tel: 0870 900 0100 www.metoffice.com

LONDON HEATHROW - October 2003 (Ref: MO19)

Report issued on Tuesday, December 16 2003 at 12:54

	Number of hours in the period 07 to 17 GMT								with Relative Humidity over 90 %
Date	with Temperature less than (deg C)								
	0	1	2	3	4	5	8	15	
1	0	0	0	0	0	0	0	2	0
2	0	0	0	0	0	0	0	4	7
3	0	0	0	0	0	0	0	0	2
4	0	0	0	0	0	0	0	10	0
5	0	0	0	0	0	0	3	10	0
6	0	0	0	0	0	0	0	9	1
7	0	0	0	0	0	0	0	10	1
8	0	0	0	0	0	0	0	3	0
9	0	0	0	0	0	0	0	1	1
10	0	0	0	0	0	0	0	0	0
11	0	0	0	0	0	0	0	4	1
12	0	0	0	0	0	0	0	6	0
13	0	0	0	0	0	0	0	5	0
14	0	0	0	0	0	0	0	5	0
15	0	0	0	0	0	0	0	6	0
16	0	0	0	0	0	0	0	7	1
17	0	0	0	0	0	0	0	10	1
18	0	0	0	0	0	0	1	7	2
19	0	0	0	0	0	0	0	10	0
20	0	0	0	0	0	0	1	10	0
21	0	0	0	1	2	2	4	10	2
22	0	0	0	0	0	0	4	10	7
23	0	0	0	0	0	0	3	10	0
24	0	0	0	1	1	2	3	10	2
25	0	0	0	1	1	2	3	10	1
26	0	0	0	0	0	1	2	10	2
27	0	0	0	1	1	1	3	10	3
28	0	0	0	1	1	2	3	10	2
29	0	0	0	0	0	0	2	10	7
30	0	0	0	0	0	1	4	10	5
31	0	0	0	0	0	0	0	10	8
MON - FRI total	0	0	0	4	5	8	27	162	50
Long-term average	0	0	0	1	1	2	11	150	37
MON - SAT total	0	0	0	5	6	10	31	193	54
Long-term average	0	0	1	1	1	3	14	185	46

Figure 2.3 (1 of 4)   Building Downtime Summary.

Building Downtime Summary (MetBUILD)

Tel: 0870 900 0100 www.metoffice.com

LONDON HEATHROW - October 2003 (Ref: MO19)

Report issued on Tuesday, December 16 2003 at 12:54

Day	Rainfall Total (mm) 00-24 (GMT)	07-17 (GMT)	Number of hours in the period 07 to 17 GMT — With Rainfall over 0.2mm	When Snow fell	Snow lying during the day 06-18 (GMT)
1	1.4	0.0	0	0	no
2	4.4	2.0	2	0	no
3	0.0	0.0	0	0	no
4	0.0	0.0	0	0	no
5	0.4	0.0	0	0	no
6	0.8	0.0	0	0	no
7	0.2	0.2	0	0	no
8	0.0	0.0	0	0	no
9	0.0	0.0	0	0	no
10	1.4	0.0	0	0	no
11	0.0	0.0	0	0	no
12	0.0	0.0	0	0	no
13	0.0	0.0	0	0	no
14	0.0	0.0	0	0	no
15	0.0	0.0	0	0	no
16	0.0	0.0	0	0	no
17	0.0	0.0	0	0	no
18	0.0	0.0	0	0	no
19	0.0	0.0	0	0	no
20	0.0	0.0	0	0	no
21	0.0	0.0	0	0	no
22	4.8	0.2	0	0	no
23	0.4	0.0	0	0	no
24	0.0	0.0	0	0	no
25	0.2	0.0	0	0	no
26	0.0	0.0	0	0	no
27	0.0	0.0	0	0	no
28	0.0	0.0	0	0	no
29	8.0	2.8	3	0	no
30	7.8	0.6	1	0	no
31	5.4	0.0	0	0	no
MON - FRI total	34.6	5.8	6	0	No. of days 0
Long-term average	37.9	16.3	12	0	0
MON - SAT total	34.8	5.8	6	0	No. of days 0
Long-term average	47.0	20.1	15	0	0

You receive this report for personal use only subject to our terms and conditions, available on request. Broadcast, publishing or redistribution prohibited.

© Crown Copyright. Met Office 2003

Figure 2.3 (2 of 4)   Building Downtime Summary.

Building Downtime Summary (MetBUILD)

Tel: 0870 900 0100 www.metoffice.com

LONDON HEATHROW - October 2003 (Ref: MO19)

Report issued on Tuesday, December 16 2003 at 12:54   Page 3 of 4

| Date | Number of hours in the period 07 to 17 GMT | | | | | | | |
| | with Mean Winds (mph) greater than | | | | | | with Gusts (mph) greater than | |
	15	18	23	26	32	39	39	46
1	0	0	0	0	0	0	0	0
2	0	0	0	0	0	0	0	0
3	0	0	0	0	0	0	0	0
4	0	0	0	0	0	0	0	0
5	0	0	0	0	0	0	0	0
6	6	6	0	0	0	0	0	0
7	9	7	0	0	0	0	0	0
8	6	4	0	0	0	0	0	0
9	5	0	0	0	0	0	0	0
10	9	4	0	0	0	0	0	0
11	0	0	0	0	0	0	0	0
12	4	0	0	0	0	0	0	0
13	0	0	0	0	0	0	0	0
14	6	0	0	0	0	0	0	0
15	8	0	0	0	0	0	0	0
16	5	0	0	0	0	0	0	0
17	3	0	0	0	0	0	0	0
18	0	0	0	0	0	0	0	0
19	2	1	0	0	0	0	0	0
20	8	6	0	0	0	0	0	0
21	0	0	0	0	0	0	0	0
22	0	0	0	0	0	0	0	0
23	10	5	0	0	0	0	0	0
24	0	0	0	0	0	0	0	0
25	0	0	0	0	0	0	0	0
26	0	0	0	0	0	0	0	0
27	0	0	0	0	0	0	0	0
28	0	0	0	0	0	0	0	0
29	0	0	0	0	0	0	0	0
30	0	0	0	0	0	0	0	0
31	0	0	0	0	0	0	0	0
MON - FRI total	75	32	0	0	0	0	0	0
Long-term average	24	8	1	1	0	0	3	1
MON - SAT total	75	32	0	0	0	0	0	0
Long-term average	29	10	2	1	0	0	3	1

Figure 2.3 (3 of 4)   Building Downtime Summary.

Building Downtime Summary (MetBUILD)

Tel: 0870 900 0100 www.metoffice.com

LONDON HEATHROW - October 2003 (Ref: MO19)

Report issued on Tuesday, December 16 2003 at 12:53   `Page 4 of 4`

Day	Air min temp (deg C)	Grass min temp (deg C)	Rainfall amount (mm)	Snow depth (cm) at 09GMT	Mean Wind speed for day (mph)	Maximum gust for day (mph)
1	11.3	8.2	3.8	0	10.0	25
2	10.7	9.4	2.0	0	5.8	15
3	12.9	11.9	0.0	0	7.7	20
4	7.9	5.5	0.0	0	11.7	30
5	3.8	1.6	1.2	0	12.4	25
6	6.9	6.9	0.0	0	16.1	33
7	11.2	9.5	0.2	0	16.6	35
8	4.8	1.2	0.0	0	11.9	36
9	11.2	9.8	0.0	0	12.3	26
10	13.0	11.1	1.4	0	13.8	30
11	7.7	3.4	0.0	0	4.0	17
12	8.4	4.6	0.0	0	11.3	28
13	10.0	7.7	0.0	0	10.6	25
14	10.5	8.5	0.0	0	13.0	30
15	8.2	5.4	0.0	0	13.5	31
16	7.5	4.7	0.0	0	11.9	29
17	6.9	3.7	0.0	0	11.2	29
18	4.4	2.2	0.0	0	10.2	24
19	6.3	3.6	0.0	0	13.7	33
20	6.1	3.9	0.0	0	14.3	31
21	1.3	-1.0	4.6	0	5.6	15
22	3.4	2.8	0.6	0	10.5	24
23	6.5	4.7	0.0	0	13.8	35
24	0.1	-2.4	0.0	0	8.9	24
25	0.3	-1.8	0.2	0	10.4	24
26	4.1	1.1	0.2	0	9.2	26
27	0.8	-5.1	0.0	0	3.6	16
28	0.4	-6.0	3.0	0	6.2	21
29	4.3	5.3	6.0	0	7.6	18
30	3.3	1.6	11.6	0	10.4	25
31	5.4	6.7	0.6	0	7.2	17

	No. of air frosts	No. of ground frosts	No. days of rain >= 1mm	No. days of rain >= 10mm	No. days of snow lying	Mean wind speed (mph)
MON - FRI total	0	4	7	1	0	10.5
Long-term average	0	3	6	1	0	7.3
MON - SAT total	0	5	7	1	0	10.3
Long-term average	1	3	7	2	0	7.3

Figure 2.3 (4 of 4)   Building Downtime Summary.

2.3.3 Weathercall

Weathercall is a joint venture between the Met Office and iTouch (UK) Ltd. offering Met Office forecasts for the next hour, day, five or ten days. These are delivered by telephone, fax, or prepaid subscription. The service offers telephone forecasts for 28 regions across the UK and fax forecasts for 9 regions.

Based upon calls from a BT landline (calls from mobiles may vary), Weathercall by telephone (09014 calls) cost 60p per minute. Weathercall five-day forecasts and additional services by fax (09060 calls) cost £1 per minute. Weathercall 10-day forecasts by fax (09065 calls) cost £1.50 per minute (prices correct at time of going to print).

For a guide to the Weathercall services please contact the Met Office (see Section 2.6).

2.4 Monitoring progress on site

A copy of the actual readings of rainfall, temperatures, wind speeds and so forth at a representative weather station can help to put the progress of a project into perspective, especially if it has been affected by adverse weather. Claims for extra time or money expended will need substantiating by an authoritative record. Met Office records provide an ideal base for resolving these questions.

2.4.1 Building Downtime Summary

The *Building Downtime Summary* is the sister product of the *Building Downtime Averages* described in Section 2.2.3. It is a four-page monthly summary of 'downtime weather', available within five days of the end of the month for each of the 50 UK weather stations shown in Fig. 2.2.

Building Downtime Summaries are available in booklet form or via e-mail.

2.4.2 NEC Monthly Summary

This is the sister product of the *NEC Planning Averages* described in Section 2.2.4 and is designed for users of the Institution of Civil Engineers New Engineering Contract. It is a one-page summary providing a comparison of the weather experienced during the month with the 1-in-10 year thresholds for that month (see Fig. 2.4). The summary is now available for about 150 locations across the UK.

2.4.3 General Monthly Summary

This is a more generalised product than either the *Building Downtime Summary* or the *NEC Monthly Summary*, available for 50 UK locations (see Fig. 2.5). It is simply a daily weather log covering maximum and minimum temperature, rainfall, mean wind speed, highest gust speed, sunshine duration and a note of any other significant weather (thunder, hail, fog, snow, gale). Wherever possible a comparison of monthly means and totals with the corresponding long-term averages is given.

NEC Monthly Summary

Tel: 0870 900 0100 www.metoffice.com

MANCHESTER (Lat=53:35 N Long=02:28 W)
November 2003 (Ref: MO19)

Report issued on Tuesday, December 16 2003 at 12:56

DATE	Daily Rainfall (mm)	Days of Rain >5mm	Minimum Air Temp (Deg C)	Days with Air Frost	Snow Depth at 0900 GMT (cm)	Days with Snow Lying at 0900 GMT
1	6.4	1	4.6		0	
2	5.8	1	5.5		0	
3	1.8		7.2		0	
4	nil		4.4		0	
5	nil		7.6		0	
6	nil		9.3		0	
7	nil		8.7		0	
8	nil		7.1		0	
9	1.2		5.1		0	
10	nil		6.6		0	
11	4.4		6.8		0	
12	1.4		7.7		0	
13	0.2		5.4		0	
14	13.0	1	7.0		0	
15	0.2		8.4		0	
16	0.2		1.5		0	
17	7.2	1	2.6		0	
18	nil		8.7		0	
19	7.2	1	11.0		0	
20	1.2		9.5		0	
21	nil		4.2		0	
22	nil		0.8		0	
23	nil		-3.2	1	0	
24	0.6		-4.1	1	0	
25	2.8		-2.3	1	0	
26	1.4		6.7		0	
27	0.2		0.6		0	
28	nil		0.7		0	
29	2.6		3.1		0	
30	3.8		4.3		0	
Total	61.6	5		3		0
Long - Term Average	78.0	6		5		0
1-in-10 Year Value	119.3	9		10		1

Figure 2.4 NEC Monthly Summary (Manchester).

General Monthly Summary

Tel: 0870 900 0100 www.metoffice.com

BIRMINGHAM - November 2003 (Ref: MO19)

Report issued on Tuesday, December 16 2003 at 12:46

Day	Temperature(C)		Rain (mm)		Wind (mph)		Sunshine (hrs)	Significant Weather
	Day Max	Night Min	00-24	06-18	Average Wind	Max Gust		
1	10.8	4.8	1.0	0.0	9	22		
2	13.6	5.4	16.4	11.6	14	38		
3	13.2	7.8	1.2	0.2	13	38		
4	13.9	4.3	tr	0.0	9	28		
5	14.8	12.1	0.0	0.0	11	28		
6	14.1	8.6	-	-	6	20		
7	13.1	5.8	0.0	0.0	6	18		
8	7.6	6.0	tr	tr	8	22		
9	8.6	4.7	3.2	3.2	7	16		
10	11.3	7.7	0.0	0.0	4	13		
11	9.6	5.6	tr	0.0	7	20		
12	12.1	9.6	3.2	0.2	8	18		
13	12.2	6.0	1.0	0.0	10	32		
14	11.4	8.6	5.2	4.0	20	50		
15	11.2	6.4	tr	0.0	7	23		
16	8.9	3.3	0.2	0.2	6	16		
17	12.6	2.3	-	2.0	9	22		
18	14.3	12.1	0.0	0.0	9	26		
19	14.2	10.3	0.0	0.0	11	29		
20	11.8	10.6	3.0	1.8	9	24		
21	7.4	3.2	0.0	0.0	2	7		
22	5.7	2.3	1.6	1.6	5	12		
23	4.6	1.1	1.0	0.8	4	10		
24	9.3	-2.3	0.0	0.0	4	16		
25	11.1	2.0	2.0	0.2	11	28		
26	9.1	7.4	17.6	4.6	11	35		
27	9.1	-0.6	0.0	0.0	5	13		
28	9.5	0.5	tr	tr	9	18		
29	12.7	7.0	3.8	3.6	13	37		
30	9.6	4.5	3.8	1.0	5	15		
Average/Total for month	10.9	5.6	64.2	35.0	8.3	23.1		
Long Term Average	9.0	3.5	60.0		10.1		64.2	

KEY: t=thunderstorm; h=hail; f=fog (@0900 GMT); s=snow (including sleet); g=gale; '-' signifies that data is not available on this occasion; tr means a 'trace' of rainfall, i.e. less than 0.20mm.

You receive this report for personal use only subject to our terms and conditions, available on request. Broadcast, publishing or redistribution prohibited.

© Crown Copyright. Met Office 2003

Figure 2.5   General Monthly Summary (Birmingham).

2.4.4 Degree Day Report

'Degree days' are an aid to the energy monitoring and management of buildings and provide an index of for how long and by how much the temperature has remained above ('cooling' degree days) or below ('heating' degree days) a given threshold value. The *Degree Day Report* confines itself specifically to the type of degree day most commonly used, heating degree days to a base temperature of 15.5 °C. The report gives an indication of the latest month's degree-day figures across the UK as represented by 18 locations. Also given is a comparison with the long-term averages for that month and a comparison against the previous year's figure. Other types of degree-day data can be produced on request.

2.5 Shoresite

Many of the comments that have been made about the services available for contracts on land hold true for contracts on the coast or at sea. An extensive computer archive of observations of weather and sea state from ships and platforms can assist with *strategy* at the planning and tendering stages and there is a wide range of site-specific weather forecast services available to help with *tactics*.

2.5.1 Planning and tendering

Among the advice available for any season or month and virtually any part of the world are the following data:

• mean values and likely extremes of any weather element
• frequencies of any weather element including wind speed and direction and wave height and direction
• average downtimes and weather windows for thresholds of interest.

Requests for information are carefully studied to ensure that specific needs are met. At coastal sites, a wave refraction study may be needed to give a meaningful picture of the wave climatology.

2.5.2 Coastal forecast services

Nearshore wave predictions are generated through the application of Met Office standard wave model products to HR Wallingford calibrated wave transformation models. The nearshore wave predictions can be further applied in calculations of surfzone parameters, or overtopping rates.

2.5.3 Offshore forecast services (see Figs. 2.6 and 2.7)

To help ensure the successful execution of day-to-day marine and offshore operations, the Met Office provides a range of site-specific forecasts covering wind, wave and weather. These forecasts, initially developed for the offshore industry,

Five Day Forecast – Example Only

Tel: 0845 300 0300 www.metoffice.com

Client (Ref: MO1)

Page 1 of 3

Forecast Issued on Monday, 21 May 2001 at 08:18

Client location

HEADLINE

Gale Warning in Force	NO	**LIGHTNING RISK:** NS: Low becoming Moderate after 1800
Sea Temp (Degrees Celsius)	8	

GENERAL SITUATION: NS: Remaining very unsettled as a series of deep Atlantic depressions dominate the northeast Atlantic and UK waters. A hint of quieter weather developing over the weekend as the main low pressure centre sinks south.

CONFIDENCE AND TRENDS: NS: High in a cold unsettled type, but medium in details of forecast beyond tomorrow. Sea and winds likely to increase from Friday with the possibility of Gales.

AT A GLANCE - VALID UNTIL 1200 Tue 22-May-2001

Phase	Wind (Mean)	Time (UTC)	Sea (Sig)	Time (UTC)
Max	20	22/0300	2.4	22/0400
Min	08	21/0600	1.4	21/0600

Cloud Height (FT) above sea level = 3/8 coverage or more below 5000ft

	Mon 21-May-2001						Tue 22-May-2001		
	06	09	12	15	18	21	00	03	06
Weather	DRY	DRY	RAIN	RAIN	RAIN	RAIN	RAIN	RAIN	RAIN
Visibility	10KM+	10KM+	10KM+	10KM+	10KM+	9KM	10KM+	10KM+	10KM+
Temp	8	8	9	9	9	9	9	9	9
Cloud	NIL SIG	3400	2200	500	200	200	200	200	200

Wind Speed in 'Knots' - Wave and Swell Heights in 'Metres' - Wave and Swell Periods in 'Seconds'

		Mon 21-May-2001						Tue 22-May-2001						
		06	09	12	15	18	21	00	03	06	09	12	15	18
Wind Dirn		S	S	S	SSW	SSW	SW	SW	WSW	WSW	WSW	WSW	WSW	WSW
10m Wnd Spd		8	13	16	19	17	17	19	20	19	16	16	15	16
10m Gust		12	18	22	26	24	24	26	28	26	22	23	21	22
50m Wnd Spd		9	14	18	22	20	20	22	24	22	19	19	18	19
50m Gust		13	20	25	30	28	29	31	33	30	26	27	26	26
Sig Wav Hgt		1.4	1.5	1.7	2.0	2.0	2.0	2.1	2.3	2.4	2.3	2.1	1.9	2.0
Max Wav Hgt		2.2	2.4	2.7	3.2	3.2	3.2	3.4	3.7	3.8	3.7	3.4	3.0	3.2
Sig Wav Prd		6	6	6	6	6	6	6	6	6	6	5	5	5
Swell Dirn		N	N	N	N	N	N	N	WNW	W	W	W	W	W
Swell Hgt		1.3	1.3	1.3	1.1	1.0	1.0	0.8	0.7	0.7	1.1	0.9	1.2	0.9
Swell Prd		7	7	8	8	8	9	9	9	10	10	10	9	9

	Tue 22	Wed 23-May-2001				Thu 24-May-2001				Fri 25-May-2001				Sat 26
	21	00	06	12	18	00	06	12	18	00	06	12	18	00
Wind Dirn	WSW	WSW	WSW	W	NNW	N	SSW	SW	SW	S	S	S	S	SW
10m Wnd Spd	15	15	18	15	10	8	7	2	1	6	11	16	16	11
10m Gust	21	21	25	21	14	11	10	3	2	8	16	23	22	15
50m Wnd Spd	18	18	23	18	12	10	8	2	2	8	14	21	20	13
50m Gust	25	25	32	26	17	14	12	3	2	11	20	29	27	19
Sig Wav Hgt	2.1	2.3	3.1	3.0	3.0	2.5	2.1	1.9	1.7	1.6	1.5	1.7	2.0	1.7
Max Wav Hgt	3.4	3.7	4.9	4.8	4.8	4.0	3.4	3.0	2.8	2.5	2.4	2.8	3.2	2.8
Sig Wav Prd	6	6	7	7	7	7	7	7	7	6	6	6	6	5
Swell Dirn	NW	NW	N	N	N	N	N	N	NNW	NNW	WSW	W	WSW	WSW
Swell Hgt	1.2	1.6	1.9	2.3	2.8	2.4	2.1	1.9	1.7	1.6	1.3	1.2	1.1	1.4
Swell Prd	9	10	12	11	8	8	7	7	7	7	7	8	9	7

Figure 2.6 Part of a 5 day forecast.

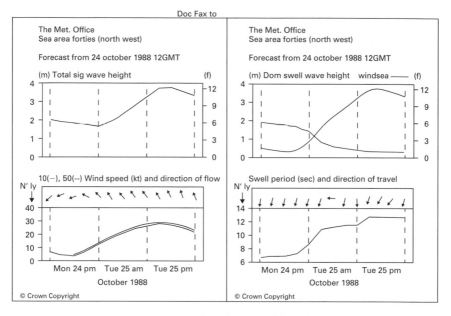

Figure 2.7 Metograph.

have many applications. They comprise mainly wind and sea conditions, with extra remarks on cloud, visibility and temperature as required.

Real time forecasts can make offshore operations cost effective and also ensure the safety of those involved. Forecasts can be issued twice daily by web delivery, e-mail, or fax. They can extend from 48 hours ahead to a 5-day forecast and can be further extended to a 6–10 day outlook forecast.

Other services available from the Marine Forecast team 24/7 include: warning services, offshore deployments, daily project briefings for planning purposes, marine consultancy, probabilistic forecasting, risk assessment forecasting and bespoke forecasting. Ensemble 5–10 day forecasts are also available.

This core product can be expanded upon for the first 36 hours or extended to incorporate details to days 3 or 5 plus. If requested, the 6 to 8 day outlook as shown in Fig. 2.8 can be provided. Whilst the cost of these and other services is small the benefits in time saved and money can be enormous.

2.5.4 Marine consultancy services

The Met Office also makes use of an extensive global wave model archive comprising data from the Global, European and UK wave models.

Resulting computations of wave and swell from the wave energy spectrum, give extensive coverage of wave characteristics for all the oceans of the world. Output from these models can either be used in forecast mode or to study wave climate at the planning stage in areas where no observations exist. These data can also be used to assist with vessel response, either in prediction mode or past event analysis.

Our suite of models includes numerical weather prediction of wind direction and speed, waves, ocean currents, salinity and temperatures, storm surges and tidal currents. Programs are in place to derive downtimes, frequency and extreme value analysis for any wave model grid point around the globe.

Metocean consultancy provides the legal and insurance sectors with retrospective marine weather analysis anywhere in the world. Expert opinion on marine weather can be provided in areas of dispute ranging from vessel performance on the high seas

FORECAST MODULES

MODULE

A	BASIC AUTOMATED PRODUCT, METOGRAPH OR METOGRAM FOR FIRST 36 HOURS
B	MODULE 'A' PLUS A GENERAL SUMMARY, WEATHER/VISIBILITY/CLOUD AND TEMPERATURE FOR THE PERIOD OF THE AUTOMATED PRODUCT. 24 HOUR TELEPHONE CONSULTANCY SERVICE.
C	(ONLY AVAILABLE WITH 'B') OUTLOOK TO 3 DAYS
D	(ONLY AVAILABLE WITH 'B') OUTLOOK TO 5 DAYS
E	(ONLY AVAILABLE WITH 'B') OUTLOOK FOR DAYS 6 TO 8
F	PERSONAL BRIEFING IN COMPANY OFFICE
G	STORM UPDATES AS NECESSARY
H	AMENDMENTS AS NECESSARY
I	OFFSHORE FORECASTER ON SITE
J	TIDAL PREDICTIONS
K	PROJECT CONSULTANCY

Figure 2.8 Offshore forecast modules.

to cargo loss. Consultants are also available to appear in arbitrations hearings and court. Support is also available to a wide range of projects in various business sectors to help manage the risks which marine weather can present. Our consultants utilise modelled, measured and observed data, sourced from an internal suite of numerical models and extensive high integrity, and often long standing data archives.

2.6   Contact points

To subscribe to any of the customised services mentioned in this chapter or to seek more specific advice, contact:

The Met Office

Tel: 0870 900 0100
Fax: 0870 900 5050
Website: *www.metoffice.com*
Email: enquiries@metoffice.gov.uk

Further reading

Standards and Codes of Practice

There are no specific British Standards or Codes of Practice dealing with the weather in relation to construction.

Other related texts

Book of the Weather: P. Eden, London, Daily Telegraph, 2002.
The Accuracy of Wind and Wave Forecasts: J. S. Hopkins, Bracknell, prepared by the Met Office for HSE, 1997.

Chapter 3 CONSTRUCTION PLANT

David Doran

Basic requirements for best practice

- **Correct selection of plant**
- **Well-maintained equipment**
- **Good safety procedures**
- **Well-trained operatives**

3.1 Plant and transport organisation

Your company may have a central Plant and Transport Department which operates a hire service for plant, commercial vehicles, small tools and radio communications. If so it would be advisable to contact that department before dealing directly with local operators. General information can be obtained from trade organisations and major plant hire companies such as Atlas Copco.

3.2 General

Plant is obviously a very wide-ranging subject; only the basic outline of some specific topics can be given here. When requesting plant, always give as much detail as possible, including the application, so that the supplier can furnish a machine which is as near as possible ideal for the job.

Site management is responsible for the supervision of plant and operators, whether hired internally or externally. It is particularly important that any operator is fully informed and has knowledge of site conditions, e.g. low capacity under bridges or low headroom over bridges, culverts, overhead power lines and underground services. It is also important to prevent machines from starting to excavate until the operator has been issued with a permit to dig and from lifting until a method statement has been provided and explained to the operator. Any loads to be lifted should always be assessed for weight before the lift commences.

See also Chapter 19 for general guidance on health and safety matters and Section 3.4.6 (Operation and maintenance).

3.3 Safety

All well-maintained and properly selected plant can be used safely in the hands of a competent and experienced operator, appropriately trained and certificated in the operation of the particular item of the plant. He should be given full details of the job to be done and be made aware of all other relevant factors concerning the material he is working on, the job site and other people and machines around. All plant can be lethal if any of the above criteria are not met. These are also the criteria which apply if you wish to get the best performance from plant.

3.4 Selection

The selection of plant is usually straightforward but the following points are relevant.

3.4.1 Cost

Hire rate

Take all factors into account if you are offered alternatives. What are the extras to the basic rate? For operated plant, these can include overtime, bonus and subsistence. Are fuel and lubricants included or excluded? Non-operated plant can sometimes be charged on a pro-rata basis for working hours above an agreed basic week (this arrangement is prevalent for diesel-driven generating sets). If you are offered a larger machine than you require, take into account any additional fuel costs (which can sometimes be significant) and check the haulage charge.

Fuel consumption

Figures for some machines may be available from your plant and transport depot.

Delivery or collection charges

These can be a significant part of the total cost, especially for short-term hire and should always be taken into account when evaluating alternatives.

3.4.2 Machine outputs

This subject would justify a separate book for each type of machine. It is not intended to cover this topic here. To assist planners, manufacturers' literature can often be helpful. Of particular use is the *Caterpillar Performance Handbook*. Liebherr and also produce useful manuals.

Note: The signaller should stand in a secure position where he can see the load and can be seen clearly by the driver and should face the driver if possible. Each signal should be distinct and clear.

Fig. 3.1 Know your crane signals.

3.4.3 Ground conditions

As a general rule, wheeled machines are preferable if ground conditions allow them to work proficiently. Under extreme conditions a low ground pressure (LGP) crawler with extra wide tracks instead of standard width (STD) tracks, may be required. Mats may be useful under crawler cranes and excavators under similar conditions. Geotextile reinforcement under a blanket of good fill should also be considered. Table 3.1 gives some typical examples of ground bearing pressures. This data however must be taken in the proper context. Ground pressures under the tracks of a crane during lifting and slewing operations will vary considerably. Should there be any doubts about the machine loadings these must be fully investigated and adequate precautions taken.

Particular care should be taken over the design, construction and maintenance of piling platforms (see Section 8.6.3).

3.4.4 Size restraints

Check access onto the site, loading on roads, traffic routes and police escorts. Are there any obstructions, low bridges, overhead power cables or other structures? Will a structure to be built as part of the contract obstruct the machine's removal? Is there room for the machine's tail-swing?

3.4.5 Space requirement for mobilisation

Crawler cranes and some large mobiles require a fairly large area for rigging (or for re-rigging if altering configuration). Craneage for the erection of tower cranes or batching plants often requires considerable space. For example, a 22RB with a 24 m boom requires a clear space of about 9 m × 32 m and a 38RB with a 30 m boom and 9 m fly-jib requires about 10 m × 50 m for efficient mobilisation.

Table 3.1 Tracked machine ground bearing pressures (kN/m²).

Crawler tractors

Dozers		Loaders		
CAT D6H STD 57 LGP 31		CAT 931B	STD 62	LGP 33
CAT D7H STD 68 LGP 42		CAT 943	STD 73	LGP 58
CAT D8N STD 85		CAT 953	STD 78	LGP 61
CAT D8L STD 101		CAT 963	STD 81	LGP 67
		CAT 973	STD 83	LGP 69

Excavators crawler

Hydraulic excavators

JCB 812	32
Liebherr R900B	37
Liebherr R942	68
CAT 245	88

3.4.6 Operation and maintenance

If in doubt about the operation of any plant delivered to site, ask for a manual and study it. If you are still in doubt, ask the supplier for a demonstration.

The hirer will be expected to carry out the basic day-to-day checks on most of the plant used on site, whether in house or hired in. Ensure that site operatives are familiar with what is required. Major services, such as oil changes, are generally carried out by the owners, but check at the time of hire for hired-in plant.

3.5 General conditions of hire

Plant is generally hired under the CPA Standard Conditions of Hire, and guidance should be given by your in-house plant specialist. A knowledge of these conditions, which often puts considerable responsibility onto site management, is important so that your contract and your company is not put at undue risk.

3.5.1 Period of hire

There will generally be a minimum period of hire, typically one week, but this can vary and is often less for small items and more for larger specialist plant. Items which require to be assembled on site can have hire commencing on delivery or on commissioning: check this, it can make a difference.

3.6 Operators

Most companies will have a group policy to cover operator competence. Typical documentation for plant operators is shown in Figures 3.2 and 3.3. This documentation should be lodged in the site office for inspection during a random visit by the HSE.

Operators of cartridge power tools (e.g. Hilti Gun) must have had training from the tool manufacturer or authorised supplier and have a Certificate of Training. Forklift and telehandler drivers have a legal requirement for training, as set out in the HSE Approved Code of Practice: *Rider Operated Lift Trucks—Operator Training*.

Your in-house plant specialist should be aware of local training facilities.

3.7 Cranes

3.7.1 General

These can be relied upon to carry out a wide variety of tasks within their capability, but do not rely solely upon the operator to ensure that the crane is kept within this capability. BS 7121 Part 1 explains the management and control of the lifting

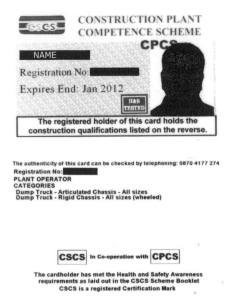

Fig. 3.2   Typical example of plant operators' authority cards.

operation using a safe system of work. This code also contains other useful information relating to safe use of cranes.

Where a banksman is used, it is essential that there is a clearly defined method of communication. Where hand signals to crane operators are used these should be in accordance with BS 7121 (Fig. 3.1).

All cranes, lifting appliances and tackle are subject to statutory regular examination and testing. The procedures should be specified in your Company Safety Instructions. The requirements are summarised in the RoSPA *Construction Regulations Handbook*. It is essential that you check that all cranes and associated equipment on site have been tested (Fig. 3.4a) and examined within the statutory periods and that section C Form F91 Part 1 (Fig. 3.4b) is completed weekly for all cranes on site.

3.7.2   Crawler

Crawler cranes are very useful where there is continuity of work. They are fairly stable and require a minimum of ground preparation. On piling works, they are still unsurpassed. Crawler cranes can be used with a plain boom or with a fly-jib. There are various heavy-lift attachments, such as American Hoist's Sky Horse, Manitowoc's Ringer and Demag's Superlift. Some larger crawler cranes can be used in a tower configuration, with a luffing jib on the vertical boom.

All common types are strut-jib with bolted or pinned additional intermediate sections to give a range of lengths.

Fig. 3.3 Plant operators' certificate of competency–National Standard may now supersede this type of documentation.

3.7.3 Mobile

There is a large variety of mobile cranes, and only the principal types used in construction are mentioned here:

Truck

Generally useful for one-off or small series lifts. Originally strut-boomed, they now largely carry hydraulically operated extending telescopic booms. In most cases, truck cranes can only lift while in a static position with outriggers jacked in place, and cannot be used for pick-up and carry operations.

F96

F.96/3R/JRT/PEF
HEALTH AND SAFETY EXECUTIVE
Certificate No. 14567

Please quote in any communication
MACHINERY REGISTER NUMBER
016666/CUK6

Factories Act 1961   ITEM No. 0730.UK
CONSTRUCTION (LIFTING OPERATIONS) REGULATIONS 1961

CERTIFICATE OF TEST AND THOROUGH EXAMINATION OF CRANE
(Prescribed by the Minister of Labour in pursuance of paragraph (1) or (2) of regulation 28)
See Note 4, overleaf, for continuation of items 10 and 11

1. Name and address of owner of crane	G. Wimpey PLC Wimpey Plant & Transport
2. Name and address of maker of crane	BPR 220 Avenue De Stalingrad Chevilly La Rue, France.
3. Type of crane and nature of power (e.g. Scotch derrick-manual; Tower derrick-electric; Rail mounted tower-electric)	Electric Rail Mounted Tower Crane (Fixed) Type GT 431D PT 7/2
4. Date of manufacture of crane	1988
5. Identification number (a) Maker's serial number	74027
(b) Owner's distinguishing mark or number (if any)	CR 0730
6. Make and type of automatic safe load indicator, if required (See regulation 30)	Richier BPR
7. Make and type of derricking interlock, if required (See regulation 22)	Not applicable
8. Date of last previous test of crane	26.1.89
9. Date of last previous thorough examination of crane	

10. Safe working load or loads

In the case of a crane with a variable operating radius (including a crane with a derricking jib or with interchangeable jibs of different lengths) the safe working load at various radii of the jib, jibs, trolley or crab must be given. Test loads at various radii should be given in column (iii) and in the case of a safe working load which has been calculated without the application of a test load "NIL" should be entered in that column

(i) Length of Jib (feet) mtrs	(ii) Radius (feet) mtrs	(iii) Test Load (tons) Kgs	(iv) Safe Working Load (tons) Kgs
55	55	4750	3800
	50	-	4250
	45	-	4800
	37	7500	6000
Height under hook 24.5 metres			
Slewing Restricted to 120° approx. due to site			
layout.			

11. In the case of a crane with a derricking jib or jibs the maximum radius at which the jib or jibs may be worked (in feet)   Not applicable

12. Defects noted and alterations or repairs required before crane is put into service. (If none enter "None")   None

I hereby certify that the crane described in this Certificate was tested and thoroughly examined on

2.6.90   and that the above particulars are correct.

Signature   *A.K.G.*   J.R. THOMSON

Qualification   Engineer Surveyor,

Name and address of person, company or association by whom the person conducting the test and examination is employed.

The Ajax Insurance Association Ltd.
Ajax House, Haslemere Road, Liphook,
Hants. GU30 7AL

Date of Certificate........ 15/6/90

4128   *For Notes see overleaf*

Fig. 3.4a Typical F96 Crane Test Certificate.

Name or title of employer or contractor .... Address of site .... Work commenced - Date		Factories Act 1961 SECTION C Construction (Lifting Operations) Regulations 1961 **LIFTING APPLIANCES** Aerial cableways, aerial ropeways, crabs, cranes, draglines, excavators, gin wheels, hoists, overhead runways, pilling frames, pulley blocks, sheer legs, winches	
Reports of the results of every inspection made in pursuance of Regulation 10(1)(c) or (2) of a lifting appliance or in pursuance of Regulation 30(1) or (2) of an automatic safe load indicator			
Description of Lifting Appliance and Means of Identification (1)	Date of Inspection (2)	Result of Inspection (including all working gear and enclosing or fixing plant or gear, and where required the automatic safe load indicator and the derricking interlock) State whether in good order (3)	Signature (or, in case where signature is not legally required, name) of person who made the inspection (4)

Fig. 3.4b Form F991 Section C-Lifting appliances and safe load indicators.

Rough terrain

As the name implies, these mobile cranes are especially suitable for site work, but generally have only limited mobility on the road. They typically have two sets of duty ratings, one in the fixed position with outriggers extended and the second free-on-wheels for pick-up and carry operations. The latter is only usable under good level site conditions, despite the name!

All terrains

These cranes are a cross between truck and rough terrain and are very useful. They combine good on-site mobility with a reasonable road speed and handling for between site moves.

3.7.4 Tower

General

For preparation for erection and dismantling, consult the HSE Guidance Note *Erection and Dismantling of Tower Cranes*. Tower cranes are structures demanding foundations which must be properly designed according to the criteria for the particular installation. Your Company Standing Instructions/Procedures should deal with this.

Tower cranes can have either luffing or fixed jibs. All tower crane operations are affected by winds; you should acquaint yourself with the maximum windspeed with which work is permissible: 70 Km/h (45 mph) is the general restriction.

Windspeed indicators, with a warning bell, are fitted to most tower cranes, and should be specified for all hired-in cranes. A continuous wind speed recorder can be fitted if required.

Static

The most common type of tower crane. Carefully consider all the factors affecting its use before deciding on the position for a static tower crane on any site. Do not forget to take dismantling into account as well as erection. If possible a tower crane should be tied into a structure. The method of doing this must be agreed with the designer of the permanent structure.

Savings can sometimes be made in the size of the crane mast sections. However, modern structural design often precludes this.

The base of a fixed tower crane is often dictated by site restrictions. However, in many cases where there is room, a ballasted travelling or cruciform base for a fixed crane is more economic than a fixed base requiring a heavy mass concrete foundation.

Travelling

Although crawler bases are available for a few types of tower crane (but not to be confused with crawler cranes in tower configuration, see Section 7.2), most travelling bases are rail mounted. The track gauge is wide, ranging from 3.8 m to 15 m or more to give stability. Tower crane rail track must be laid level, and there can be no exception to this. Large radius curves can be negotiated.

Climbing

Most tower cranes can be *climbed*, but the present economics of mobile crane hire for maximum height erection, as against labour for periodic climbing, generally favours immediate maximum working height installation.

Internal stairwell/lift-shaft climbing tower cranes are useful in certain cases and can have the advantage of a much shorter jib while still reaching all points of the building. On high-rise work, there can be a considerable saving on mast sections. Any proposal to use this type must be discussed with the engineer responsible for the design of the permanent structure.

3.8 Excavations

3.8.1 Selection

For bulk excavations involving continuous production work it is usual to select the largest machine that can be given a continuous workload. However, where there is a limited amount of work to perform, study other factors carefully and select a machine accordingly. A larger machine than necessary generally incurs more cost.

Loose material takes up far more room once excavated. See Table 3.2 for some useful bulking factors.

3.8.2 360° Excavator

The 360° crawler-propelled excavator is the standard bulk excavation tool. However, wheeled types are often preferable where travelling is involved if ground conditions permit their use or where an existing finished surface must be protected.

Wheeled excavators are generally found in medium sizes, but rubber crawlers on some midi and mini excavators allow similar use. Within their capabilities mini excavators are extremely popular. They are economic on fuel, cheap to transport, and especially useful on restricted sites.

3.8.3 180° Excavator or digger loader

The 180° excavator of the JCB 3CX Sitemaster type is extremely useful for general site duties having the ability to perform rehandling, loading and fork-lift duties in addition to its capability as an excavator. With four-wheel drive, it has good tractability in most site conditions. Almost all these machines currently available in the UK have an offsettable excavator attachment, however there are some with a fixed centrepoint which limits excavation close to walls or obstructions.

3.8.4 Depth of dig

Manufacturer quoted depths of dig are absolute maximum figures. If there is a very little excavation at maximum dig, then a machine near its limit is a practical

Table 3.2 Densities and bulking factors.

Material	Bulk density (kg/m₃)	Bulking factor
Cinders	860	1.50
Clay	2000	1.23
Clay & Gravel	1750	1.18
Earth	1980	1.25
Loam	1540	1.23
Granite (broken)	2730	1.64
Gravel – Pitrun	2170	1.12
– Washed—dry	1800	1.12
– Washed—wet	2260	1.12
Limestone (broken)	2610	1.70
Sand—dry	1600	1.12
Sand—wet	2080	1.12
Sandstone	2520	1.67
Shale	1660	1.33
Topsoil	1370	1.43

Note: The bulking factor is the reciprocal of the swell factor.

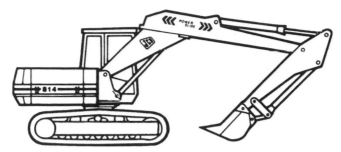

Fig. 3.5 Typical 360° excavator.

Fig. 3.6 Typical 180° excavator or digger loader.

application, but for any quantity of excavation it is generally more economic to use a machine with a slightly greater digging depth than that required. This makes it easier to obtain a full bucket and does not require frequent repositioning of the machine, thus impeding output

3.8.5 Lifting

Excavators are extremely useful for lifting but the regulations associated with lifting are restrictive. An excavator which is not subject to regular statutory inspection *cannot be used for lifting under any circumstances*. Machines without both a safe load indicator (SLI) and a load radius indicator (LRI) are restricted to their worst position lifting capacity up to one ton and one ton only. This applies even if their nominal capacity is greater, unless the work is directly associated with excavation being undertaken by that particular machine at that time, in which case the machine may be rated at higher loads under a *Certification of Exemption* if it is fitted with check valves. If an excavator is required to do any lifting, ensure that the supplier is aware of this and furnishes a suitable machine. *Form F91, Part 1, Section C* must be completed weekly (see Fig. 3.4b).

Table 3.3 shows typical lifting capacities. However, the figures shown must not be taken as accurate for any individual machine. Model variations, boom and dipper sizes, track width and length, etc. can make for a large spread of ratings for any one basic model. Check both machine markings and test certificate in every case.

3.8.6 Attachments

Breaker

There are several types available, with Montabert being the most common. They are extremely useful for general demolition and breaking out concrete or other hard materials. Be sure that the breaker is matched to the excavator for size and hydraulic flow/pressure characteristics.

Grapple/Grab

As well as the normal bucket almost any type of hydraulic grapple can be fitted to 360 hydraulic excavators. The finger or claw type is extremely useful for the accurate placing of large material.

3.9 Dumpers

The dumper remains the economic method of moving small to medium quantities of material around site. They are almost invariably diesel driven, and the majority are four-wheel drive with articulated steering. Two-wheel drive rigid models are, however, much cheaper and can be used on good surfaces. A wide range of capacities from 0.5 to 8 tonne is available with 2 to 3 tonne being most popular. The basic model has a forward tipping muckskip, hydraulically operated except for the very smallest which may be manual. There are rotaskip variations which allow for side tipping.

The high-lift type has a much higher skip pivot level and is useful used in conjunction with a special concrete pouring throat for the delivery of concrete to skips and low shutters. From 8 tonnes upwards more sophisticated, fully cabbed,

Table 3.3 Typical hydraulic excavator lifting capacities (tonnes).

	Lift with LI and LRI	Lift without SLI and LRI	Lift in connection with excavation work
JCB 3 CX Sitemaster (dipper arm fully retracted)	N/A	1.0 t	1.0 t
JCB 805 B (2.30 m arm)	5.1 t	1.0 t	1.1 t
JCB 812 (2.30 m arm)	4.3 t	1.0 t	1.2 t
Liebherr 900 (2.2 m arm)	5.3 t	1.0 t	2.0 t
Liebherr 902 (2.2 m arm)	7.3 t	1.0 t	3.0 t
CAT 215 B LC (1.8 m arm)	4.6 t	1.0 t	2.2 t
CAT 225 B LC (1.98 m arm)	8.4 t	1.0 t	3.3 t
CAT 245 (2.59 m arm)	16.7 t	1.0 t	6.8 t

Maximum lift capacity, no attachments fitted. Based on BS 1757 recommendations of 71.4% full circle tipping load.
Note: For additional information please see *The Reference Manual for Construction Plant.*

articulated dumptrucks are available. Under typical site conditions, these can be used for general haulage operations, but, if bulk production is required, it is recommended that a well-maintained haul road is laid down.

3.10 Pile hammers

3.10.1 General

There has been enormous development in recent years. Diesel and air hammers have been largely ousted in favour of vibratory hammers (usually hydraulically driven, but occasionally electric) and hydraulically operated drop hammers. This is due to the comparatively low weight, efficiency and relative environmental friendliness of these systems. There are also proprietary systems, usually noise reduced, involving rope-operated drop hammers and hydraulic jacking systems.

Pile leading and guiding systems are as important as the hammer itself and must be carefully evaluated for all applications.

3.10.2 Diesel hammers

These are very reliable and can be used in conjunction with all types of material. They are, however, relatively noisy, even when fitted with hushpacks, and do not have the degree of control available with hydraulic and electrically driven hammers.

Diesel hammers can occasionally be difficult to start. Be extremely cautious in the use of starting aids, especially ether. It is not unknown for a piston to be fired right out of the cylinder and land some distance away!

3.10.3 Air hammers

These are similar to diesel hammers in many respects. They generally have exceptional reliability, but consume large quantities of air, thus requiring large compressors. Many are reversible and can be used for pile extraction.

3.10.4 Hydraulic hammers

These are generally very effective but tend to be more noisy and slower than vibrators. They are, however, good in clay and will penetrate harder materials, although drilling or other preliminary work is necessary in many hard rocks. Hydraulic hammers cannot usually be reversed for extraction.

3.10.5 Vibratory hammers

These offer many advantages when driving piles in suitable materials. They are generally unproductive in clay or very hard materials. They are relatively quiet. Most can be used as extractors for piles driven in almost any material with a crane of adequate line-pull. The effect of the vibrations can affect surrounding buildings. Preliminary surveys and movement monitoring may be necessary.

3.10.6 Pile rigs

Smaller hydraulic rigs, mounted on hydraulic excavator chassis, are extremely useful for sheet piled walls, with pile lengths above 15 m now feasible. Such rigs can operate without any temporary works, and can have relatively high rates of output if the job is set out with good runs of work, with piles distributed along the line of work.

Some rigs can be fitted with a drilling attachment that can be useful for many site operations.

3.11 Water pumps

Pumps for moving large volumes are generally of the centrifugal type, with diaphragm-type pumps useful for smaller quantities. They can be powered by petrol or diesel engines, by electric motor or by compressed air, the latter two types usually being submersible pumps. For the largest flows against high heads, electrically driven submersible centrifugal pumps are usually most efficient. Electric submersible pumps can be set up for automatic operation when used in conjunction with float switches. They are also available with special high head impellers which sacrifice flow volume for pressure, making these ideal for large level differences. If very high heads or long distances are involved, booster pumps can be inserted along the line.

Small petrol-driven centrifugal pumps are portable and extremely useful for pumping out small quantities of water. Diaphragm pumps are useful where solids handling is involved, although modern centrifugal pumps can handle high solids ratios. Table 3.4 indicates some examples.

3.12 Electric generators

3.12.1 General

Electric generators are used where it is impossible or uneconomic to acquire a mains supply. They can be from 0.5 kVA to over 5000 kVA, and may be driven by

Table 3.4 Water pumps.

Description	Type	Power	Flow (m³/h)	Head (m)	Weight (kg)
50 mm Selwood 50A	Centrifugal	Petrol engine	30	24	18
50 mm Honda	Centrifugal	Petrol engine	24	16	18
40 mm Wickham	Diaphragm	Petrol engine	11	18	29
50 mm Sykes	Centrifugal	Diesel engine	34	25	230
100 mm Sykes	Centrifugal	Diesel engine	152	36	530
150 mm Sykes	Centrifugal	Diesel engine	212	55	860
100 mm Flygt	Submersible	Electric	108	20	
100 mm Flygt	Submersible	Electric	218	91	
100 mm Flygt 2201	Submersible	Electric	545	38	
62 mm CP10	Submersible	Air	603	316	
62 mm CP20	Submersible	Air	733	328	

petrol or diesel engines, or, for the very large sizes, by gas turbine. They are generally rated in kVA, but may be quoted in kW by some suppliers who have allowed
for the Power Factor in accordance with the manufacturers' specification.

3.12.2 Small petrol engine generators

Up to about 4 kVA generators are usually petrol driven and hand transportable with
single phase alternating current outputs of either 110 or 240 V, but can be dual volt-
age 110/220 V. Such generators can quickly be moved along as work progresses,
but can be very noisy especially in an enclosed space, unless fitted with an acoustic
enclosure and silencer.

3.12.3 Diesel engine generators

Diesel engine generators are used from 5 kVA to 1000 kVA. These are generally
very rugged, and have the advantage of long engine life and low fuel consumption. Larger single phase sets, up to about 12 kVA, tend to be only 240 or 110 V
as specified, whereas sets of above 12 kVA are generally skid mounted and diesel
driven with three phase four wire system, i.e. 415/240 V outputs. Unless required to
be installed in a purpose built structure, generators should be obtained either complete with weather protection enclosure or acoustic chamber. Some generators have
small fuel tanks, so larger or additional tanks should be specified to provide fuel
capacity to suit requirements, especially if unattended night running is involved.

3.13 Electric tools and transformers

Portable electric tools for use on construction and building sites are generally at
110 V AC. They should be used strictly in accordance with BSEN50144 particularly in respect of earthing requirements.

The use of all electrical tools should be carried out in accordance with all relevant electricity supply regulations, and temporary electrical installation requirements. The *Electricity at Work Regulations 1989* place responsibilities on suppliers
and users of electrical equipment for a safe system of working, use of tools within
its strength and capability and protection of electrical equipment from adverse or
hazardous environment in and out of service. *Guidance Note PM32* on the safe use
of portable electrical apparatus gives safety advice on installation, use and maintenance.

3.14   Compressed air equipment

3.14.1   Compressors

For most applications these are diesel driven, but for large jobs of longer than usual
duration, there is a case for electric units, especially where noise and exhaust fumes
are likely to give problems. Their usual operating pressure is 7 bars (100 psi) which

corresponds to 6 bars (90 psi) design working pressure of many air tools, but 10 bars pressure (150 psi) is a common option. Portable compressors are available in a wide range of sizes ranging to beyond 474 l/s (1000 cfm).

For economic working the compressor size should be matched to the air consumption. Table 3.5 shows the air consumption of commonly used air tools. As many of these tools are of American origin the Imperial ratings are also shown. Approximate conversion factors are:

- 1 bar = 14.5 pounds per square inch (psi)
- 1 litre per second (l/s) = 2.12 cubic feet per minute (cf/m).

3.14.2 Compressed air tools

Hand-operated tools are available for almost any purpose, from drills to chainsaws. They are generally durable and can be safer than electric tools, especially where water is present. They also have the security advantage of being relatively unattractive for domestic use.

The most popular tools are hand-held breakers, with the CP117 size being useful for general breaking out and the much smaller CP9 being useful for light work. Compressed air is also commonly used to power water pumps.

Always ensure that an air bottle is in the line to lubricate tools and that it is regularly serviced. It is essential that airways are kept clean at all times. This is particularly important with immersed concrete vibrators, with which extra care should be taken, especially when they are temporarily put to one side during a pour.

Air losses and tool efficiency must be taken into account as these will affect the size of the compressor required.

Table 3.5   Air consumption at 6 bars (90 psi).

Model	l/s	cf/m
Consolidated Pneumatic		
CP9/RR	10	22
FL22F Pick/C.Dig	22	46
222 Pick	21	44
117 Demo Tool	29	62
32A Rock Drill	47	100
CP10 Pump	54	115
CP20 Pump	59	125
1½ Vibrator	9	20
190 Vibrator	27	60
3250 Vibrator	42	90
CP6 Scabbler	14	29
23 inch Danarm Chainsaw	35	75
3270RW800 Woodborer	52	110
Errut		
U3 Floor Scabbler	52	110
U5 Floor Scabbler	75	160
U7 Floor Scabbler	99	210

3.15 Compaction equipment

The general principle is to use vibration for granular materials and pad or sheep-foot rollers (which may or may not be vibratory) for cohesive materials.

For small amounts of material, single- or twin-drum self-propelled 'walk-behind' rollers are generally effective. Larger quantities and areas can be dealt with by use of heavier 'ride-on' varieties. The three-point roller still has some applications in asphalt finishing but has been largely superseded by tandem vibratory rollers, either with twin or single vibrating drum. For fill and sub-base, single drum self-propelled vibratory rollers with pneumatic-tyred drive wheels are generally used. A more accurate finish may be obtained by a model having the vibrating drum also driven in traction. Some rollers do not have this facility. For rock fill, or where usage is intermittent, towed vibratory rollers hitched to a crawler or wheeled tractor are appropriate.

On granular fill and non-cohesive soils, vibratory plate compactors are useful in confined areas. Reversing models are more productive but more expensive.

Petrol engine or compressed air powered rammers are also useful in confined areas and are suitable for most materials.

3.16 Forklift trucks and telescopic handlers

The delivery to site of most materials in packages has made forklifts and telescopic handlers basic tools on many construction sites.

Forklifts are useful for loading and off-loading vehicles with suitably packaged items, and can place items at a height where they can approach directly to the off-loading point. They can be fitted with many attachments, including a front bucket, although the latter is of very limited use.

Telehandlers can generally perform as a forklift, with the advantage of being able to reach forward over obstructions, or ground that should not be disturbed. They are more expensive to hire than forklifts but can use attachments more efficiently, especially a front loading shovel. As forklifts and telehandlers are able to deposit packaged loads of two tonnes or more on a scaffolding, scaffolding load capacity should be checked before arranging material distribution.

Fig. 3.7 Typical forklift telescopic handler.

Construction machines of these types are almost invariably diesel powered and whether two- or four-wheel drive, are, for stability, usually rigid rather than articulated. Many have very good rough terrain capabilities, and are mobile in poor ground conditions. They should, however, be driven very slowly over uneven ground or when climbing curbs.

Loaded forklifts and telehandlers should always face uphill and be driven in reverse when descending. Empty machines should always face downhill. Forklift and telehandler stability is greatly reduced under cross fall conditions.

3.17 Site communications

Good communication is essential on construction sites to achieve the most efficient use of manpower, materials, machines and transport. Your in-house plant specialist should be able to offer advice on the most suitable equipment. A wide range of VHF, UHF, band 3 radio and mobile telephones are available. These include battery-operated handsets for individual use, mobile equipment for fitting to vehicles and base stations. Intrinsically safe equipment is available for use in hazardous areas such as oil refineries, nuclear power stations etc. A radio licence is required for the operation of all VHF and UHF radios.

Further reading

Standards and Codes of Practice

BS EN 13000: 2004 *Specification for Power-driven Mobile Cranes*.
BS 4363: 1998 *Specification for Distribution Assemblies for Electricity Supplies for Construction and Building Sites*.
BS 7121: 1997 to 2006 *Code of Practice for Safe Use of Cranes*.
Note: There are several parts to this BS dealing with such topics as inspection, testing, mobile cranes, lorry loaders, tower cranes and recovery vehicles. When completed, BS 7121 will replace CP 3010.
BS EN 50144: 1996 to 2006 *Safety of Hand-held Motor Operated Tools. General Requirements*.
Note: This Standard is in several parts dealing with drills, wrenches, sanders, knives, hammers, spray guns, shears, nibblers, tappers and planers.

Other related texts

Notes: The RoSPA [Birmingham] produce a considerable volume of publications related to construction. For Health and Safety regulations see Chapter 19 and www.hse.gov.uk. The undermentioned companies produce useful handbooks:

- Caterpillar: *Performance Handbook*.
- Liebherr: *Technical Handbook Earth Moving Product Lines*.

The Reference Manual for Construction Plant: ICES, Cheshire UK, ICES, 2001.
Note: This comprehensive survey of construction plant quotes dimensions and capacities for very many types of plant.
The Operation and Maintenance of Bridge Access Gantries and Runways: IStructE, London, SETO, 2007.

Operator's Safety Code for Rough Terrain Lift Trucks: BITA, Ascot UK, BITA, 2002.

Crane Stability on Site: CIRIA, London, CIRIA Report SP131, 1996.

Tower Crane Stability: CIRIA, London, CIRIA Report 1891, 2003.

Code of Practice for the Safe Use of Lifting Equipment, 5th ed.: LEEA, Bishops Stortford UK, LEEA, 2001.

Management of Construction Equipment, 2nd ed.: F. Harris *et al.*, Hampshire, MacMillan, 1991.

Introduction to Civil Engineering Construction, 3rd ed.: R. Holmes, Reading, College of Estate Management, 1999.

Construction Methods and Planning: J. R. Illingworth, London, Spon, 1993.

Plant Engineers Reference Book: D. A. Snow, Oxford, Butterworth Heinemann, 2002.

Design and construction of deep basements including cut-and-cover structures: IstructE, London, IstructE, 2004.

Guide to inspection of under water structures: IstructE, London, IstructE, 2001.

Heavy Construction—Planning, Equipment and Methods: J. Singh, Taylor & Francis, London, 2006. [A new edition is in preparation.]

Management of Off-Highway Plant & Equipment: D.J. Edwards, F.C. Harris and R. McCaffey, E & Routledge, London, 2003.

Chapter 4 SETTING OUT

Martin Sheward

Basic requirements for best practice

- Correct choice of instruments
- Well-trained staff
- Good maintenance of equipment
- Correct procedures
- Good coordination with workforce

4.1 Introduction

Setting out probably carries more risk than any other task a young engineer is likely to be asked to do. The cost of mistakes in both time and money can be horrific. Errors can have major implications.

In spite of this it is usually treated as a necessary evil which takes an engineer's time away from the more 'important' tasks. Getting help or understanding to overcome the physical problems of setting out can be received with little sympathy.

When setting out, a risk assessment should be carried out. The greater the effect of wrong or inaccurate setting out, the more time, effort and care should be taken. What could start out as a quick job that anybody can do, may actually, in some months down the construction process, involve a project-critical dimension. It may only be a line of sheet piles in a muddy field, but it may form the basement of a building that already has too tight a space to allow for setting out inaccuracies.

Setting out is the fastest way for a new engineer to make an impression. Make it a good one!

4.2 How accurate do you need to be?

It is necessary to balance the need for setting out to be accurate enough for the task, against spending valuable time on being more accurate than necessary. It must also be born in mind that a line set out on a slab done once, accurately, will save you doing it several times to varying degrees of accuracy. Approximate setting out lines used in the early stages of a job should be obliterated to avoid misuse at a later stage. The line may also survive longer than the engineer's recollection that the line was only a rough grid, it may get used for something that needs to be positioned carefully. If a line can be made to survive by permanently marking the structure in some way it is worthwhile spending time on all lines put down. This, as the job goes on will be a valuable investment in a control framework, which can be used again for a variety of purposes.

Modern construction methods mean that often a large part of a structure is manufactured off site. After an *in situ* foundation and basement levels, for example, the structural steel frame and cladding, as well as the lifts and a good deal of the services, pipes etc. will be brought to site already made, and they will be expected to fit. In addition, each of the separate manufacturing processes has its own production tolerances, plus their own installation tolerances. For example when a window comes to site, it is desirable that the window fits the opening that has been provided for it. Not only does it have to fit the opening, it has to fit with just the right amount of space between it and the wall it fits into. The gap at the edge needs to be small enough to be neatly sealed. That opening would have a specified size as well as a plus or minus tolerance, i.e. big enough to fit the window, but not so big you cannot seal it and it looks unsightly. Engineers should be aware of the sometimes detrimental effect of a succession of positive (or negative) tolerances.

A simple example, but now think of the same principle and apply it to a twenty storey building or a bridge which has to have its beams fit across the span. Often there will be the coming together of several trades each with their own tolerance regime. No excuse is acceptable when the first train through a station hits the signal gantry. So accuracy of setting out has to be good enough to allow the person it is intended for to do his job, and the finished item be within tolerance either as stated in the specification or as dictated by the following trade. That usually means the setting out has little latitude for inaccuracy.

That said, a common sense view needs to be taken of each item to be set out. It is perfectly acceptable to be 'near enough is good enough' for a stage one strip of a site, or the setting out of a base on a dig to be blinded to ensure the dig is sufficiently deep.

4.3 Tolerances and permitted deviations

The contract documents should contain a section on tolerances. These are often lifted cold from BS 5606 *Accuracy in Building*. The section will lay down the criteria for accuracy. For example if it states that the position of a pile shall not deviate from its intended position by more than 75 mm, and your setting out is only accurate to 25 mm, it can easily be seen that poor setting out will quickly erode the permissible tolerance. Although it is the responsibility of the design team to define tolerances, it will often be necessary to inspect drawings and interface details to see how accurate each stage of the construction process needs to be. Tolerances may need to be defined where none is given or where the BS tolerance is too generous or restrictive.

On finished work, particularly that which is seen by the public or building user, tolerances are not so much a structural or fit requirement, but what an architect deems to be acceptable to the eye. This can be a tough test to pass. A contractor may occasionally decide to demolish work that is within tolerance but is not pleasing to the eye. With much visible work, *if it looks right it is right.*

Checklist

- Do you have the latest drawings and information?
- Have any design variations been issued since the drawing was last revised?
- Have any checks been made on the quality of the information?
- How critical is the item to be set out?
- What will happen if it is not set out correctly?
- Are the calculations correct, and have they been checked?
- What accuracy does the item to be set out require?
- How much of the construction tolerance is going to be taken up by inaccurate setting out? Is it an acceptable figure?
- Are the instruments, methods and assistance of a sufficient quality to achieve the required accuracy?
- Have the tradesmen been consulted for their requirements?
- Does the setting out look right?
- Have there been sufficient checks made on the setting out to flag up an error should one exist, or is it perfectly correct but in the wrong field!
- Will the setting out lines last, and is the marking out sufficiently clear and unambiguous?
- Has the end user been shown around what has been done and is he satisfied?
- Have you checked to ensure the setting out is being used correctly?
- What is the total out of position due to summation of plus and minus tolerances?

4.4   Avoiding mistakes

A man who never made a mistake never made anything. (anon.)
Most setting out and out-of-tolerance errors come into this category and can be avoided by good practice.

Types of errors:
- Gross errors—These are the mistakes that we all make. With good practice and cross checking they are noticed.
- Systematic errors—These are usually small and of the same sign, but if unchecked can accumulate to an unacceptable level. These are often caused by poorly calibrated equipment.
- Random errors—These are may be positive or negative and occur randomly. This is the residual error when all others have been dealt with. Repetitive observations should be taken to reduce random errors.

4.4.1 Good practice in the office

Setting out starts in the office. Before embarking on a task:

• Check the information you are about to use. Is it the latest drawing? If you have already worked to a drawing and it has subsequently been superseded, do the revisions affect your work and have all the revisions been noted?
• Check the quality of the information. Do the dimensions make sense? Do they agree with the other drawings (i.e. structural/architectural)?

   Calculating the setting out information can often take longer than the setting out itself. Take care over it. If it is an essential piece of setting out, get a colleague to produce independent calculations. He may interpret the raw data in a different fashion (maybe correctly) and end up with a different answer. This is a better procedure than getting calculations checked as the checker may follow your calculations in detail without noticing errors of principle. Do final figures look right? Make rough scale checks off drawings as a way of checking gross errors. If you are using some software to do all your thinking for you, do you know how it works? Just inserting figures into a program will give you no idea of whether they are right or not. There is little on site that is so complex that it cannot be worked out using a calculator: this will give you a better feel for what is going on. (See Fig. 4.1 for a summary of the formulae useful for solving setting out problems.)

4.4.2 On-site control

Primary baselines and benchmarks should be regularly checked for accuracy. Approximate setting out for excavation may not be sufficiently accurate for precise concreting work.

   Concrete monument stations may be damaged by earthmoving and other equipment and should be regularly monitored. When working from stations marked on a concrete slab it is advisable to check that the correct station is being used. This may be done by measuring back to an adjacent station or by simple observation of surrounding landmarks.

4.4.3 Equipment and staff

Now that you have the calculations and the right control, ensure you have the equipment and assistance (chainman) adequate to achieve the required accuracy (see also Section 4.5).

4.4.4 Checks

It is essential to include cross-checks as you proceed. We have all made mistakes, the crime is not being the first one to spot them!

   A simple example would be in setting out a box: to pre-calculate the diagonals as part of the setting out calculations is easy enough, and it provides a quick and easy check on accuracy. You will also need to have worked out the rejection criteria. This can give you confidence in your own work. If queried, you can say with confidence that the setting out is to the required standard.

Basic formulae—Almost every setting out problem can be solved by breaking it down into basic trigonometry solutions. Complex formulae are not necessary. With the formulae below and basic right angle trigonometry there will be little that cannot be solved.

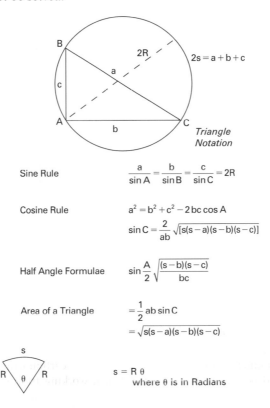

Sine Rule	$\dfrac{a}{\sin A} = \dfrac{b}{\sin B} = \dfrac{c}{\sin C} = 2R$
Cosine Rule	$a^2 = b^2 + c^2 - 2bc\cos A$
	$\sin C = \dfrac{2}{ab}\sqrt{[s(s-a)(s-b)(s-c)]}$
Half Angle Formulae	$\sin\dfrac{A}{2}\sqrt{\dfrac{(s-b)(s-c)}{bc}}$
Area of a Triangle	$= \dfrac{1}{2}ab\sin C$
	$= \sqrt{s(s-a)(s-b)(s-c)}$

$s = R\,\theta$
where θ is in Radians

Figure 4.1 Simple formulae.

Setting out is an ongoing process, so cross-checks to already constructed work should be carried out. This can take the form of reworking the setting out of existing work. It can highlight problems early on if checks do not tie in. It is tempting to set out new work from existing features. To do this will compound errors in setting out tolerances and construction errors from new and existing work. It is easy to see that errors would rapidly accumulate when doing this. To avoid the accumulation of errors, work should be set out 'from the whole to the part'. Always set out from the main control to the part to be built to ensure that each part has approximately the same accuracy, and errors do not accumulate.

The same can be said when it comes to establishing control. It is best to set secondary control from the main control and not have a series of add-ons each one further away from its intended accuracy. A simple example of this is the old story of setting out a housing estate by going from house to house. By the end there is not enough room for the last house. The correct way is to set out each house from the main control. That way each house is in the correct place, within permitted deviations. The same principle applies to all setting out tasks.

The final checks on setting out should be the taking of a moment to stand back and see if it looks right. Check that it eyes through with existing features, pace it out as a rough check, does it look sensible? A common sense last look can save the day.

4.4.5 Marking the ground

The best setting out in the world is of no avail if it does not last the first operation or if nobody can understand it. Before starting, ask the person who is going to use your setting out what he wants, and how he is going to go about the job. That done, make sure what you have done is clear and easy to understand.

Do not be afraid to write notes and sketches next to your lines on walls and floors (provided they are not finished surfaces) as to what the lines mean. Try to keep offsets in sensible dimensions, e.g. 500 mm 1000 mm etc. Tape tags can be affixed to a variety of objects that are otherwise hard to mark, such as pins in the ground, reinforcement cages etc. These can be labelled 'Top of Blinding' etc. You cannot give too much information. When finished, take time to explain it to the end user. Remember, you know what you have done—all they are presented with is the fruits of your labour, and it may not be as obvious to them as it is to you. Keep an eye on it, you may need to explain it twice!

4.4.6 Use of grid offsets

Building grids will need to be offset from actual gridline position so lines of sight are not obscured. In a simple square grid structure with columns on gridline inter-sections, the grid should be set out as an offset to the true position. Offsets should be large enough to allow for column size and formwork. The offset should be as small as practicable so that errors in out-of-square taping do not become too critical.

For example, a 500 mm square column will overhang the gridline either side by 250 mm. Formwork and its support will add another 150 mm or so all around, mak-ing about 400 mm. Therefore a 500 mm offset line would mean the line of sight clearing the column formwork by about 100 mm. When checking columns for pos-ition, the tape reading will be small at about 100 mm. With an offset this close to the columns, if the tape is not held square to the column face when checking, there will not be any significant error. If a larger offset was chosen, there is more chance of a clear line of sight past building debris etc. but a poorly held staff or tape off the column face will produce apparent errors in position for the column. Whatever is chosen for the offset it should be the same for the whole site to avoid errors. If it is unavoidable to have a different offset, then the offset should be an obvious differ-ence of half a metre or a metre, changes of a small amount will be assumed to be the same as the main grid whereas large changes will be more obvious.

If space permits, the grid for a simple framed structure would be marked as an offset all around the site with survey monuments or nails in the road. When the base slab is cast this would then be transferred to the main slab as marks at each end of the gridline, as well as intermediate marks for ease of use by tradesmen.

These marks are then replicated on each floor (see Section 4.10 Control of verticality) in the same pattern as the structure goes up. These are then used by all trades. Therefore when deciding on a grid layout, some advance thinking is required to ensure grids are suitable for all trades. With everybody working from the same

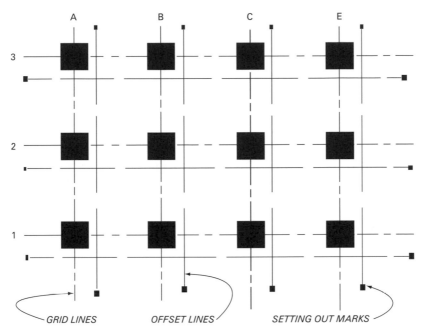

Figure 4.2 Example of a basic grid referenced and offset.

reference line, problems of fit will be minimised, and any problems will not be due to grid discrepancies.

4.5 Instruments

All instruments contain errors, nothing is perfect, but we can aim to have our instruments perform within acceptable limits by enforcing a checking and calibration programme. This is often laid down in the company quality assurance plan. We can further aim to reduce these errors in the way in which we use instruments by following observation routines. Observation routines, as well as cancelling out many errors present in instruments, can also show their magnitude. This not only gives ongoing checks that the calibration of the instrument is within acceptable limits, but also helps to give some confidence in the equipment and results obtained in using them.

A good way of ensuring that an instrument remains in adjustment is to look after it. Most instruments look fairly robust and many are even waterproof. What's more, if you drop one not too many bits fall off. However, an instrument contains delicate and fragile components capable of achieving a high degree of accuracy. Remember, a good total station costs more than the annual salary of most junior engineers! So here are a few guidelines for looking after them:

- Before removing an instrument from its box make sure you know how to return it. If the lid does not close easily, you have got it wrong! Forcing the lid back on a box is likely to damage the instrument.

- When taking an instrument from its box, close the lid. An open box fills with water, dirt and dust. This may be transferred to the instrument when it is returned to the box.
- Clamp the instrument to its tripod as soon as it is removed from the box. Do not wander about with it. Do not put an instrument on a tripod without securing it as it may get knocked off its perch.
- In moving from one set-up to another, always take a theodolite or total station off the tripod, put it in its box, and carry it in its box to the new position. Carrying an instrument on a tripod will quickly put it out of adjustment and you run the risk of collision damage. When setting out you need all the help you can get to achieve accuracy. Maltreatment of an instrument will quickly erode tolerances as the instrument will be out of calibration.
- Never leave an instrument set up and unattended. You may return to find someone has pulled a power lead, which has ruined your instrument. It will not be his fault! If you need to leave an instrument, unclip the tribrac and retain the set up. Alternatively, an instrument can be returned to its tripod once removed. As long as the legs have not been disturbed, all you need to do is level and re-centre the instrument. Better to leave an instrument on the floor rather than risk it being knocked from a desk.
- Try not to set up over air lines or electrical site power lines. These may be wrenched away and in doing so may move the instrument.
- Although many instruments claim to be waterproof, some are only showerproof. Many are neither—especially so as they age. It is therefore advisable when setting out in the rain to try to keep the instrument dry. A wet instrument must be dried before returning it to its box. A wet instrument left in a box will sweat causing water vapour to damage the instrument.
- Instruments such as automatic levels may be stored upside down in their box. This is in order for the instrument compensator to sit in its 'parked' position when inverted and to avoid damage and maladjustment when being transported. Think how instruments work and treat them sympathetically.
- It is bad practice to stand on an instrument box to achieve a better view. Such boxes are expensive to replace. Should the box contain an instrument, flexing of the lid might damage the instrument.
- Calibrating instruments on site it is often a simple but delicate task. This should be carried out at leisure and only by an experienced person. It is very easy to end up with an instrument further out of adjustment than before you started. The inexperienced should be wary that adjustment screws are very small and easily sheared off.
- Tripods also need care and attention. Assembly bolts will need periodic tightening. If these are allowed to loosen too much the tripod will cease to be a stable platform, and will produce observation errors.

4.6 Levelling

Levelling has been in use since early Egyptian times. When constructing the pyramids it is said they filled the excavations with water and fat. This left a level 'tidemark' around the walls of the trench as a datum. It is not necessary for us to flood the site these days. Our fat line can be achieved with an optical level, which gives a level reference plane. The height of the fat line, or *height of collimation*, is measured by reading a staff from a *Bench Mark* (BM), or *Temporary Bench Mark* (TBM) of known value: this is known as levelling, and is the means by which we control the height of any construction project.

All levels are measured from a datum level. The datum usually used is the ordnance datum (OD) which is derived from mean sea level as measured at Newlyn in Cornwall, UK. A local site datum or a GPS derived datum may also be used. Datum values measured to the ordnance datum are noted as above ordnance datum (AOD). You may also see 'RL' meaning reduced level. In mining and tunnelling, in order to avoid negative values, the datum value will have, for example, 100 m added to it. On site, the site datum will be established as part of the site control. If that task falls to you, it will be necessary to find at least two OS (Ordnance Survey) datums and level from them to the site and back. It is always advisable to find more than one as a check. This will confirm you have found the datum to be correct and will identify any relative error between them. Should the error between the chosen datums be large it would be advisable to check against a further datum. If your site is to tie in to another (e.g. a motorway contract split into sections), you will also need to check compatibility between sites.

4.6.1 Equipment

Levelling by conventional means is usually carried out using a dumpy type level, tripod and levelling staff. The level will probably be automatic i.e. with a compensator which keeps the line of sight horizontal once the instrument has been levelled using its circular bubble (see Fig. 4.3a and Fig. 4.3b).

Other levels in common use

Digital electronic levels that read a bar-code staff. These operate using the same principle as optical levels, and can be used in either digital or optical mode. When used in the digital mode a bar-code staff is read to 0.1 mm, or better depending on the instrument, and stores the results in a level book format also recording distances to the staff. Results, as well as being calculated by the level as work proceeds, can be downloaded into a computer. This type of level can be very efficient for taking many readings e.g. when monitoring or surveying. Downloading into a computer also saves time and reduces input errors. These instruments are not suited to all conditions, particularly when there are obstructions to a clear line of sight. Being an electronic instrument, they also require a battery.

Water levels, although very low-tech, can be very useful when working in confined spaces or to allow tradesmen to transfer their own levels, e.g. brick courses or footing levels. They can also be used where other instruments may not be suitable e.g. when transferring a level through an aperture between two tunnels about to connect, or on a scaffold which is too unstable as a base for an instrument. Water levels are most suitable for transferring a level to another point, e.g. finished floor levels around a room.

Rotating laser levels are often used by tradesmen. A laser level produces a level plane of laser light (see Section 4.11 Construction lasers). Levels are then transferred in the same fashion as with optical levels. Once set up (and occasionally checked by an engineer) they can easily be used by tradesmen without constant intervention by an engineer. They are of particular use to bricklayers, concretors and those installing

Figure 4.3a   Engineer's automatic level.

Figure 4.3b   Standard E pattern staff.

false ceilings who can use the laser plane as a benchmark from which to measure. Additionally engineers working alone can make use of such devices.

4.6.2   Basic principles

An optical level provides a level line of sight to an accuracy that is related to the quality of the level. A level plane can be achieved by rotating the level. In transferring a level from the site datum the level needs to be set up in a position from where the datum is visible. A set-up approximately half way between the datum and the point to be levelled is ideal as this will minimise instrument errors.

Knowing the instrument height, the level of other unknown points can be recorded. The staff reading at point A to be surveyed is 1.003. If this is deducted from the height of the line of sight (i.e. the height of collimation), the level of point A is calculated as 11.234 − 1.003 = 10.231 m RL (see Fig. 4.4).

If another point is required further away, knowing the height of point A, the instrument can be moved to set-up 2 and the procedure repeated.

Now that point A is of a known height, the procedure can be repeated to establish the height of point B. This time the reading onto the known point (called the backsight) is 0.457 so the height of plane of collimation (HPC) is 10.231 + 0.457 = 10.688. The staff is now placed on point B and read. This is called the foresight. The foresight reads 1.538 so the reduced level of B is 10.688 − 1.538 = 9.150 m RL (see Fig. 4.5).

This procedure can be repeated for as many points as necessary. To transfer a datum it may be necessary to have several set-ups. When the points levelled are used only to transfer the level over a distance of several set-ups, and are not required in their own right, these are called change points. For every reading that is taken there will be a small amount of error will be incurred. The more set-ups taken away from the benchmark the more error will creep in. For this reason it is necessary to 'close'

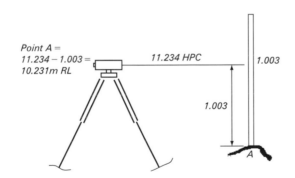

Figure 4.4 Height of collimation.

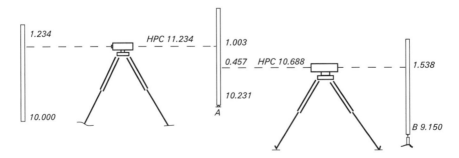

Figure 4.5 Levelling calculation.

the levels back to either the start point, or another point of known value. This acts as a check on the accuracy of the levelling, and ensures that no gross errors have been incurred e.g. misreading the staff by 100 mm. An acceptable closing error will depend on the task in hand.

4.6.3 Intermediate sights

A series of readings taken from a single set up, are called intermediate sights. Each reading is taken in turn from the HPC (height of plane collimation).

4.6.4 Inverted staff readings

A levelling staff can be used to obtain readings from above and below the HPC (height of collimation). For example, to take the level of a ceiling soffit, invert the staff so the zero is on the point to be measured, and read as before. When calculating levels, inverted readings are treated as negative values, i.e. you add the staff reading to the HPC to get the reduced level of the point. When booking these readings it is essential to identify that they are an inverted reading so that it is not forgotten when it comes to calculating the result. Conventional practice is to put a ring around the reading.

4.6.5 Booking readings

It is essential to book readings correctly for the following reasons:

• When working at speed and under pressure it is important to have a correct booking format, as this is when a mistake is most likely to occur.
• It is easier to calculate large numbers of readings at one sitting.
• If an error occurs, the levelling can be re-checked in a logical fashion.
• Others can follow your work in the event of a dispute or an audit.
• There are two main booking formats for levels on site, the 'Rise and Fall' method and the 'Height of Collimation' method. As the height of collimation is the most commonly used method (and being slightly easier to grasp), this method is covered here.

4.6.6 Common sources of error

It is assumed that you have correctly levelled the instrument on firm ground and that the instrument has not moved during reading. However:

• the instrument may be out of adjustment
• the staff may not have been held sufficiently upright
• the staffman (often somebody close to hand) may not have held the staff on the correct point (this type of error can lead to the levelling loop having an acceptable closing error but the whole loop being at the wrong level)
• the staffman may have failed fully to extend the staff.

Most of the errors come from the staffman, so it is essential to ensure the staffman is correctly briefed and a firm eye is kept on what he is doing. Always ensure the staff is plumb (see Fig. 4.6).

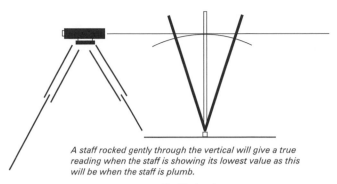

*A staff rocked gently through the vertical will give a true
reading when the staff is showing its lowest value as this
will be when the staff is plumb.*

Figure 4.6   Staff plumbness.

4.7   Use of theodolites and total stations

Many types of theodolite and total stations are available, no two are alike. Electronic instruments are just an electronic version of older (now scarce) mechanical equipment. Although each one is different in appearance and software, they are all basically very similar (see Fig. 4.7a–c). Some manufacturers ensure that software is similar, not only between models of instrument in the range, but also similar to compatible systems such as GPS. A theodolite measures both horizontal and vertical angles. A total station measures horizontal and vertical distances as well as angles. Total stations usually incorporate useful software and most can receive data from a computer. Once set up they are capable of calculating their position from known points and, in addition, can calculate bearings and distances for points to be set out.

Some instruments have motor-driven directional telescopes that can if required lock onto the prism pole and take a reading. For example, this could be employed on a survey of a wall into which your building has to fit. After the start point and stop point have been entered, the instrument will then survey at your specified intervals and record all readings. Readings can then be downloaded into a CAD package for further plotting work. Technology advances at a startling rate. Whilst this relieves the hard-pressed engineer, it is important to understand the basic principles of changing techniques. A blind acceptance of the capabilities of a new instrument or piece of software is an invitation for trouble. Inevitably there will be a time when for some reason it gives you results that are not correct. Huge errors are easy to spot. The nearly right results are the hardest to find, and those most likely to go unnoticed until its too late.

When using a total station for the first time, do not be daunted by its high-tech capability. Remember it is just a theodolite which measures distances, and it has a built-in calculator. Until used to it and all its functions, re work calculations longhand as a check. This will help you to understand the way in which the equipment is meant to work. Without experience it would be possible, for example, to mistake the order of the Eastings and Northings on the screen.

Setting out with an instrument on the ground is similar to setting out a technical drawing. To commence you need a reference line from which to work. The coordinates of the points you wish to set out can then be calculated relative to this reference line. The angle and distance on the page would be measured with a protractor and

Figure 4.7a   Views of a Leica Total Station with reflectorless EDM.

a scale; on the ground they would be measured with a theodolite and a tape or by electronic distance measurement methods (EDM).

4.7.1   Electronic distance measurement

EDM operates by sending an electronic beam from a telescope to a reflector or reflective surface, which then returns the beam back to the instrument so that it can calculate the distance to the centre of the prism pole or surface concerned.

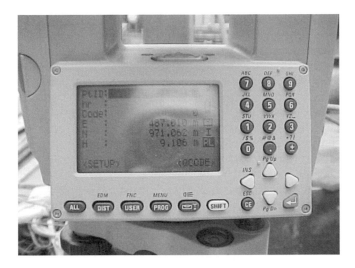

Figure 4.7b Leica Total Station keyboard and display. The display can be changed to show a variety of information i.e. angles and distances. Other makes are similar.

Figure 4.7c Leica target incorporating Tribrac (bottom), optical plummet section. In this instance it is a laser optical plumment which shines a laser onto the floor, prism and target plate.

How accurate is it? Most, but not all, instruments display distances to the nearest millimetre. An EDM will have a quoted accuracy of, for example, 3 mm ± 3 ppm. That indicates that distances will be accurate to 3 mm plus another 3 mm for every 1000 m (or part of) of displacement from the instrument. This accuracy can be rapidly degraded if good practice is not followed. For example, if a prism pole is held out of plumb, then tolerances will be quickly eroded. If an EDM signal is too close to an obstruction the signal can be disrupted giving poor results. EDM is also sensitive to temperature and pressure changes. Current values need to be input to ensure the instrument applies the right corrections to observed distances: the longer the line the more adverse the effects become.

Prisms are not necessarily interchangeable: some prisms have a different prism constant. Using a substitute prism may give incorrect readings causing a constant but systematic error. If in doubt carry out a check over a known distance.

The instruction manual will identify how a particular instrument works. Many instruments also have a calibration routine, which allows you to follow a set procedure that identifies and corrects for instrument errors. The more basic instruments will need to be calibrated by using traditional calibration tests. Older mechanical instruments can sometimes be adjusted quite simply once the calibration error has been identified. Instruments should not be randomly adjusted and such adjustments should only be attempted by qualified personnel. Instruments are precise pieces of equipment, in the hands of an inexperienced operator they will give inaccurate results. Calibration screws are fragile and may be easily damaged, so should be handled with care.

4.7.2 Instrument errors

To minimise instrument errors, when reading angles always read on both faces of the instrument. Observe your backsight (your reference target or base line end) and record the angle, e.g. 00 00 00. Then, observe the foresight (the target to be observed or unknown point) and record the angle, e.g. 90 18 04.

To change face you flip the telescope over, and turn the instrument 180° to point back to the foresight, re-observe the target (your reading should be approximately 180° different to the first reading) and record the reading, e.g. 270 18 10. Then, point back to the backsight; observe and record the reading (which should also be about 180° different to the first reading), e.g. 180 00 08.

The value for the angle will be:

Face left: 90 18 04 – 00 00 00 = 90 18 04
Face right: 270 18 10 – 180 00 08 = 90 18 02
The two readings are then averaged to give the angle: 90 18 03.

The amount the instrument differs from face left to face right for the two readings indicates the size of the face error (in this case about 6–8 seconds). Some of this error might be due to poor setting up. For accurate work several *rounds* (angles read on both faces) of angles need to be taken.

The following will also minimise instrument errors:

- Read distances to EDM targets on both faces when doing work of a extreme accuracy. An instrument resolves the slope distance observed to a horizontal distance using the vertical angle. If a large face error is present, this can have an effect on observed distances.
- Check the optical plummet regularly. The optical plummet is situated in the top portion of the instrument, and can be checked by setting over the mark, turning the instrument through 180° and looking through the optical plummet again. You should be still over the mark, if not, the distance off the point is twice the error. Half this error can then be taken out by repositioning the instrument over the point halfway towards the point over which you are centering.

- Ensure the instrument is level. Level the instrument using the long plate bubble, turn the instrument around 180° and recheck the bubble. The bubble should be in the same place, if it is not, the displacement represents twice the bubble error. Re-level the bubble eliminating out half the error.

4.8 Linear measurement

In order to set out it is necessary to measure distances accurately.
Methods available include the use of:

- pocket tape 3 m or 5 m—steel class II
- 30 m tape—steel class I or II (fibreglass or fabric)
- land surveyor's chain—for surveys only
- steel band
- total station or theodolite mounted EDM
- hand held EDM.

It is important to choose the correct equipment for the task and to be aware of its limitations. Although EDM is an excellent technique, it is not suitable for measuring very short distances: these are better measured using a tape. Likewise a tape is not the best equipment for measuring 300 m across a littered construction site.

4.8.1 Pocket tapes

A pocket tape will probably be no better than class II. The class is usually identified near the end of the tape by a roman II ringed by an oval. Do not use anything less than a class II tape for setting out (see Fig. 4.8).

Figure 4.8   Tape showing class 2 (II) mark and temperature at which the tape is designed to be correct.

4.8.2 30 m tapes

For accurate measurement, use a class I tape. As with the pocket tape, this is identified by a roman I inside an oval near the start of the tape. Class I tapes have a blank end section so the zero mark can be lined up with the point to be measured. This prevents errors generated by not picking the correct starting point for tapes, which are graduated from the end. Ensure that your chainman or assistant correctly holds the end of a tape to remove the possibility of error. Class II tapes are identified in the same manner but have their zero at the end of the tape rather than a blank section. Some engineers prefer to work from the 1 m mark to avoid *end* errors. If in doubt, double check.

Unclassified tapes should not be used for accurate work. Fibreglass and fabric tapes should really only be used for very approximate work.

Steel bands, although not commonly used, are capable of high degrees of accuracy.

4.8.3 Sources of error

Likely sources of error include:

- Reading errors.
- Tape calibration errors—tapes may be stretched or damaged; steel tapes bend and kink easily. All these factors may affect on the length of a tape.
- Temperature effects—a tape is generally calibrated at 20 °C. Should temperatures in use vary considerably form the calibration temperature then suitable corrections will need to be made. The length of a steel tape will vary by 0.1 mm/m per 10 °C.
- Tension—too much or too little tension will give incorrect readings. The calibrated tension is marked next to the class of the tape and is applied with a spring balance.
- Slope—you know, from basic trigonometry, that if you are measuring on a slope, you will not be measuring a true plan distance. You can correct for this by measuring the angle of the slope and calculating the horizontal distance. Alternatively you can measure the slope in a series of short horizontal steps. Complicated taping exercises should be avoided, if possible, due to the potential for errors.
- Measuring distances by tape should not be attempted for distances of 10–30 m, depending on the surface measured over. Correct tension should be applied.
- EDM is very useful for measuring most distances on site, but when measuring close to the instrument it can be inaccurate due to the steep inclination of the telescope. Short distances should be checked with a tape.

4.9 Control of verticality

Control of verticality will be necessary in a number of situations. For example in setting out vertical formwork or transferring levels up successive floors of a multi-storey building. There are several methods for dealing with this, which include:

Plumb-bob

A string or cord line attached to a heavy weight will, under gravity, hang vertically and provides a quick and reliable method of checking or transferring a plumb line. Such a technique is applicable to plumbing a lift shaft or in other indoor situations where the performance of the line will not normally be influenced by environmental factors such as wind. Moderate oscillations of the plumb line can be counteracted by damping the plumb-bob in water or oil. If oil is used it should not be disposed of into the drainage system.

Where the height to be plumbed is considerable, wire may be substituted for cord. Care should be exercised when lowering a plumb line that the safety of those working below is not compromised. It may be prudent to use a small weight to initially establish a line and then replace the weight by a heavy bob for final plumbing.

Spirit level

This is only suitable for small-scale work, for example in checking door frames or concrete formwork. Where a spirit level has been used for an approximate check, more sophisticated methods should be used for a final check.

Theodolite

This instrument can be supplied with a diagonal eyepiece to enable the user to look vertically upwards. With such an attachment, the theodolite can be used for checking the alignment of columns and in many other similar applications. By using a tape in conjunction with the theodolite it is also possible to measure the extent to which a member is out of plumb. The ease with which the telescope of a theodolite can be set to move in the vertical plane makes it a very useful instrument for use in this type of work.

Optical plummet

An optical plummet is an instrument which sights directly up or directly down. It houses an automatic compensator (similar to an automatic level) which gives the instrument greater accuracy than is achieved using a theodolite with a diagonal eyepiece. Designed for the purpose, the optical plummet will achieve excellent results (see Fig. 4.9).

4.10   Coordinate calculations

Coordinate references, although on the face of it quite complex, are in reality very easy to use. On some sites the site grid will be tied into the Ordnance Survey (OS) national grid. This may be due to it being a road, railway or some other large project that needs controlling on a large scale both before and after the construction phase of the project. The grid on the site may be little different from map grid references used when reading an ordinary OS map. More often than not though, the site grid will be a local grid designed to be on the same orientation as the main column layout or some other major feature of the project. This is done for ease of use for dimensioning the works. It is useful to note that jobs designed using a CAD system, are able to provide coordinates at the click of the button. On jobs where, for example, you have 500 piles to set out all referenced from some obscure point on a skew grid, it is worth making a formal request for the information in coordinate form. Armed with coordinates, a site control network with coordinated values and an instrument

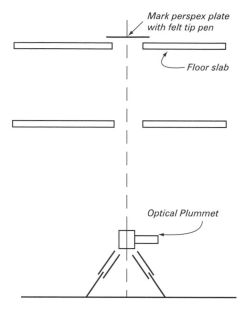

Figure 4.9   Optical plummet.

which calculates setting out, you can make an immediate start and set out points just by typing figures into the total station.

Principles

The zero of the coordinate system is always on the bottom left of the map. With site grids, it's the same principle, although architects GAs can be drawn in another orientation altogether, often just to make it fit the page.

Eastings

These are all points of equal 'eastness' and run along the bottom of the page. It is customary to quote the Easting before the Northing. However American practice is to quote Northings first. It is advisable to note 'E' or 'N' after written figures to avoid misunderstandings.

Northings

These are all points of equal 'northness' and run up the page. In practice, avoid having grid origins so close to any work to preclude negative values. Figures should be to three decimal places indicating the desirable level of accuracy for most tasks.

Bearings

To work with coordinates we don't use angles but bearings from grid north. This is because to calculate angles may lead to errors, e.g. which side of the line was the angle calculated from? (particularly difficult to discern if the angles are very small).

4.11 Construction lasers

The idea of using a laser usually invokes thoughts of a high-tech site using cutting-edge technology. This is not entirely true. Lasers have been available for use for a long time. A laser is basically just a visible line of sight, a string line of light. A laser is classed by its power output and may be class 1, 2 or 3.

Class 1 lasers

Invisible to the eye, needs a laser detector and is the safest class of laser.

As Class 1 lasers need a laser detector or laser eye, these are generally restricted to laser level type of use. The laser is then used in the same fashion as a level, the laser eye can be attached to a *boning-rod* or staff. This can be given to the operative who is controlling the excavation, levelling a line of wall copings or other operation. They can also be used in machine control systems e.g. they can be used to control the blade of a bulldozer. The engineer will however still need to set up the instrument and calculate rod heights etc.

Class 2 lasers

Visible line of light. These are eye-safe in so far that the eyes' natural blink reflex time is sufficient to protect the eye. The beam must not be observed through an instrument, or stared at.

Safety precautions: terminate beam at the end of its useful length (not a rule but advisable); avoid where possible setting up at eye-level.

Class 2 lasers emit a visible line of sight, which make them very useful for a variety of tasks. These also come in the form of laser levels, but some can also be used to form vertical and inclined planes. This can be useful for setting wall lines, or ramps. A staff or stick used in conjunction with a laser can be marked with a variety of points to indicate, for example, the bottom of a dig followed by the top of the bedding and then the top of the paving. The advantage over the Class 1 laser is that a laser detector is not required. It is also possible to get a laser eyepiece attachment for some instruments. This can be used to transform a visible line of site for taking measurements and setting out. An example of this would be to set an instrument on the centre-line of a tunnel. A laser can then be shone through the instrument to give the tunnelling gang a line to follow. With a string line, this would require regular resetting. Class 2 lasers are also used in pipe lasers for setting out drainage. The laser is set up in a section of pipe (at a manhole) and the line and grade set for the drainage gang to follow.

Class 3a lasers

A visible line of light but of higher output than Class 2 lasers. These should not be used where a lower class of laser will suffice due the increased risk to the eyes.

Safety precautions are as Class 2 lasers, except that the termination of the beam at the end of its useful length is a requirement.

Class 3a lasers have the same function as Class 2 lasers but are more powerful. In some instances (such as on long lines of sight, in poor dust laden atmosphere or where a screen to detect the laser is used), more power may be necessary.

There are higher classes of laser (for example Class 3b) but these are too power-ful for normal use on site due to the risk of eye damage. Consequently the strict safety measures that need to be in place hamper their use.

4.12   Setting out simple buildings

Simple buildings can be set out in a variety of ways. Consider, for example, a simple rectangular domestic house. The first priority will be to set out the foundations. It may be sufficiently accurate to set out basic excavation dimensions by taping from exist-ing work. However, it is preferable to carry out a dry run using an accurate method of setting out. This will flag-up any unexpected calculation errors or problems in obtain-ing lines of site. A satisfactory sequence of events might be as follows:

- Mark out, from the main site control, the four superstructure corners of the build-ing. This can be done using coordinates or grids depending on the type of site control in use.
- Carry out the necessary checks on diagonal dimensions. If unacceptable errors are found then re-work the original setting out until sufficient accuracy is established.
- Make a by-eye check to see that the general location is correct.
- Recognise that the corner marks will soon be lost in construction and that offset lines need to be established and identified by pegs or concrete monuments. These lines should be close to the building lines so that tradesmen can measure short distances for their work.
- Once a concrete floor or raft has been established, setting out lines can be trans-ferred to the concrete surface using lines marked on spray paint.
- Where concrete strip footings are being used, check that the subsequent masonry is properly positioned on the footing. Failure to do this may result in foundation malfunction due to an eccentricity of the load.
- From existing setting out lines, provide bricklayers with setting out for base courses. A theodolite may facilitate this work.
- Use recommendations in Section 4.6 if it is necessary to provide setting out data related to floor levels.
- At all stages carry out simple checks to confirm that setting out is correct.
- Bear in mind that once the lower courses of masonry have been constructed, follow-ing tradesmen will probably use this to set out the remainder of the construction. It is, therefore, essential that the initial setting out is to an appropriate standard.

When setting out rows of houses it is essential to work *from the whole to the part*, (see Section 4.4.4) i.e. from gridlines, not house to house. Where engineers have worked from *one end* it has been reported that the last house in a row could not be accommodated on the remaining plot!

4.13   Satellite positioning

A 1960s surveying text book consulted in 1980 would reveal little change in twenty years. That is not true today, with modern technology, systems and software are being continually updated. Nowhere is this more obvious than with satellite posi-tioning. It has exploded onto the construction market changing some operations beyond all recognition. As technology improves, accuracy increases and costs come down it becomes more economical to employ it on smaller and smaller jobs.

What is it?

Satellite positioning is the determination of the position of a point using a satellite receiver. Satellite positioning is generally known as GPS or *global positioning system* after the American military system, which was first available for public use. Unlike most surveying and setting out tasks, the skill required of the operator is minimal. The skill with GPS is with the management of the system's input and output data. The satellite receiver does all the work in gathering the data and outputting or storing it as required. With setting out it can provide the operator with predetermined setting out coordinates.

How accurate is it?

Accuracy depends on the methods employed and the equipment used. For construction setting out centimetre level accuracy is achievable. This makes it suitable for many setting out tasks. Unlike traditional survey methods, each point is independent of the points around it, and therefore each point is of a similar accuracy. Degradation of accuracy (due to creep) with distance from the main station is no longer a problem. If used in unsuitable conditions, accuracy may be compromised. An error in one point is not passed on to adjacent points.

What are the advantages?

When used for setting out, a single engineer with a setting out pole equipped with a satellite receiver can set out points almost as fast as he can mark them. With a road centre-line for example, the operator can walk the route and mark centre-line points at whatever frequency is required. The setting out information can be taken straight from the design on disk without the need to input a mass of figures. Work is unaffected by weather or daylight or a lack of it. Visibility between points is not required, so local obstructions (shrubbery, mechanical plant, low buildings, walls etc.) do not hinder the process. Productivity increases are considerable. As well as giving plan coordinates (Eastings and Northings), it will automatically provide heights as a matter of course. Satellite systems can also be integrated into computer-controlled plant, in which, for example, a grader has the road design in its memory. The grader blade is automatically adjusted to give the correct earthwork profile. This eliminates the need for a complete setting-out team along with their instruments, forest of timber-work, chainmen and their transport.

What are the disadvantages?

Cost is always an issue, but this has to be balanced against productivity. GPS is not suited to all locations. Due to the fact the position of the receiver is derived from observing a number of satellites, a clear view of the sky is necessary. This may make GPS unsuitable for city centre sites shielded by adjacent tall buildings. A received signal may give inaccurate results if deflected off the side of a building.

GPS is not suitable for tunnelling work. However GPS can be used very efficiently to establish a control either side of an obstruction under which tunnelling is required.

GPS does not work well in tree-covered areas, again due to the need for a clear line of sight to the sky.

The height element of the output is of a lower order of accuracy than the plan coordinates. Additionally, heights given are not above mean sea level (as with traditional levelling), but above the mathematical model of the Earth, WGS84 (World Geodetic System 1984). Unfortunately, for Europe this does not run parallel to mean sea level. However the GPS output can be configured to give correct information.

GPS is not sufficiently accurate enough to obtain the 1 mm precision that can be achieved with a theodolite.

Checklist

- Is the site suitable for using GPS?
- Are there obstructions to the horizon?
- Can design information be obtained on disk?
- Can the GPS system accept design information in the given format?
- Is the output configured in such a way as to give heighting information that is correct for the site? If necessary check known benchmarks using GPS.
- Will GPS give the required accuracy for the task?

Further reading

Standards and Codes of Practice

BS 5964: 1990 *Building Setting Out and Measurement. Methods of Measuring, Planning and Organisation and Acceptance Procedure.*
Note: This Standard is in several parts and deals with such topics such as measuring stations, targets and checklists for the procurement of surveys and measurement services.
BS 7334: 1990 *Measuring Instruments for Building Construction. Methods for Determining Accuracy in Use.*
Note: This Standard is in several parts and deals with such topics as theory, measuring tapes, optical (including plumbing) instruments, theodolites, lasers, procedures for setting out using a theodolite and steel tape, and the use of electronic distance measuring equipment up to a range of 150 m.
BS 5606: 1990 *Guide to Accuracy in Building.*
BS 6954: 1988 *Tolerances for Buildings.*
Note: This Standard is produced in several parts and deals with such topics as basic principles, recommendations for statistical basis for predicting fit between components having a normal distribution of sizes, and recommendations for selecting target size and predicting fit.
BS EN 30012–1: 1994 *Quality Assurance for Measuring Equipment.*

Other related texts

Setting Out on Site: D.W. Quinion, Civil Engineer's Reference Book 4th Edition, Butterworth, 1989.
ICE design and practice guides: *The management of setting out in construction*: J. Smith (ed.), London, Telford, 1997.
Setting Out on Site (Video): CIRIA Report SP86, London, CIRIA, 1991.
Surveying and Setting Out: F. Bell, Aldershot, Avebury, 1993.

Chapter 5   FLEXIBLE ROADS AND OTHER PAVED AREAS

Robert Langridge

Basic requirements for best practice

- Adequate site investigation and specification
- Adequate drainage gradients
- Competent site supervision
- Correct aggregate properties
- Independent on-site quality control
- Trial areas of work
- Systematic rolling patterns
- Good compaction
- Accurate compacted layer thickness

5.1   Introduction

This chapter provides an insight into the construction of flexible roads and other paved areas. The aim is to provide information for people concerned mostly with job specification and site management, and to help avoid some common problems that may be identified during and after the construction of flexible roads and other paved areas.

Asphalts and macadams used in the construction and maintenance of flexible roads and other paved areas consist of a mixture of aggregate and binder. The mixtures to the required specification are usually produced in batches in a manufacturing plant. One or more batches of mixture form a lorry load. Lorries transport the mixture to the point of laying.

For new flexible construction and for the maintenance of roads and other paved areas, some or all of the basic requirements listed above may need to be considered by the specifying authority:

5.2   Flexible construction

In many countries, asphalts and macadams using bitumen or modified bitumen binders and often referred to as bituminous mixtures, have been in increasingly widespread use for roads and other paved areas since the latter part of the
nineteenth century. During this period, much research into asphalts and macadams and their constituent aggregates and binders, and into their uses, has been
undertaken. An example of this work has been the development of stone mastic
asphalt.

Stone mastic asphalt had been developed in Germany and was in use in that country for many years before its introduction into the United Kingdom in the1980s. At
first, in the United Kingdom a number of road trials were carried out. In these trials
the gradings of the mixtures had sometimes been modified, for example to improve
texture depth. Different types and grades of binder and additives were tried. Stone
mastic asphalt can be laid in thin layers, for instance a 15–20 mm compacted thickness is possible with 6 mm size aggregate grading for roads carrying a low volume
of traffic.

It can be very durable and resistant to age hardening. It can have a low void
content and the added fibres or polymers permit a relatively thick bitumen film on
the particles of aggregate. A stable structure of many interlocking aggregate particles enables stone mastic asphalt to be very resistant to deformation. Stone mastic
asphalt has only a small reduction in volume during rolling, so it may be laid over
rutted or uneven surfaces which are otherwise sound.

On roads, a very important advantage of stone mastic asphalt, is the marked
reduction of tyre noise. There are some disadvantages to stone mastic asphalt and
these include—where good skid resistance is required—a high quality aggregate
for the full depth of the layer; the need for additives; high bitumen content, and a
high mixing temperature. However, the advantages of stone mastic asphalt would
in many instances outweigh the disadvantages significantly. Proprietary mixtures of
stone mastic asphalt are available from a number of UK manufacturers of surfacing
materials.

When bituminous mixtures for surfacing roads and other paved areas have been
well specified and correctly manufactured, and the workmanship on site is to a high
standard, they should give long trouble-free service. The potential for a long service
life of roads and other paved areas is, for example, a very important environmental
consideration.

Flexible roads and other paved areas are constructed in layers. To perform
properly, each layer should be fully compacted to the specified thickness and be
to the specified levels. Unbound materials, for example Type 1 sub-base, may
require added water to assist compaction. For hot laid bituminous mixtures, proper
compaction is only possible when they are generally above the minimum rolling
temperature for the particular mixture and are workable. Compaction should commence as soon as possible, but without undue displacement of the mixture. Proper
compaction of warm or cold laid bituminous mixtures is also important. No roller
marks should remain in the compacted mixtures.

Layers of bituminous mixtures should be well bonded to each other. The layers
of construction and the sequence in which they are laid may vary.

The following is a guide:

Capping layer

A capping layer might be required where the subgrade has a California Bearing Ratio (CBR) of less than 5%. The CBR test is described in BS 1377.

Sub-base

At formation level, which may be the top of the subgrade or the top of any capping, the sub-base is laid. This may be an unbound granular material, either Type 1 or Type 2. Type1 sub-base is a crushed rock or other crushed hard material complying with certain grading requirements. Type 2 sub-base may consist of sands and gravels and other permitted materials. The grading of Type 2 material may be smaller sized and finer than Type 1. For Type 2 material, the plasticity index on the fraction passing the 425 micron BS sieve, when tested in accordance with BS 1377, may be up to 6%. For the sub-base of some less heavily trafficked roads and paved areas, comparatively lower quality materials might be acceptable if they meet certain criteria, and these include, for example, quarry waste, hardcore and hoggin. As an alternative, a soil cement sub-base might be acceptable. Where the subgrade is suitable this may be stabilised using a mixed *in situ* method, or ready mixed material may be brought to site. The sub-base has several functions, three of these are to provide some protection to the subgrade; form part of the structural design, and be a working platform on which to lay the remainder of the construction.

Roadbase laid on the sub-base

Bitumen bound roadbases should be manufactured and laid in accordance with either BS 594 *Hot Rolled Asphalt for Roads and other Paved Areas*, or BS 4987 *Coated Macadam (asphalt concrete) for Roads and other Paved Areas*. Base mixtures should be strong, dense, able to resist cracking, flexible and have good load spreading properties.

Binder course laid on the base

The binder course is bitumen bound and provides a surface on which to lay the surface course. It provides part of the structural strength of the construction. If the surface course is porous asphalt, water should drain across the surface of the binder course.

Surface course

This has to perform many functions. Generally, it must be:

- skid resistant
- resistant to abrasion
- resistant to indentation
- resistant to deformation

- resistant to water
- able to retain good riding quality.

5.3 Some design and specification considerations

Bituminous mixtures when laid and compacted, and during use require good support. A site investigation survey should be made. The types, thicknesses and number of layers of construction required will depend on the type of subgrade and the purpose of the finished work. For high-grade roads such as motorways or trunk roads the Highways Agency (HA) will provide very specialised specifications.

Freedom from pools of water on the finished work is likely to be very important. Requirements for the values of gradients are given, for example, in BS 4987 and BS EN 12056 Part 3. Small deviations in level during construction or, subsequently, minor settlement may lead to large areas of standing water.

Drains should be located at the low points of gradients. On existing surfaces which are to be overlaid, a level survey should be made at sufficiently close centres to enable the gradients to be calculated and compared with the relevant British Standard. Where required, any reduction in the level of a surface, or making up with a regulating course, should be undertaken before the new surface course is laid.

Will the construction materials be hand laid or machine laid? Bitumen bound mixtures should, whenever possible, be machine laid. After spreading, particularly by hand, hot mixtures may cool to an unacceptable extent before compaction is complete. The rate of cooling will depend, for example, on the thickness of the layer and on the weather conditions. To facilitate hand laying, a volatile flux, generally kerosine, to temporarily soften the binder is sometimes added to the mixture. Not very much of the flux is likely to evaporate during cold weather. The rate of evaporation of the flux from a close graded or dense mixture is likely to be slow at any time, and can result in the laid material becoming soft in warm weather.

Where kerbs are installed before any of the road construction, it is important that these are to the correct line and level. Kerbs may be used to determine the levels and profiles of the layers of road construction.

Weed killers may be required, possibly on existing surfaces and should be applied in accordance with the manufacturer's instructions. Weed killers are toxic and should not be allowed to contaminate surfaces that are to remain exposed.

Mixtures made with bitumen may be attacked by oil. Some bitumens may be oil resisting. For oil-resistant surfacings bitumen/epoxy resin binders are available. This type of mixture may have a high bitumen content and be very impermeable to water.

Where resistance to skidding is important, an aggregate with the required minimum polished stone value and suitable in all other respects should be selected.

Coloured pigmented surfacings are available. Those having a bitumen binder are generally restricted to shades of red or green. Lighter coloured mixtures may require a clear resin binder. Unless a suitably coloured aggregate is used, the colour may be lost when the coloured binder film has worn away.

5.4 Requirements for good quality work

- At the outset, allow for the cost of independent specialist advice, followed by frequent independent routine checking and testing of the work for compliance with the specification.
- Ensure that the supervising staff have been suitably trained and are competent.
- Ensure that drawings are complete and show correct levels to enable the required gradients to be installed.
- Specify *Hold Points* to enable inspections of the work to be made jointly with the sub-contractor at least once per day. An example of a checklist for inspection of sub-base, base, binder and surface course which may be used is given in Section 5.6 (Fig. 5.1).
- On new work, gradients should be formed in the subgrade.
- Ensure that surfaces (substrates) are suitable to receive each layer of construction. The following should be ensured:
  - the formation is to the correct levels and properly compacted—any soft places have been excavated and made good
  - the capping layer, if any, is suitable material and complies with the specification, and has been laid and compacted in accordance with the specification
  - the sub-base is the correct type, and has the required grading and moisture content to allow proper compaction
  - all gullies and manholes etc. should be in place and to the correct levels—however, for practical reasons it is sometimes necessary to complete the installation of these items after the surface course has been laid
  - good access.
- Where the binder course is not newly laid, the surface should be thoroughly cleaned if necessary using water jetting, followed by a tack coat of BS 434 K 1–40 bitumen emulsion. The tack coat should be applied evenly at a rate of between 0.3 to 0.4 litres per m^2. Avoid ponding in hollows. Allow the emulsion to turn from brown to black before laying the surface course.
- All vertical faces abutting the surface course should be cleaned and painted with a thin uniform coating of 50 pen or 70 pen hot bitumen or cold thixotropic bitumen compound of similar grade.
- For the surface course, unless the *in situ* material is at working temperature or has been suitably heated, all lane and transverse joints should be cut to form a vertical face. All loose material should be removed, if necessary using an air line. The cleaned vertical face on the surface course should then be painted with a thin uniform coating of hot 50/70 pen bitumen or cold thixotropic bitumen compound of similar grade.
- Bitumen may be damaged by overheating or prolonged heating. If hot bitumen is used for painting joints, it is suggested that the temperature of the bitumen should not exceed 190 °C. An example of a record of temperature readings is given in Section 5.6 (Fig. 5.2).
- For the surface course, all joints should be well compacted, flush and of neat appearance.
- Overpainting of finished joints with bitumen should not be permitted.
- The surfacing contractor should supply a project-specific test and inspection plan detailing how they will ensure that the materials and workmanship on site comply

with the specification. Details of supervisory staff, including their experience, job description, qualifications and frequency of proposed site visits should also be supplied.

• At the commencement of the surfacing work, lay a trial area which includes all type conditions. The location of the trial area should be agreed with the main contractor at the pre-order meeting and be marked on a site plan. If the trial area is acceptable, it may be included in the works.

• Full implementation of required health and safety procedures.

5.5   Quality control

Good on-site quality control of the surfacing and associated work is essential. It should help, for example, with keeping to the programme by avoiding as far as possible any need to redo work and with achieving long-term client satisfaction. Quality control should be the responsibility of all concerned with the works. On every aspect of the work, close cooperation between the designers and the main contractor, and between the main contractor and the surfacing and groundworks subcontractors and consultants should be established as soon as possible.

Bituminous mixtures may be damaged, for example by overheating or prolonged heating. Conversely, if the mixture is not hot enough, proper spreading and compaction may not be possible. The temperature of the mixture should be checked and recorded within 30 minutes of the arrival of the delivery on site. An example, of a record sheet: 'Record of delivery and rolling temperatures for base and surfacing mixtures' is given in Section 5.6 (Fig. 5.3). The thermometer to be used for checking the temperature should be of the electronic type and have a probe length of at least 300 mm. Take approximately six temperature readings, in as much of the load as practicable. Allow time for the cold thermometer to warm up by moving it from one place to another in the load before taking the first temperature reading. There will be some variation in the temperature of the mixture throughout the load: average the results. The specified limits for the temperature reading will depend on the grade of bitumen and the type of mixture.

If samples of bituminous mixtures are required for laboratory tests, samples should be taken in accordance with BS 598–100.

5.6   Examples of quality control record sheets

Checklist for inspection of sub-base, base, binder and surface course

Contract:_____ Date:_____ __

Inspection made by:_____ _____ Grid ref:_____

1. Weather conditions:_____

2. Gradients correct: Yes/No

3. Material in situ: sub-base/base/binder course/surface course*

4. Material to be laid/being laid: sub-base/base/binder course/
    surface course*

5. Rollers (type and model):_____

6. Surface levels within tolerance?   Yes/No

7. Compaction satisfactory?   Yes/No

8. Measured compacted thickness of layer:_____mm to_____mm

9. Evidence of diesel or other oil spillage?   Yes/No

10. Substrate clean and dry before overlaying?   Yes/No

11. Coated chippings rate of spread and embedment satisfactory?   Yes/No

12. Cooled joints in surface course properly cut back and
     vertical faces painted with hot bitumen or other approved?   Yes/No/NA

13. Freedom from pools of water?   Yes/No

*Delete where not applicable.

Signed:_____ Date:_____

On behalf of subcontractor:_____

The completed form should be handed by the subcontractor to the site agent at the end
of each day's work.

Figure 5.1   Checklist for inspection of sub-base, base, binder course and surface course.

Record of temperature readings taken in hot bitumen compound ex bitumen heater

Contract:_____ Date:_____

Temperatures recorded by:_____ Grid ref:_____

_____

	Time	Reading °C*
8 a.m. – 9 a.m.		
9 a.m. – 10 a.m.		
10 a.m. – 11 a.m.		
11 a.m. – 12 noon		
12 noon – 1 p.m.		
1 p.m. – 2 p.m.		
2 p.m. – 3 p.m.		
3 p.m. – 4 p.m.		
4 p.m. – 5 p.m.		

*Readings to be recorded once each hour during the working day.

Signed:_____ Date:_____

On behalf of subcontractor:_____

The completed form should be handed by the subcontractor to the site agent at the end of each day's work.

Figure 5.2   Record of temperature readings taken in hot bitumen compound ex bitumen heater.

Record of delivery and rolling temperatures

Contract:_____ Date:_____

Temperature readings recorded by:_____

Delivery ticket no	Type of bituminous mixture	Temperature of mixture in lorry within 30 minutes of arrival on site (°C)	Temperature of mixture immediately prior to rolling (°C)	Chainage/ grid ref

Signed:_____ Date:_____

On behalf of subcontractor:_____

The completed form should be handed by the subcontractor to the site agent at the end of each day's work.

Figure 5.3 Record of delivery and rolling temperatures.

Further reading

Standards and Codes of Practice

BS 434 1 & 2: 1984 *Bitumen Road Emulsions (anionic and cationic)* and *Code of Practice for use of Bitumen Road Emulsions.*

BS 594 1 & 2: 2003 *Hot Rolled Asphalts for Roads and other Paved Areas.* Specification for materials, transport, laying and compaction of rolled asphalt.

BS 598: 1996 Sampling and Examination of Bituminous Mixtures for Roads and other Paved Areas.

Note: This BS is in several parts dealing with such topics as methods for sampling for analysis; methods of test for the determination of density and compaction; methods for determination of the condition of the binder on coated chippings and for measurement of the rate of spread of coated chippings; and methods for the assessment of the compaction performance of a roller and recommended procedures for the measurement of the temperature of bituminous mixtures.

BS 1377: 1990 *Methods of Test for Soils for Civil Engineering Purposes.*

Note: This BS is in several parts dealing with such topics as classification tests; chemical and electro-chemical tests; compaction-related tests; compressibility; permeability and durability tests; consolidation and permeability tests in hydraulic cells with pore pressure measurement; shear strength tests(total and effective stress); and *in situ* tests.

BS 4987 1 & 2: 2003 *Coated Macadam (asphalt concrete) for Roads and other Paved Areas.* Specification for constituent materials, mixtures, transport, laying and compaction.

BS EN 12056–3: 2000 *Gravity Drainage Systems inside Buildings.* Roof drainage, layout and calculation.

Other related texts

Manual of Contract Documents for Highway Works, Volume 1: Specification for Highway Works; Volume 2: Notes for Guidance on the Specification for Highway Works: DoT, London, DoT/HA.

Bituminous Mixes and Flexible Pavements, An Introduction: BACMI/QPA, BACMI/QPA, 1992.

Development of a Performance-based Surfacing Specification for High Performance Asphalt Pavements: D. Weston *et al.*, Crowthorne UK, TRL Ltd, 2001.

Asphalt Surfacings: a Guide to Asphalt Surfacings and Treatments used for the Surface Course of Road Pavements: E. Nicholls (ed.) London & New York, Spon, 1998.

The Structural Design of Bituminous Roads: W. D. Powell *et al.*, Crowthorne UK, TRRL, 1984.

Highways: J. A. Turnbull, Oxford UK, Butterworth Civil Engineer's Reference Book 4th ed., 1989.

Standards Related to Road and Paving Materials: ASTM, Philadelphia USA, ASTM, 2000.

Chapter 6 DRAINS, SEWERS AND SERVICES

Les Richardson

Basic requirements for best practice

- **Clear drawings and specifications**
- **Proper selection of materials**
- **Close attention to bedding and haunching**
- **Properly made joints**
- **Adequate testing before back-filling**

6.1 Materials

Most underground drainage systems will be constructed using one of the following materials (note European standards increasingly apply):

- vitrified clay (to BS EN 295: 1995 or BS65: 1997)
- concrete (plain, reinforced or prestressed to BS EN 1916: 2002 and BS EN1917: 2002)
- grey iron (to BS EN 877: 1999)
- ductile iron (to BS EN 545: 1995, BS EN 598: 1995 and BS EN 969: 1996)
- glass fibre reinforced plastics (GRP) (to BS 5480: 1996)
- steel (to BS EN 1123: 1999)
- unplasticised PVC (PVC-U) (to BS EN 1401: 1998)
- polyethelene (to BS 6572: 1985, BS 6437: 1984, BS 3284: 1984 and BS 6437: 1984)
- polypropylene (to BS EN 1852–1: 1998).

The design of any underground drainage system should have taken account, amongst other things, of the following:

- ground conditions (including possible corrosion of pipes from surrounding ground and proximity of significant vegetation)
- construction method
- expected design life of system
- economics
- aggressive nature and temperature range of effluents in pipes.

It is preferable for any system to be of one type of material and to a consistent design. However if it is necessary to use more than one material, then suitable transition sections may be available from some manufacturers. Such specialist components may also be necessary to connect an old, Imperial measurement system, to a metric system. Alternatively, repair and adaptor couplings are now readily available.

6.2 Definitions

Key terms are defined as follows:

Benching: A surface at the base of an inspection chamber or manhole to confine the flow of sewage to avoid the accumulation of deposits and provide a safe working surface. The surface is sloped so that any surcharge flow runs off it.

Blinding: Material that will fill interstices, irregularities and excavated soft spots in the exposed trench bottom and, when adequately compacted, will create a firm uniform formation on which to place the pipe bedding material.

Note: Hoggin, sand, gravel, all-in, graded or single-sized aggregate or lean-mix concrete may all be used. Care needs to be taken with the use of any concrete in the base of the trench in order to avoid the creation of outstanding hard spots.

Branch drain: A pipeline installed to discharge into a junction on another pipeline or at a point of access such as an access junction, inspection chamber or manhole.

Branch vent: A ventilating pipe connected to one or more branch drains.

Cesspit: A pit formed of concrete or brickwork in which sewage is allowed to collect. The pit will be periodically emptied using purpose-made hydraulic equipment.

Crown: Highest point of the external surface of a pipe at any cross section.

Deep manhole: A manhole of such depth that an access shaft is required in addition to the working chamber.

Drop-pipe connection: A vertical connection to or near to the invert level of a manhole from a sewer or drain at a higher level.

Flexible pipe: Pipe that deforms to a significant extent before collapse, e.g. plastic.

Flexible pipeline of rigid pipes: A line of rigid pipes with flexible joints.

Formation: The finished level of the excavation at the bottom of a shaft, trench or heading, prepared to receive the permanent work, such as the pipe bedding.

Foulwater: Water discharged after being used in households or in a process. Wastewater is now the international definition.

Inspection chamber: A covered chamber constructed on a drain or sewer to provide access from the ground surface for inspecting, testing or the clearance and removal of obstructions, and usually situated in areas subjected to light loading only.

Invert: The lowest point of the internal surface of a drain, sewer or channel at any cross section.

Inverted siphon: A portion of a pipeline or other conduit in which sewage flows under pressure due to the soffit of the sewer dropping below the hydraulic gradient and then rising again.

Junction: A fitting on a pipeline to receive a discharge from a branch drain.

Lamphole: A small shaft, constructed of pipes, for the purpose of lowering a lamp into the sewer to facilitate inspection and to indicate change of direction in a sewer between manholes.

Main: Either a pipe or cable—a primary distribution system, normally located beneath an adaptable area such as a footway or service zone.

Manhole: A working chamber with cover constructed on a drain or sewer within which a person may inspect, test or clear and remove obstructions.

Nominal size (DN): A numerical designation of the size of a pipe, bend or branch fitting. Confusingly it may refer to either the internal or external diameter of the pipe according to the standard for the material.

Rigid pipe: Pipe that fractures before significant deformation occurs, e.g. clayware and concrete.

Rigid pipeline: A line of rigid pipes with rigid joints. Rarely laid now but found in drain and sewer construction prior to the 1960s.

Semi-rigid pipe: Pipe that behaves either as a rigid or flexible pipe according to diameter and thickness, e.g. fibre (asbestos)–cement, ductile iron and steel.

Septic tank: A purifier for sewage where no sewer is available. It is a tank through which sewage flows slowly enough for it to decompose and be purified. It is divided into two or more chambers separately by scum boards.

Service: A pipe or cable connection from the main to the building, usually within the curtilage of that building.

Sewage: Water-borne human, domestic and farm waste. It may include trade effluent, subsoil or surface water.

Sewerage: A system of sewers and ancillary works to convey sewage from its point of origin to a treatment works or the place of disposal.

Shallow manhole: A manhole of such depth that an access shaft to the chamber is unnecessary.

Soffit: The highest point of the internal surface of a pipe or conduit at any cross section.

Stormwater: Surface water from heavy rainfall combined with wastewater diverted from a sewer by a stormwater overflow.

Storm overflow: A device, on a combined or partially separate sewerage system, introduced for the purpose of relieving the system of flows in excess of a selected rate. The size of the sewers downstream of the overflow can thus be kept within economical limits or the flow to a sewage treatment works limited, the excess flow being discharged to a convenient watercourse.

Surface water: Water that flows over, or rests on, the surface of buildings, other structures or the ground.

Trade effluent: The fluid discharge, with or without matters in suspension, resulting wholly or in part from any manufacturing or specialist process.

6.3 General

6.3.1 Setting out

The main contractor is normally responsible for setting out the works and for the correctness of the position, levels and dimensions of the work. Check the specified requirements on your particular contract. Any site variations should be agreed with the person or authority responsible for design and recorded.

6.3.2 Safety

Comply with your company safety instructions and the Contract Health and Safety Plan. These will normally refer to the Health and Safety at Work Act, 1974 and relevant Regulations, in particular:

- Health and Safety Executive Guidance Note HS (G) 47: Avoiding danger from underground services
- Construction (Health, Safety and Welfare) Regulations, 1996
- Construction (Design and Management) Regulations, 1994
- Safety regulations of the various utility companies
- Safe Working in Sewers and at Sewage Works published by the National Joint Health and Safety Committee for the Water Service.

A fuller list of relevant regulations and advice is found in the further reading list and also Chapter 18.

In particular, the CDM Regulations, 1994, require that health and safety is taken into account and managed through all stages of a project, from conception, design and planning through to site work and subsequent maintenance and repair. Designers need to be aware that it is best to prevent any hazard and if necessary alter the design to avoid the risk.

Entry into inspection chambers, sewers and similar confined spaces is always hazardous due to possible accumulation of toxic gases or oxygen deficiency. It must always be assumed that such dangers exist and when it is necessary to enter, detailed procedures must be established (see references in further reading). Water companies in England and Wales and Water Authorities in Scotland and Northern Ireland have their own procedures which must be adhered to when entering their systems.

Typically these procedures should include initial testing of the atmosphere followed by continual monitoring during the course of the work. It is most important to maintain a safe atmosphere by providing constant ventilation. In addition to the foregoing precautions, a suitable safety harness and line must be securely held by a person stationed outside the workings and be capable of being quickly attached to a purpose-made rescue winch. Self-contained escape breathing apparatus should be carried by operatives within the chamber or sewer and they should have been trained in its correct use. It is essential that all persons in charge of the work are thoroughly trained and operatives fully instructed.

If such entries are to continue over a period of time, a permit to work system should be operated, detailing responsibilities, equipment to be used and work procedures to be adopted. If in doubt, contact your health and safety advisor.

6.3.3 Dilapidation survey

Before commencing any work, a detailed inspection should be made of any roads, buildings, fences or land in the vicinity of operations and a record kept of their conditions with particular regard to any defects. These records, supported with photographs where appropriate, should be agreed with the owner, client or Local

Authority as a true record before any works commence. It may, for example, be necessary to affix tell-tales across cracks in adjacent structures to determine if the works caused subsequent movement.

6.3.4 Record of existing mains, services and sewers

Prior to commencing any work on site, the contractor should obtain a plan from each utility company (water, gas, electricity, telecom etc.) indicating the location of any plant or apparatus they have in the vicinity. (In some cases it may be necessary to contact the National Grid operator but the local electricity company will advise on this.) These plans should not be accepted as complete and accurate records, as they are often only diagrammatic (see Fig. 6.1).

Before excavating it is essential to establish the position of the services as accurately as possible. This is best achieved by using cable and pipe locating devices used in connection with available plans or other information. Such devices should always be used in accordance with the manufacturers' instructions and be regularly checked and maintained in good working order. The tools are commonly referred to as CATs

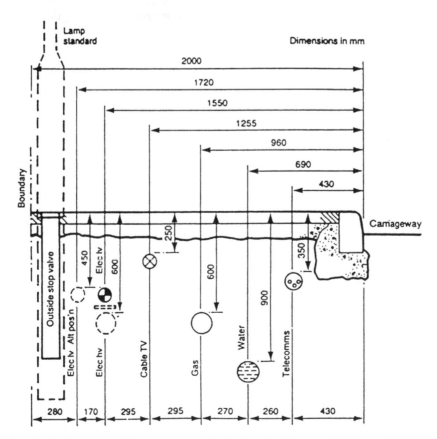

Figure 6.1 Recommended arrangement of mains in a 2 m footway including cable TV duct.

(cable avoidance tools). Using them correctly requires training followed by practice. It should be appreciated that such devices will not detect the presence of plastic pipes. Once determined, the line of any identified services should be noted and
marked with waterproof crayon chalk or paint on paved surfaces, or wooden pegs
in grassed or unsurfaced areas. Steel pins, spikes or long pegs which could damage
the services should not be used. Once the location device has been used, hand dug
trial holes may help with depth location.

Public utilities do not always record the location of existing services to individual houses, and therefore on redevelopment sites proceed with caution when
excavating within the original footways or immediately behind them. Live services
could remain.

If in doubt—stop work and have it checked by the appropriate utility companies.

6.3.5 The Party Wall etc. Act 1996

There are certain legal requirements when excavating close to a boundary of adjoining property which are covered by this Act. A useful free booklet explaining these
requirements may be obtained from the office of The Deputy Prime Minister.

6.4 Trench excavation

6.4.1 General

BS EN 1610 now provides the specification for the construction and testing of
drains and sewers. This begins with trench excavation and the provision of adequate
trench support where necessary to ensure stability and safety. Care should be taken
to limit loads close to the edge of the trench, such as excavated material, which
should be no closer than 1.5 m to any trench. Particular ground conditions may
require detailed consideration as described below. For example, in unstable ground
such as running sand or silt, additional measures such as dewatering operations or
consolidation by freezing or chemical means may be necessary.

In wet, fine grain soil such as soft clays, silts or fine sands, suitable blinding or
other stabilising material should be placed on the virgin soil immediately after the
last cut and before any traffic is permitted on the trench bottom to prevent disturbance and softening of the foundation. The use of a filter fabric is also effective in
preventing movement of fine material, both on the trench bottom and around pipe
bedding, as described in the section on waterlogged ground.

Where the formation is low and does not provide continuous support, it should
be brought up to the correct level by placing and compacting suitable material.

6.4.2 Trench support

The main hazard associated with excavation work is ground collapse; no soil can be
relied upon to support its own weight for any length of time, a factor that becomes
increasingly important as additional loads are applied, such as plant and materials.

Even a minor collapse of ground can cause serious injury—1 m^3 of earth weighs
approximately 1.3 tonnes and gravel and clay soils are even heavier.

If an excavation cannot be battered or sloped back to a safe angle, the sides will require support to prevent the possibility of collapse.

Excavation collapses may be prevented if:

- workers are trained in the safe installation of basic systems of support
- the basic support equipment is provided on site, and is used
- adequate and competent supervision is provided to ensure that correct safety precautions are observed, and assessments are made of changes in conditions that might necessitate changes in levels of safety.

The soil on each side of a trench excavation is usually supported by a structure that is held in position by forces transmitted through waling, and struts from opposite faces. Provided the support system is strong enough, an *equilibrium* is set up that keeps the structure, and hence the excavation, stable. As the structure relies heavily upon secure strutting, it is very important that the struts are securely fixed in place to prevent *accidental* removal.

Trench shields or drag boxes provide an alternative to traditional trench sheeting but are provided mainly as protection for workers in the excavation rather than for excavation support. Side sheets are kept apart by struts that span the width of the device, thus producing a rigid box. As work advances the excavation machine pulls the box forward to the next work location. This can lead to severe disturbance of the trench walls and can affect the integrity of pipe bedding material if they are not moved progressively with bedding and backfill construction.

The Construction (Health, Safety and Welfare) Regulations, 1996, cover the safety of excavation works. An important provision in the Regulations concerns the inspection of the excavation support structure. It must be thoroughly examined by a competent person. The working part of an excavation must be inspected once a day or, if greater than 2 m depth, before each shift. *No person* shall work in an excavation before it has been completely examined and approved.

Materials, plant or equipment must *not* be sited near the edges of excavations in a manner likely to endanger persons working in the excavation. When material or equipment is being placed into an excavation it must be done in a safe manner.

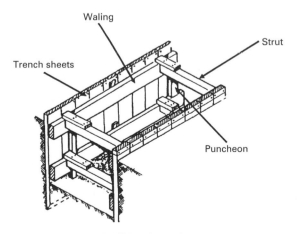

Figure 6.2 Traditional trench support system.

A safe means of access and egress must be provided for all excavations and ladders provide the main method of achieving this. Care should be taken to see they are in good condition. They must be suitably secured to prevent undue movement and extend above the excavation to give the necessary height required for a safe handhold. To allow for adequate means of escape in an emergency, it is considered that one ladder every 15 m is an average to work to, but more may be required depending upon the number of workers and the potential risk, e.g. where there might be a possibility of flooding from a rain event.

Handrails must be provided around every accessible part where a person is liable to fall. Gangways across excavations should have guardrails and toe boards.

6.4.3 Soft spots

Pockets of peat, chalk slurry or unconsolidated ground occurring below formation level should be removed and replaced with suitable well-compacted material.

In mixed non-uniform soil containing soft spots, rock bands or boulders, the foundation should be tested at suitable intervals and soft spots hardened by tamping in suitable material. Hard spots should be generously undercut and large boulders removed. The resulting holes should be filled with suitable material, to make the foundation as uniform as possible.

6.4.4 Rock

In rock, the bottom should be trimmed and screeded either with concrete or with not less than 200 mm of bedding material so that there is no rock projection that could damage the pipes. Where first-class workmanship and supervision can be guaranteed and where the variation in depth of bed does not exceed 25 mm and the socket depression 25 mm, the minimum depth of granular bed may be 100 mm.

6.4.5 Soft ground

Where the trench formation has little bearing strength and will not support pipe bedding material effectively, it is necessary to provide a stable formation before pipelaying. Such conditions most commonly occur in peat, silty ground, soft to very soft alluvial clays, running sand or in artificially filled ground.

Although trench formations are sometimes stabilised with concrete, this is unlikely to assure long-term stability in all cases, and a form of flexible bedding construction is the preferred method of dealing with this situation.

The trench formation and manhole base should be over-excavated by 600–800 mm, depending on the bearing strength of the ground. Gravel reject material or small hardcore, less than 75 mm, is then compacted in layers to form a firm trench bottom. A 50 mm thickness of lean-mix concrete is then placed as blinding. The pipe is then laid on granular bedding material.

The pipe bedding construction requirements are calculated in the normal way. It is important that 'wide trench' design criteria are used because 'narrow trench' conditions cannot be guaranteed in this situation. The extra depths of granular bedding material (Fig. 6.3); 150 mm for sleeve-jointed pipes and 200 mm for socketted

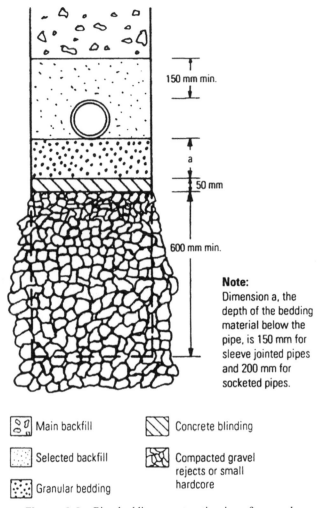

150 mm min.

a

50 mm

600 mm min.

Note:
Dimension a, the
depth of the bedding
material below the
pipe, is 150 mm for
sleeve jointed pipes
and 200 mm for
socketed pipes.

Main backfill   Concrete blinding

Selected backfill   Compacted gravel
                                 rejects or small
Granular bedding   hardcore

Figure 6.3 Pipe bedding construction in soft ground.

pipes, rather than the usual 50 mm and 100 mm respectively, are required because of the hard nature of the constructed trench bottom.

For a Class F bedding, selected backfill material is then placed to 150 mm above the pipe and compacted before the main backfill is placed. Where Class B or Class S beddings are required, additional bedding material will either partially or wholly replace the selected backfill material.

Where groundwater exists at a level above the interface between the rejects and the new trench bottom, the procedure detailed for waterlogged ground should also be applied. The geotextile should surround both the material in the base of the trench as well as the pipe bedding material. The use of a geotextile around the compacted material in the base of the trench will also assist compaction in exceptionally soft ground conditions, as well as limiting the movement of fines.

6.4.6 Waterlogged ground

Moving groundwater at a level above trench formation in fine grained soils can reduce the strength of pipe beddings. Granular bedding material encourages water movement and this washes fines out of the surrounding ground, causing a loss of support to the bedding and pipeline. This may occur particularly in peat, silty ground, soft to very soft alluvial clays, running sand or artificially filled ground.

The traditional method of dealing with this problem was to include a proportion of coarse sand in the bedding material in order to fill the interstices that might otherwise take up the fine material from around the trench. This limits the movement of fines, but the bedding material requires much more compaction energy than if it were single sized or graded.

A more effective method is to wrap the whole of the bedding construction, including any additional compacted material in the trench bottom, in a geotextile as detailed for soft ground. The grade of material should be chosen to allow the movement of water through the bedding material, but to prevent the movement of fine material, and retain it in the ground around the trench. In such conditions, measures are also needed to prevent similar movement of fines under manholes. The geotextile construction should be continued around the outside of the manhole excavation and under any manhole bedding material. The specification for the geotextile, particularly the pore size, should be related to the nature of the fines in the ground, and specialist advice may need to be sought on this.

Prior to commencing pipelaying, it is essential to satisfactorily dewater the trench formation. Any well point dewatering must also be suitably filtered to prevent continuous removal of fine sands and silts. Sump pumping from the end of the trench is not recommended even when filtered, as instability of the formation can arise because fine material may be pulled from the trench formation.

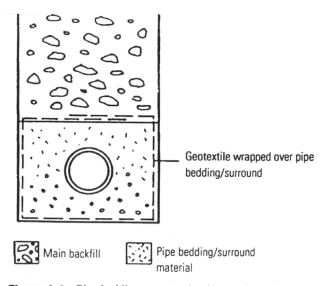

Figure 6.4 Pipe bedding construction in waterlogged ground.

'Drag-box' or similar trench support systems should not be used in waterlogged fine-grained soils because the pipe and bedding are likely to be disturbed when the support is removed. Steel or timber sheet trench support systems are recommended. They should not be driven or extracted with vibratory hammers and backfilling should proceed progressively as the support system is removed. Further information is contained in Chapter 7 Earthworks and Excavation.

It should be noted that if a utility company is being used to lay services, agreement on programming the work is essential to provide continuity, if not they may pull off site and cause delay due to the difficulty in getting them back.

6.4.7 Trenchless excavation

Trenchless excavation is sometimes carried out for the installation of services. There is a range of different systems which are appropriate to various ground conditions and pipe or cable types.

Micro-tunnelling

This method can be used on pipelines from 150 to 900 mm diameter and uses a steerable remote controlled tunnelling machine. The pipes are jacked into place behind the cutting head either during or after the tunnelling process.

Pipe-bursting

This method uses the existing line of pipes to pull the new pipeline through behind a head which bursts the old pipes aside. Some upsizing of the pipe size is possible, depending on the power available at the head and the surround to the original pipes.

Directional drilling

This technique involves drilling a small diameter pilot hole in a shallow arc underneath a major obstacle such as a river or railway. The pilot hole is enlarged using back reaming techniques until large enough to receive the pipeline.

Flow Mole

The Flow Mole system uses a remote-controlled and guided boring tool known as the Guide Drill, which incorporates high pressure slurry jet nozzles for cutting soils to provide a pilot bore. The bore produced is impregnated and lined with slurry helping to stabilise it before insertion of the final pipeline or cable.

Narrow trenching

Although not strictly a trenchless construction method, the width of trench excavated is little more than 50–100 mm greater than the pipeline. The equipment used incorporates a circular cutting wheel with chain type trenchers. Often used for shallow services in roads as it requires a minimum of backfilling and resurfacing.

Mole ploughing

A traditional method used mainly for installing land drainage in open fields but can be adapted for installing service pipes where ground conditions permit. The plough is towed and its blade forms a subsurface void into which the pipeline or cable is inserted immediately following the plough blade.

Specialised contractors operate these trenchless systems, which can provide savings in time and labour over traditional methods, particularly where surface disturbance needs to be minimised, such as in busy roads.

6.5   Drain and sewer installation and testing

6.5.1   Prestart

- Check that invert levels on drawings are correct to gradients and to any existing invert levels.
- Check service drawings to establish if any existing services cross proposed drain runs.
- Calculate the position of manhole centre lines for setting out purposes.
- Calculate gradients if laser is to be used.
- Draw a sketch of proposed drain run (ideally in a permanent duplicate book), showing the following information:
  - manholes with identification
  - diameter of pipe
  - type of pipe e.g. clay, concrete, cast or ductile iron, plastic
  - junctions on pipe run
  - diameter of junction branches
  - distance to junction from centre of manhole
  - manhole chamber diameter or other size
  - number of rings required
  - connections into manholes with diameters.
- Check trench width for size of machine bucket.
- Check trench depths for planks or sheets and strutting.
- Check type of pipe bed material.
- Check type of initial backfill material.
- Check type of main backfill material.

6.5.2   Setting out

- Set out centre line of manholes and provide offsets.
- Erect profile boards at manholes (where a laser is used, this is only for a check).
- Mark on the ground any existing services lines crossing trench.
- Set up pipe boning rods or set laser gradient as appropriate.
- At each manhole, set out pegs for directions of connections.

Note: Class 3a visible red beam pipe lasers may only be used by trained site engineers. The use of hired or subcontractors' pipe lasers on site requires the appointment of a laser safety officer.

6.5.3 Pipelaying

- Check pipeline is running in correct direction.
- Check trench width is correct.
- Ensure that machine driver and banksman are aware of any service crossings.
- Check planks or sheets and strutting or other means of trench support are being used correctly.
- Check that correct pipe bedding is being used.
- Check pipeline by means of an air test to ensure that they are correctly jointed. Test to be witnessed by appropriate inspector and client's site representative and to be recorded on a check sheet.
- Check with level staff that pipes are at correct level.
- Measure distances to junctions and record in duplicate book.

6.5.4 Pipe bedding

Defined as Bedding Construction Type 1 in BS EN 1610, the granular bedding material should be placed to invert level and should usually extend to the full width of the trench. Gradings for the granular materials to be used for pipe bedding are specified in tables B.15 to B.17 of Appendix B to BS EN 1610, which also specifies the standards that define the materials themselves. These may be BS 882, BS 1047 or BS 3797, although these are now replaced by BS EN 13242.

Placing bedding material

The bedding material for the specified bedding class is placed as required in the lower bedding, upper bedding, sidefill and initial backfill areas (see Fig. 6.2 and Fig. 6.3).

Socket holes should be formed at each joint position for socketted pipes. These should be deep enough to prevent the weight of the pipe and the load upon it bearing on the socket or coupling and should be a minimum of 50 mm deep, leaving a minimum depth of 50 mm of bedding material beneath the joint. For clay sleeve jointed pipes and couplings (not socketted) this permits a minimum bedding depth construction of 50 mm under couplings and pipes to be specified.

Placing sidefill material

Sidefill of either granular or selected backfill material, depending on the class of bedding, should be placed evenly on either side of a rigid pipe taking care not disturb the line and level. Bedding material should not be compacted in the socket holes.

The sidefill should be placed to the top of the pipe and hand tamped in 150 mm layers, ensuring that the line of laid pipeline is not disturbed. The initial backfill should be placed to 150 mm above the crown of the pipe and this layer hand tamped.

Selected fill, whether selected from locally excavated material or imported, should consist of uniform, readily compactible materials, free from vegetable matter, building rubbish, frozen material and materials susceptible to spontaneous combustion. It should also exclude clay of liquid limit greater than 80 or plastic limit greater than 55 and materials of excessively high moisture content. Clay lumps and stones retained on 75 and 37.5 mm sieves respectively should be excluded from the fill material.

It should be noted that the side support for flexible pipes depends upon the extent and the degree of compaction of the bedding and the nature of the surrounding soil. For this reason, flexible pipes should be installed with specified bedding materials which allow appropriate compaction.

Placing backfill material

As soon as the pipes are correctly jointed, bedded, tested and approved, selected material and backfill should be placed. The minimum thickness of the initial backfill over the pipe barrel shall be 150 mm with a minimum of 100 mm over the joint.

For clay pipes with class D, N, F or B bedding, selected material from the trench excavation can be used as sidefill above the bedding, or for the layer directly above the pipes, provided that it is readily compactable. It should exclude stones retained on a 40 mm sieve, hard lumps of clay retained on a 100 mm sieve, timber, frozen material and vegetable and foreign matter. For class D, N or F bedding, the selected material should be placed under the haunches of the pipe, care being taken not to displace the pipe from its correct line and level.

Mechanical compaction equipment should not be used until there is a minimum of 450 mm of hand-compacted material above the crown of the pipe. The need for effective compaction is important at all stages to minimise subsequent settlement.

6.5.5 At manhole

- Check centre line of manhole against offset pegs.
- Check invert level of manhole.
- Ensure that all required connections are positioned and are the correct diameter.
- Check that benching is constructed correctly.
- Check that the correct rings are being used where appropriate.
- Check height of manhole against finished levels.

6.5.6 Joints and jointing

For the materials dealt with in this chapter, most pipework will be flexibly jointed with either a sleeve or spigot and socket joint. Also in replacing short sections of pipe it may be more practical to use flange connections. There are two principal types of joints:

- flexible joints
- rigid joints.

The following guidance should be observed:

Flexible joints

This type of joint is most commonly used and provides limited articulation, longitudinal travel and ease of jointing. Jointing systems are available for all the materials in Section 6.1 above. *It is essential that the manufacturer's instructions are closely followed.* A well-constructed flexible joint will permit an angular tolerance of up to 3° between consecutive sections of pipe.

Jointing surfaces and sealing rings should be clean and, where specified, a lubricant should be used. To allow for longitudinal (telescopic) movement a small gap should be left between the spigot and the shoulder of the adjoining socket, or between pipes in a sleeve joint.

Depressions in the bedding material should be carefully fashioned to accommodate pipe sockets. This is essential to ensure that the pipe is evenly supported along its barrel.

Small bore pipes can be joined by hand. Care must be taken when using mechanical equipment for large diameter work that winch anchorages are placed remotely from constructed pipework to avoid damage.

Rigid joints

If rigid joints are being used, they will need to be protected against ground movement. *As with flexible joints the manufacturer's instructions must be rigorously followed.*

Flanged joints require special protection to the bolts to avoid corrosion. Depressions in bedding material should be fashioned to accommodate flanges. This is essential to ensure that the pipe is evenly supported along its barrel.

6.5.7 Backfilling

- Ensure that all trench support is removed in stages appropriate to the ground conditions.
- Ensure that a marker is positioned at junctions.
- Ensure that the pipe has been surrounded correctly.
- Ensure that all backfill material is compacted as instructed, with no mechanical compaction applied before there is at least 300 mm cover to the pipes.
- Reinstatement of any road surface affected by the works should be to the HAUC specification and final reinstatement is usually carried out by the relevant authority or its own contractor.

6.5.8 Testing

Generally, drainage works are inspected and tested in three stages: during construction prior to backfill; after backfill, and upon completion prior to handover. The works should be protected during all stages of construction, and the entry of foreign matter into any part of the system prevented.

The current specification for testing is found in BS EN 1610 clause 2.12. Earlier details are set out on BS 8000 Part 14.

6.5.9 Inspection and test plan

All tests must be formally recorded on an appropriate form and witnessed, if necessary, by the client's representative or the building inspector. Check your project specification, Building Regulations and Project Quality Plan for the test requirements (see Fig. 6.5). Reference should also be made to BS 8000 *Workmanship on Building Sites* parts 13 and 14.

DRAIN AND SEWER TESTING

Company	Contract		Document Ref. ....
			Page Ref. ....
			Date ....

Length Ref.	Check	Timing	Carried out	Witnessed	Result
A1 – A2	Pipe material	When laid		-	
	Bed material	When laid		-	
	Levels	When laid		-	
	Cleaning	Pre-test		-	
	Inspection	Pre-test		-	
	Air test	On laying			
	Air test	After backfill			
	Infiltration	After backfill			
	Water test	When Required			
	CCTV	Post-test		-	
MH A1	MH material	Construction		-	
	Backfill	Construction		-	
	Cleaning	Pre-test		-	
	Inspection	Pre-test		-	
	Infiltration	After backfill			
	Water test	When required			

Figure 6.5 Sample inspection and test plan.

6.5.10   Test procedures

Construction stage testing

Before any test, check for any damage and clear out any debris. Ensure that plugs are clean and in good condition before they are fitted.

Tests should be carried out to locate and remedy any defects in the soundness that may exist during construction. Such tests should take place immediately before side-fill is applied and the work is covered up to facilitate the replacement of any faulty pipe or fitting or to rectify any joint defect.

Tests may include visual inspection for line and level, joints, damage or deform-ation, connections, linings and coatings. The standard specifies that the vertical change in diameter of flexible pipes shall be checked for compliance with the design. Inspection of the pipeline will also reveal any defects in the support and bedding. Testing for leaktightness includes the testing of connections, manholes and inspection chambers.

Final testing

Testing and inspection should take place immediately before handover when all relevant works have been completed.

A water test is generally preferred since it relates very closely to the conditions found in practice. It does, however, suffer the disadvantage of using large quantities of water.

An air test is easier to apply but the results can be affected by small changes in temperature. A change in temperature of 1 °C will result in a corresponding change in air pressure of about 38 mm water gauge. Therefore, if the pressure drop is above that specified, the pipeline should not be deemed unsatisfactory on the air test alone.

The number of air tests allowed to be carried out is not restricted but where recourse is made to the water test, it alone shall be decisive.

6.5.11   Tests for straightness and obstruction

Tests for line, level and freedom from obstruction should be applied before commenc-ing and immediately after backfilling. A straightness test can be carried out by means of a mirror at one end of the pipeline and a lamp at the other or by laser instruments.

6.5.12   Air test

The air test (specified in BS EN 1610 13.2) is carried out by inserting plugs or inflat-able bags in the upper and lower ends of the pipeline and pumping in air under pres-sure. A glass U-tube gauge (manometer) and the air pressure source are connected. Air pressure is then applied by mouth or hand pump to achieve the required test pressure.

The test time for the air test varies according to the pipe size as defined in the table below. The test method (LA) is the same as that previously specified in the UK, except that the test period is increased for pipe sizes above DN200.

It is expected that this will remain the preferred UK method for the future, avoid-ing the additional safety precautions needed when using the higher test pressures favoured by parts of continental Europe.

Table 6.1   Test pressures, allowable pressure drop and test periods for the air test

Test method	Test pressure mbar (kPa)	Allowable drop mbar (kPa)	Test period in minutes				
			100 mm	200 mm	300 mm	400 mm	600 mm
LA	10 (1)	2.5 (0.25)	5	5	7	10	14

The pressure is first brought up to approximately 10% more than the test pressure and held for 5 minutes. It is then adjusted to the test pressure and the drop in pressure measured at the end of the specified test period. If the measured drop is less than the allowable drop, then the pipeline passes the test. The test equipment is required to measure with an accuracy of 10% of the allowable pressure drop.

As with the water test, there are possible contributing factors that could cause an apparent failure in the air test. These are: temperature changes of the air in the pipe due to direct sunlight or cold wind acting on the pipe barrel, these may show up during the 5 minute stabilisation period; dryness of the pipe wall and leaking plugs or faulty testing apparatus.

6.5.13 Water test including manholes and inspection chambers

The water test (specified in BS EN 1610 13.3) is applied as an internal pressure-head of water above the pipe at the high end of the line to be tested. A maximum limit is placed on the head at the lower end to prevent damage to the pipeline. Steep drains and sewers may need to be tested in short sections in order to avoid too great a head at their lower end. The test is carried out to a maximum pressure of 50 kPa ($\approx$5 m head at the lower end of the pipeline) and a minimum pressure of 10 kPa ($\approx$1 m head) measured at the top of the section of pipeline under test. The pressure is applied by filling the test section up to the ground level of the upstream or downstream manhole or inspection chamber as appropriate. Pipes are strutted as necessary to prevent movement.

After the test section has been pressurised, the pipeline should be conditioned by allowing to stand full for at least one hour. A longer time may be needed in dry climatic conditions.

The test is carried out over a period of 30 minutes during which the pressure is maintained within 1 kPa of the test pressure by topping up with water. The test requirement is satisfied if this amount of water does not exceed the following amounts per square metre of internal wetted area:

0.15 l/m^2 during 30 minutes for pipelines
0.20 l/m^2 during 30 minutes for pipelines including manholes
0.40 l/m^2 during 30 minutes for manholes and inspection chambers

For various pipe diameters this rate of top-up over the test period may be expressed as shown in the Table 6.2.

Table 6.2   Water test: allowable
rate of loss in litres per metre over
30 minute test period

Pipe size	Pipelines
DN100	0.047
DN150	0.071
DN225	0.106
DN300	0.141
DN400	0.189
DN500	0.236
DN600	0.283

The volume of water added may be measured directly or calculated by measuring the drop in level in a standpipe or chamber.

Even though there may be no visible leakage, an apparent loss of water may exceed the permitted figure at first, due to the effect of continued absorption by the pipes or air trapped in the joints being dissolved. In such cases the line should be allowed to stand for a further period until conditions have stabilised. The test should then be repeated.

Unless there is an actual leak, the rate of apparent loss will decrease rapidly with time. Other reasons for apparent failure may include changes in ambient temperature, trapped air or leakage past plugs or stoppers.

6.5.14   Closed circuit TV surveys

Water companies now require a closed circuit TV survey of any sewerage system prior to accepting it for adoption. This will be carried out by specialist subcontractors employed by the authority or by the company.

In view of the detailed defects that can be highlighted by this method of inspection a high standard of workmanship is required throughout the construction operation if expensive remedial works are to be avoided. It is essential that the drainage system is completely cleaned out before the TV survey commences. Cleaning of drains and sewers usually involves jetting operations at high pressures. The WRC sewer jetting Code of Practice provides for a maximum pressure of 340 bar (5000 psi) for clay and concrete pipes, whereas only 180 bar (2600 psi) is allowed for plastic pipes.

6.6   Location of services

New and existing service ducts and pipes on site should be marked and recorded to avoid subsequent damage as well as making it easier to find them when required at a later stage in the works. This is particularly important with regard to ducts under highways where the depth should also be recorded and the position related to drainage or a permanent feature in the carriageway.

Recommendations for positioning of utilities are given in NJUG No 7. The recommended standard is shown in Fig. 6.1. It will be seen that the street lighting column is shown on the property side of the footway rather than the kerb side to conform with the requirements of BS 5489.

The relative depths of lay required for the various mains argue powerfully in favour of the lateral dispositions illustrated and they are therefore recommended as standard locations. The lateral clearances between adjacent utility mains should be considered as the minimum and represent the best use of the limited space available. It should also be noted that an allocation of footpath/service strip space is shown for cable television.

6.7 Maintenance

Responsibility for the maintenance of the works is normally with the main contractor. In the case of adopted sewers this is normally 12 months from practical completion. Many structural problems showing up during the maintenance period are related to connections. To alleviate these, the following procedure should be adopted:

- The drain or sewer should be inspected before and after connection is made and a check should be made that the connection is to the correct pipeline.
- The connection should be made using a preformed junction. The use of repair couplings when making post connections ensures watertight, flexible joints.
- The system should be checked for watertightness after the connection is made.
- It is necessary to ensure that the drain or sewer is not weakened or damaged by the connection and that no operational problems are caused by the connection.

6.8 Records

Records must be kept of the sizes and positions of all pipes, ducts and cables and of the tests required by the contractual specification, company procedures and project quality plans.

Further reading

Introduction

The design and construction of sewers and drains in the United Kingdom has been largely controlled over recent years by the provisions of BS 8005, the Guide to new sewerage construction and BS 8301 together with Approved Document H to the Building Regulations and the equivalent Building Regulations documents for Scotland and Northern Ireland. The other principal reference documents in these fields are the Civil Engineering Specification for the Water Industry and Sewers for Adoption, the Water Industry design and construction guide for developers.

The publication during 1997 and 1998 of new European standards covering the design and construction of drains and sewers has somewhat changed this picture. Both BS 8005 and BS 8301 have been replaced by the provisions of BS EN 752; BS EN 1295–1 and BS EN 1610. These are now the references to be used in specification documents and with this, some of the procedures to be adopted have been changed.

In areas where existing national methods needed to be retained, specific national annexes are provided to the standards. For the British editions of these standards, the national annexes largely contain the familiar information from BS 8005 and BS 8301, brought up to date where necessary. In this way, most pre-existing design and construction practices will not need to change with the advent of the European standards. Two exceptions to this are BS EN 752–4 covering hydraulic design and BS EN 1610.

Alongside these Code of Practice standards, many products used for the construction of drains and sewers are now covered by their own European standards which should be referred to when specifying materials.

The following provides a range of relevant references:

Standards and Codes of Practice

BS EN 295–1, 2 & 3: 1991 *Vitrified Clay Pipes and Fittings and Pipe Joints for Drains and Sewers.*

BS EN 752 *Drain and Sewer Systems Outside Buildings*: Parts 1–3: 1996, 4–7: 1998.

BS EN 877: 1999 *Cast Iron Pipes and Fittings, their Joints and Accessories for the Evacuation of Water from Buildings.* Requirements, test methods and quality assurance.

BS EN 1295–1: 1998 *Structural Design of Buried Pipelines under Various Conditions of Loading.*

BS EN 1610: 1998 *Construction and Testing of Drains and Sewers.*

BS EN 12056–2: 1998 *Gravity Drainage Systems Inside Buildings*: Sanitary pipework, layout and calculation.

BS EN 545: 1995 *Ductile Iron Pipes, Fittings, Accessories and their Joints for Water Pipelines.* Requirements and test *methods.*

BS EN 598: 1995 *Ductile Iron Pipes, Fittings, Accessories and their Joints for Sewerage Applications.* Requirements and test *methods.*

BS 65: 1997 *Specification for Vitrified Clay Pipes, Fittings and Ducts, also Flexible Mechanical Joints for use Solely with Surface Water Pipes and Fittings.*

BS 882: 1992 *Specification for Aggregates from Natural Sources for Concrete.*

BS 1047: 1983 *Specification for Air-cooled Blast Furnace Slag Aggregate for use in Construction.*

BS 3797: 1996 *Specification for Lightweight Aggregates for Masonry Units and Structural Concrete.*

BS 8000: 1989–2002 Workmanship on Building Sites. Part 13: Code of Practice for above ground drainage and sanitary appliances; Part 14: Code of Practice for below ground drainage.

Note: See also Chapter 1.

BS 3868: 1995 *Specification for Drainage Stack Units in Galvanized Steel.*

BS 4660: 2000 *Thermoplastics Ancillary Fittings of Nominal Sizes 110 and 160 for Below Ground Gravity Drainage and Sewerage.*

BS EN 1401–1: 1998 *Plastics Piping Systems for Non-pressure Underground Drainage and Sewerage.* Unplasticised poly(vinylchoride) (PVC-U).

BS EN 1852–1: 1998 *Plastics Piping Systems for Underground Drainage and Sewerage (Polypropylene PP).*

BS EN 1916: 2002 *Concrete Pipes and Fittings, Unreinforced, Steel Fibre and Reinforced.*

BS EN 1917: 2002 *Concrete Manholes and Inspection Chambers, Unreinforced, Steel Fibre and Reinforced.*

BS 65: 1991 *Specification for Vitreous Clay Pipes, Fittings and Ducts, also Flexible Mechanical Joints for use with Surface Water Pipes and Fittings.*

BS EN 12056: 2000 *Gravity Drainage Systems Inside Building. General performance requirements.*
Note: This BS EN is in several parts dealing with topics such as sanitary pipework, layout and calculations, roof drainage, wastewater, installation and testing.
BS EN 13242: 2002 *Aggregates for Unbound and Hydraulically Bound Uses.*
BS EN 13252: 2001 *Geotextiles and Geotextile Related Products.* Characteristics required for use in drainage systems.
BS 5911: 1982 to 1994 *Precast Concrete Pipes, Fittings and Ancillary Products.*
Note: This Standard is produced in several parts.
BS 437: 1978 *Specification for CI Spigot and Socket Drain Pipes and Fittings.*
BS 534: 1990 *Specification for Steel Pipes and Specials for Water and Sewage.*
BS 5480: 1990 *Specification for GRP Pipes, Joints and Fittings for use for Water Supply or Sewerage.*
BS 2760: 1973 *Specification for Pitch-impregnated Fibre Pipes and Fittings for Below and Above Ground* Drainage.
Note: Although this Standard has been withdrawn it is useful in repair work.
BS 4660: 1989 *Specification for PVC-U Pipes and Plastics Fittings of Nominal Sizes 110 & 160 for Below Ground Gravity* Drainage *and Sewerage.*
BS 5481: 1989 *Specification for Unplasticised PVC Pipe & Pipe Fittings for Gravity Sewers.*
BS 3506: 1969 *Specification for Unplasticised PVC Pipes for Industrialised Uses.*
BS 5955: 1980 to 1990 *Plastics Pipework (Thermoplastics Materials).*
Note: This Standard is produced in several parts.
BS 6572: 1985 *Specification for Blue Polythene Pipes up to Nominal Size 63 for Below Ground use for Potable Water.*
BS 6437: 1984 *Specification for Polyethelene Pipes (type 50) in Metric Diameters for General Purposes.*
BS 3284: 1967 *Specification for Polythene Pipe (type 50) for Cold Water Services.*
Note: This Standard is obsolescent but is useful for repair work.
BSCP 302: 1972 *Small Sewage Treatment Plants.*
BS 5489: 1992 to 1996 *Road Lighting.*
Note: This Standard is produced in several parts.
BS 8005: 1987 to 1990 *Guide to New Sewerage Construction.*
Note: This Standard is produced in several parts.
BS 8301: 1985 *Code of Practice for Building Drainage.*

Other related texts

Note: For HSE documents see Chapter 18.
Provisions of Mains and Services by Public Utilities on Residential Estates: NJUG Report No. 2, London, NJUG.
Cable Locating Devices: NJUG Report No. 3, London, NJUG.
The Identification of Small Buried Mains and Services: NJUG Report No. 4, London, NJUG.
Model Guidelines for the Planning and Installation of Utilities Supplies to New Building Developments: NJUG Report No. 5, London, NJUG.
Service Entries for new Dwellings on Residential Estates: NJUG Report No. 6, London, NJUG.
Recommended Positioning of Utilities Mains and Plant for New Works: NJUG Report No. 7, London, NJUG.
Civil Engineering Specification for the Water Industry, 5th ed.: WRc, Swindon UK, WRc, 1998.
Sewers for Adoption—Design and Construction Guide for Developers, 5th ed.: WRc, Swindon UK, WRc, *2001.*
Sewer Jetting Code of Practice: WRc, Swindon UK, WRc, 1997.

The Building Regulations 2000: Approved document H: Drainage and Waste Disposal:
HMSO, London, HMSO, 2002.
Building Standards (Scotland) Regulations 1990: Part M: Drainage and Sanitary Facilities:
HMSO, London, HMSO, 1990.
*The Building Regulations (Northern Ireland) 1990: Technical Booklet N: Drainage and Waste
Disposal:* HMSO, London, HMSO, 1990.
Manual of Contract Documents for Highway Works Vol. 1–3: London, HMSO.
*The Specification, Design and Construction of Drainage and Sewerage Systems using Vitrified
Clay Pipes:* CPDA, Chesham UK, CPDA.

Building Advisory Service

Construction Safety Manual: Section 23: Confined Spaces.
Safe Working in Sewers and at Sewage Works: NJH&S Committee for the Water Service.
Specification for the Reinstatement of Openings in the Highway: HAUC.
Water Supplies: B. H. Rofe, Oxford, Butterworth Civil Engineer's Reference Book 4th ed.
(L. S. Blake ed.), 1989.
Irrigation, Drainage and River Engineering: W. Pemberton *et al.*, Butterworth Civil
Engineer's Reference Book 4th ed. (L. S. Blake ed.), 1989.
The Law of Sewers and Drains, 8th ed., J. F. Garner *et al.*, Crayford UK, Shaw, 1995.
Testing of Drains and Sewers: Water and Air Test: CPDA, Chesham UK, CPDA, 1990.
Sewerage and Sewage Disposal: Watson Hawksley, Oxford, Butterworth Civil Engineer's
Reference Book 4th ed. (L. S. Blake ed.), 1989.
Buried Pipelines and Sewer Construction: D. J. Irvine, Oxford UK, Butterworth Civil
Engineer's Reference Book 4th ed. (L. S. Blake ed.), 1989.
Concrete Pipe for the New Millennium: ASTM, Philadelphia, ASTM, 2000.

Chapter 7   EARTHWORKS AND EXCAVATION

Michael Tomlinson

Basic requirements for best practice

- **Clear drawings and specifications**
- **Good site investigation**
- **Well-chosen and maintained plant**
- **Well-trained staff**
- **Good safety practices**

7.1   Site investigations

7.1.1   General

Be fully aware of all information and requirements contained in site investigation reports and elsewhere before work starts on site. As far as possible ensure that the depth of investigation is greater than any required excavation. A desk-top study may be available; if so make sure it is read. Particularly in mining areas, a good knowledge of underground workings is essential before excavation can commence.

There are certain legal requirements related to The Party Wall etc. Act, 1996 when excavating adjacent to adjoining property. A useful *free* booklet explaining these requirements may be obtained from the Office of The Deputy Prime Minister.

A ground investigation can vary from a visual inspection of, trial pits or open trenches to a full investigation with boreholes to considerable depth. Similarly, the soils testing can vary from simple index tests to the very detailed.

Every excavation on site is in effect a trial pit and every opportunity should be taken to observe the soils. This information can be useful to gain a general picture of the soils underlying the site. Site investigations are normally carried out in accordance with BS 5930.

7.1.2   Contaminated ground

It is becoming increasingly necessary to develop land contaminated by the residues from the industrial activities of the past 200 years. Many such sites contain a variety of dangerous chemicals. It is, therefore, essential to obtain as much information as possible regarding the use to which the site was previously put and dependent upon this, organise a detailed survey by a qualified laboratory. Such a survey will highlight potential dangers (e.g. the presence of methane, radon or other gases) thereby enabling a safe system of work to be established, advice on which should be sought from the local health and safety advisor.

7.1.3 Code of Practice

BS 8004 should be considered as the 'bible' for reference during both the planning and working stages of a project involving earthworks, excavations and foundation construction.

7.2 Safety

7.2.1 General

No excavated ground can be assumed to be safe and all ground can collapse without warning.

The conditions likely to be met in excavations will vary widely. These notes cannot, therefore, be entirely specific in relation to the support of excavations, but guidance is given as to when expert advice should be sought.

These notes only relate to excavations utilising open cut or temporary support methods. Specialised methods such as diaphragm walls, bored pile walls and ground stabilisation by injection, freezing and dewatering are excluded, as their use must always relate to designed conditions or specialist advice.

The support of excavations divides into two categories:

- where so-called standard solutions are used:
    sloping side (open cut) excavation
  - standard details from BS 6031 or CIRIA Report 97
  - proprietary systems.
- purpose designed for the particular situation by persons competent to do so.

Note: See also Chapter 19 for general advice on health and safety.

7.2.2 Requirements

It is mandatory for a competent person to be made responsible for the supervision of excavation under the general direction of the site manager or other person having overall responsibility for the contract. This applies whether standard or designed solutions are used.

The management team of any project involving excavations deeper than 1.2 m should include the necessary requirements in their company standing instructions or procedures. Such requirements usually include the following:

- All excavations exceeding 1.2 m deep, to be supported or excavated to a stable slope.

- All excavations exceeding 1.2 m, if intended to be left unattended, to be protected by a rigid barrier not less than 1 m in height—or be covered by robust material.
- Provision of safe means of access into excavations.
- Provision of barriers or stop blocks where cranes, dumpers, lorries etc. are required to manoeuvre close to the edge of the excavation.
- Excavations to be inspected by a competent person daily or at the beginning of each shift. In addition, this person must carry out and record in the official register HSE Form (see Fig. 7.1) the results of thorough examinations at least every seven days or after the use of explosives or where an unexpected fall of rock, earth or other material has occurred that is likely to have affected the stability of the timbering or other supports.
- Provision of adequate supplies of trench sheets, timbering and propping materials of a suitable quality must always be available and to hand before excavation starts.
- Location of spoil heaps at a safe distance from the top of an excavated slope or trench.

7.2.3 Standard solutions

The use of standard solutions should only be applied in the following circumstances:

1. double-sided, narrow trench support, not exceeding 4 m deep, in dry ground
2. shallow pits, not exceeding 4 m deep
3. where water problems have been eliminated by other means, e.g. well pointing, but within the limitations of 1 and 2 above.

- When deciding the slope of open cut excavations, proper account must be taken of the material in which the excavation will take place.
- Do not assume that excavation in rock is necessarily stable.

Name or title of employer or contractor	Factories Act 1961		SECTION B
....	Construction (General Provisions) Regulations 1961		
Address of site	**EXCAVATIONS, SHAFTS, EARTHWORKS,**		
....	**TUNNELS, COFFERDAMS AND CAISSONS**		
Work commenced—Date			
Reports of the results of every thorough examination made in pursuance of Regulation 9(2) of an excavation, shaft, earthwork or tunnel or in pursuance of Regulation 18(1) of a cofferdam or caission.			
Description or location (1)	Date of Examination (2)	Result of thorough examination State whether in good order (3)	Signature (or, in case where signature is not legally required, name) of person who made the inspection (4)

Figure 7.1 Reprint of HSE Form No: F91 (Part 1).

- Where support (timber, trench sheeting etc.) is to be used, recognised good practice must be followed in detail.
- Installation of support must maintain safe conditions at all times.
- Where proprietary systems are adopted, installation must be strictly in accordance with manufacturers' instructions.
- All decisions in relation to the foregoing must be determined by the person on site responsible for ground works, under the direction of the site manager or other persons responsible for the contract as a whole.
- Persons erecting or removing supports must always work within the protection of some form of existing support or from outside the excavation, and be under immediate supervision.
- The system of work adopted must be explained in detail to all persons engaged in placing or removing supports, and others who will be working in the excavation.
- Written work instructions must be provided if safe working cannot be maintained without them. The method statement and instructions given must be strictly adhered to. If doubt arises the nominated competent person must be consulted.

7.3 Geotechnical process of dewatering and treatment of substrata

7.3.1 General

These processes cover the application of a number of established methods to facilitate foundation construction in difficult ground conditions, and include: groundwater lowering, injection of grouts into the ground to change its physical characteristics, compaction and freezing the groundwater.

7.3.2 Exclusion of water

Methods:

- open drains with sumps and pumps to keep out surface water
- cut-offs to limit water entry
- well points
- ground freezing
- deep wells
- injection of chemicals.

7.4 Removal of water from excavations

When soil conditions are such that the sides or base of the excavation remain stable, usually the least expensive method is to pump the water out. Plan for the correct pump capacity, or drainage (with the necessary local authority agreement).

If, however, there is a possibility of instability of unsupported slopes or softening or heave of the bottom of the excavation, it is advisable to remove the water from the ground before it reaches the excavation by means of an external groundwater lowering installation.

The resulting changes in groundwater levels, and the re-established water levels afterwards, may cause settlement of adjacent structures. Monitoring procedures may need to be undertaken as outlined in Chapter 8. There is the additional danger of loss of fine material through pumping.

The requirements of the local authority for the discharge of pumped water into sewers or watercourses should be followed. This will probably require provision of settlement tanks for removing sediment before discharge.

7.5 Ground treatment to change the physical properties of the ground

A number of methods are available as follows:

- Removal and replacement of the weak strata with strong compact material: it is necessary to check the compatibility of the proposed fill material with the existing chemical ground conditions.
- Compaction of soils *in situ* by vibro-compaction or dynamic compaction.
- Pre-consolidation of soil by preloading the site with imported fill: check compatibility of fill material as above.
- Injection of grout into the ground.
- Construction of *in situ* diaphragms.
- Removal of water by sand drains and drainage wicks.

All these are operations normally carried out by specialist subcontractors. See company standing instructions or procedures for recording work in progress.

7.6 Regrading

This is a term loosely applied to all operations involved in adjusting ground levels and gradients. All work is required to be in accordance with the drawings and specification but any changes which site management think may be appropriate should be brought to the attention of the engineer for his consideration at the earliest opportunity.

Note: The engineer must approve any proposed changes before they are carried out.

7.6.1 General

Regrading entails the movement of considerable quantities of earth before the construction of roads or buildings commences and can be categorised into two sections:

1. earthworks involved in major civil engineering contracts, e.g. motorways, railways, open-cast mining
2. on site regrade for estate roads, housing and commercial developments.

The requirements for the first category are so wide ranging and subject to specific requirements regarding method of working, plant operation and outputs that they are not covered specifically here, although the comments do apply in general terms.

The following information is for the assistance of the engineer supervising the necessary regrade for the second category. General guidance is given in BS 6031.

7.6.2 Embankments and filling

The checks that need to be made before work starts include the stability and levels of the original ground or foundations at the base or top of embankments, selection of suitable fill material and the stability and profiles of side slopes.

Note: The stability of slopes to cuttings and embankments (in all weather) is critical to safety on site. **If in doubt, ask**.

If a full soils report was not available at the time of design, certain assumptions may have been made regarding the ground conditions. It is essential to verify these before commencing regrade and, if proved to be incorrect, inform the engineer immediately.

Note: Know your soil conditions before starting any large-scale earth moving operations.

7.6.3 Existing services

Underground services can cause major problems on site especially when their location is in doubt. The correct procedure for dealing with some existing services is detailed in Chapter 6: 'Drains, Sewers and Services'.

Overhead services can also restrict the operation of crane jibs, tipping wagons, mechanical excavators etc. Follow company safety instructions and consult the appropriate service authority before any operations commence. When working in London (and other areas where Tube or similar underground lines exist) there may be statutory requirements to be met. A close working relationship with London Underground Ltd or similar authority is essential.

7.6.4 Filling material

Suitability depends on composition, moisture content and what the material is to be used for, therefore it is not easily defined. It is easier to define materials that are generally unsuitable for reuse, such as:

• material from swamps, marshes and bogs
• peat, logs, stumps and perishable materials
• materials susceptible to spontaneous combustion
• material in a frozen condition
• soft clays
• material having a moisture content greater than that required to give the specified standard of compaction
• rock material which swells when exposed to air or moisture
• chemically contaminated material (e.g. industrial waste)
• some demolition material.

Other factors which need investigation:

- the chemical characteristics and compatibility of the soils when they are in contact with foundations of buildings, roads, sewers and services so that the appropriate precautions can be taken
- sulphates (see BS 8004)
- frost susceptibility, especially related to sub-grade for road construction or exposed ground-bearing slabs
- industrial waste.

7.6.5 Stripping

Keep stripping of topsoil to a minimum and only remove as necessary, taking into account the nature of the subsoil and the use of special earth moving plant. For example, clay subsoil deteriorates when exposed to wet weather and easily becomes unworkable.

The location and size of spoil heaps should be carefully planned considering:

- the need to keep the topsoil on site
- is the topsoil reusable?
- where the topsoil is to be replaced
- period of time before topsoil is to be reused
- working areas for construction
- stability of slopes.

Note: Topsoil may be suitable when stripped but can deteriorate if kept in a heap for a long time: weathering causing segregation; excessive height of the heap causing changes in the internal water pressures in the soil. Also remember that double handling of soil is expensive.

7.6.6 Stability of slopes

The safe angle selected for a slope is dependent on the nature of the material under the worst conditions in the short or long term. The engineer should have indicated on the drawings the position of the top and bottom of any permanent slope after taking into consideration the necessary slope angle.

Where softer material overlies rock, excavate the former at a flatter slope than the latter and leave a berm or horizontal ledge at the junction of the two materials.

With cuttings in chalk and other friable rocks, allow for seasonal surface disintegration caused by weathering. Where necessary, safeguard the public and workforce from the falling, or accumulation, of debris.
Note: If in doubt seek instructions from your engineer.

7.6.7 Slippages

These can occur on slopes without any apparent warning. Some causes are:

- excavation or erosion at or near the toe of a slope
- additional loading at the top of a slope caused by spoil dumps, traffic, building, plant tipping etc.
- increase in weight of soil by taking up water

- steepening of slope
- increase in internal water pressure
- softening of clay soils
- surface drying of clay soils causing shrinkage cracking with consequent water penetration
- increase in water table increases buoyancy of granular soils
- removal of vegetation binding the surface soil.

Note: Changes in weather conditions can cause slips to occur.

7.6.8 Filling

Before starting, consider the stability of the material on which the filling is to be founded—the additional weight may cause problems. In general, topsoil must be removed before filling commences.

The choice of suitable fill materials depends on:

- the purpose of the embankment
- availability of local material
- consolidation and settlement properties of the material
- wet weather working
- type of plant to be used.

Ditches, land drains or French drains are generally located at the toe of finished embankment slopes depending on the topography of the site. Ditches may also be required to drain the working area as the filling proceeds, especially if an existing watercourse or drain is disrupted when a permanent diversion has to be provided.

7.6.9 Compaction and testing

Correct compaction of embankments is essential and must be strictly in accordance with drawings and specification.

Choice of the most suitable compaction plant and careful planning are essential for the most economic results. Refer to the Highways Agency Specification for Highway Works and the contract specified requirements.

For general fill areas, the specification will usually require that the *in situ* density is a percentage of maximum dry density, usually 95%. The strength of the formation for road construction will usually be specified as a percentage California Bearing Ratio (CBR). The CBR tests can be carried out in the laboratory but some Highway Authorities may require *in situ* CBR tests. See also BS 1377.

7.6.10 Quantities and planning

The design should have ensured the most economical balance between cut and fill, but this may have been based on a small amount of information of the soil types to be encountered and also there may have been only little knowledge of the local cost and source of fill material to be used. Therefore if site conditions vary greatly from the information available at the design stage, advise the engineer of these changes so that the need for any redesign can be considered.

If the designed levels are the most economical, the locations of the cut and fill areas should be available and careful planning of haul routes and working methods should be completed before starting work.

Note: Do not rush. Careful planning can save time and money.

Check any quantities provided and check whether or not they include construction thicknesses of any surfacing or sub-base.

Full details of the quantities are required to plan the most economic operations. Just the total quantities of cut and fill in cubic metres are not of great use without knowing the average depths and areas involved. This especially applies to topsoil quantities over large areas—a small increase in depth can greatly increase total volume. This is important when considering how much site control is required in supervising topsoil strip. In handling any quantity of soil, allowance must be made for bulking which is usually in the order of 10%.

Before commencing work on site, record a grid of existing ground levels and compare with the original survey. A long period might have elapsed between original survey and start on site and any discrepancies between the two sets of levels must be brought to the attention of the engineer before work commences.

7.6.11 Drawings

The regrade information will generally be shown on the following drawings:

- Road longitudinal sections augmented by the road construction details showing the construction depths.
- Regrade layout that normally shows the finished levels as contours superimposed on the existing ground contours. Spot heights may clarify areas not covered by regrade contours. The layout will also show any features closely linked with regrade, e.g. retaining walls, banks or steps.
- Site cross sections showing proposed and existing levels.
- Cut and fill drawing with areas of cut and fill areas delineated by colour. If this drawing is hand drawn, it merely denotes areas of cut and fill with no consideration given to the depth. However, computer plotting can show a range of depths denoted by density of shading.
- Quantities produced using computers.

Note: Before starting regrade, study carefully all drawings and information.

7.7 Minor regrading or landscaping

This is the process of re-profiling surface areas between structures and surrounding constraints such as roads and boundaries. This minor regrade is often carefully designed to minimise in order of priority: retaining walls, tanking, under-building, banks and soil movement. However, draw the engineer's attention to any opportunities to reduce these items while maintaining acceptable bank slopes, gradients for vehicles and pedestrians etc. Do not make unapproved changes.

Plan operations to reduce double handling of material and to prevent unacceptable contamination by builders' rubbish. Where excavated material is to be used as fill elsewhere, plan to excavate, transport and fill in one operation whenever feasible. Do not disturb roots of trees to be preserved nor place excessive fill around such trees.

The maintenance of finished ground level at least 150 mm below any DPC is essential. When turfs are to be laid over regraded areas, ensure that sufficient allowance is made close to buildings.

Where banks and slopes are to be formed, ensure that the minimum widths of flat areas adjacent to buildings (1 m) and footpaths (0.75 m) are maintained. Also, where rotary clothes driers are provided in private development there must be no more than 300 mm fall in any direction over the area covered by the drier.

Surplus and poor quality fill may often be used in open space areas provided that the final surface is adequately drained and stable.

Where there is a deficit of fill material to complete the designed regrade, notify the engineer as soon as possible so that redesign can include the largest possible area. Beware of bridging DPCs if working close to buildings.

7.8 River bank retention

There are numerous methods available varying greatly in cost. Temporary works may be required in order to complete the permanent works. Such temporary works must be designed by a qualified engineer.

The final design can be influenced by degree of retention required, design life, cost, approval of statutory authority, hydraulic characteristics of a watercourse and ease of maintenance. The main types are:

Bagwork

Hessian or porous plastic bags filled with dry concrete or sand and cement. These are laid to the design profile in interlocking courses (like stretcher bond brickwork) and are only suitable for shallow depths with relatively tranquil flow.

Concrete lining

This can be reinforced or un-reinforced with movement joints and may or may not include concreting the bed of the watercourse. Where the bed is not concreted, protection against scouring and undermining of the banks must be considered.

Interlocking steel sheet piling

This is suitable where vertical banks form a continuous wall. Where scour of the bed is likely, protection will be required to prevent the piles being undermined. Where the retained height is shallow the piles can act as cantilevers. For greater retention heights, the tops of the piles are anchored back to king piles or concrete gravity blocks.

Interlocking concrete blocks

These come in many shapes, sizes and finishes depending on the service requirements. The method of construction and use will depend on the type specified. Manufacturers' instructions, design drawings and specification will need to be closely followed.

Stone and block pitching

This is similar to interlocking concrete systems, except natural stone or cut blocks are used. Joints may be pointed with sand/cement or asphaltic mastic to improve stability. Asphaltic mastic is preferred to allow small movements without cracking.

Gabions

Steel mesh baskets or mattresses are filled with natural stone or broken concrete. For increased stability a pourable asphalt mastic may be used to fill the joints (see pro-prietary literature).

Geotextiles

There are many recent advances in the use of geotextiles with grass and fascine work (follow current technical literature).

7.9 Retaining walls

All retaining walls must be properly designed and construction details provided by the engineer. Construction must be in accordance with the specified requirements. Take particular care with backfilling and back drainage which can cause severe instability if not carried out properly. The wall must not be surcharged during or after construction unless so designed.

An outline of the principal forms of retaining walls is given below:

Reinforced concrete

Generally this is expensive for low wall situations, reinforced concrete is therefore normally used for high structures or where there are problems with limited space for the works.

Reinforced concrete walls rely for stability on adequate soil foundation and the vertical loading onto the base slab from the retained material. Alternatively a forward projecting base slab can provide stability. Piled foundations can be used in weak ground.

Mass concrete

These rely on the soil foundation and their own mass to retain the soil behind. They are infrequently used nowadays as cheaper alternatives are available.

Brickwork

There are significant differences between low brick retaining walls and higher walls with heavy loading. Generally it is cheaper to construct a high reinforced concrete wall with a brick facing and therefore massive brick retaining walls are no longer built.

Reinforced blockwork

This is an economical form of construction for low- to medium-height retaining walls. Care in construction to tie in reinforcement properly and to infill with concrete at specified levels is most important. Specified heights must not be exceeded by adding an extra course of blocks.

Crib wall

This is made from interlocking precast concrete or treated timber components which form a three-dimensional grid. This is then infilled with non-cohesive soil or stone or both to form a mass-type retaining wall. Various sizes are available for different heights and the width of the wall can be varied with height by reducing the number of grids. Although these types of walls are easy to assemble and modify, it is essential that they are built in accordance with a proper design and specification.

Reinforced earth

This technique is used where the material to be retained is yet to be positioned. It is ideal for reinforcing a road embankment where a battered slope would occupy too much space. It is constructed by placing a row of interlocking pre-cast concrete panels on a strip foundation, galvanised metal strips are then bolted onto the back face of the panels and compacted in with the layers of fill material. The wall height can be increased by adding successive panels, strips and fill material (typically a sandy gravel). A composite structure is formed and remains stable due to the friction between the reinforcing strips and the fill material being greater than the outward force the fill material exerts on the rear of the panel.

Geotextile

Systems such as the 'Tensar Geogrid' use a polypropylene mesh interwoven between successive layers of soil: this acts as reinforcement between the soil layers producing a composite mass. Vegetation will grow on the exposed face to bind the soil further.

Paving slab

The paving slab on edge is a very cheap way of retaining up to, say 500 mm where space is too restricted for a bank. Where the retained height varies or runs down a slope, the top edge of paving should be cut to the slope and not stepped, which gives an unsightly appearance. In certain circumstances it may be necessary to provide damp-proofing arrangements.

7.10 Excavations and stability of ground, slopes and rock faces

For open excavations, two main factors govern the angle of slope. First, consider the type of ground, and second, the possible damage caused by a slip. Flow of ground-water or interference with the flow could initiate a slip. Drainage may be required to direct surface water away from the slope.

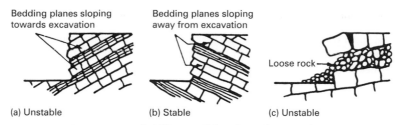

Figure 7.2   Stability conditions in rock excavations.

If, for example, important property is close to the top of the excavation, there obviously must be no risk of a slip and a high factor of safety must be adopted.

Mass movements of the ground can occur due to mining subsidence, swallow holes, land slips on unstable slopes, rock falls, creep on clay slopes and frost or water damage to newly exposed faces. The sides of rock excavations must not be assumed to stand vertical safely. Their stability depends on the angle of bedding planes and degree of shattering of unsound rock. Bedding planes sloping steeply towards an excavation provide dangerously unstable conditions, especially when groundwater lubricates the bedding planes. Only when the bedding planes are horizontal, or sloping away from the excavation, may stable conditions be assumed. An apparently soundly bedded rock face can conceal a mass of shattered rock fragments behind, which are liable to fall out causing undermining and major collapse of overlying sound rock.

In certain types of soil it may be necessary to take care of heave effects on surrounding structures (see Fig. 7.2).

7.11   Steel sheet piling

7.11.1   Method and safety

- Piling contractors must be requested to provide a written method statement as appropriate for the piling operations. Find out what induction training and information specific to a method statement is provided by immediate site supervisors to piling operatives.
- Note that cranes must be selected and used in accordance with CP3010 and with the Construction (Lifting Operations) Regulations 1998. A firm level base of adequate bearing value must be provided, or crane mats used.
- Check that cores of pendant/bridle ropes are not fractured.
- Any crane used for raising or lowering operatives must be fitted with a dead man's handle and the descent must be effectively controlled; the latter may be achieved by power lowering. Properly constructed man-carrying cages, which are unable to spin or tip, must be used. The cages should be regularly and carefully inspected.
- All lifting appliances and gear must carry appropriate certificates of test and examination, and must be adequate for the job, paying particular attention to the risk of damage to gear by sharp edges.

- All personnel working on piling operations must wear safety helmets (helmets appropriate for piling work are now available with retaining strap and smaller peaks). Ear and eye protection should be provided.
- Piling machine operators must be at least 18 years of age, trained, competent, medically fit and authorised by site management to operate the machine.
- When piling from a pontoon or adjacent to water, personnel must wear life jackets. Rescue equipment (e.g. a safety boat and lifebuoys with lifelines attached) must be kept ready for immediate use and enough operatives must know how to use it.

7.11.2 Materials handling

- When splitting bundles of sheet piles, use chocks. If large quantities of piles are handled, the use of purpose-made strops and grips is advised.
- Piles should not be stacked too high or in a cantilever position. Use spacers and chocks where necessary. Tubular piles should not be stacked more than four high and should be properly chocked.
- When lifting piles or piling hammers, use hand lines to control the load. Give due consideration to wind speed during the operation (see Section 7.11.4 Pitching piles).
- Check the dimensions and alignment of clutches. If necessary perform trial clutching of piles. This is advisable for Z-shaped pile sections where the clutch is less *positive* than on trough sections (e.g. Larssen piles).

7.11.3 Gate systems

See Fig. 7.3.

- Positioning (pitching) and driving sheet piling is usually done by using a temporary supporting structure (gate). This is made up of heavy steel or timber H-frames supporting horizontal-heavy H-beam guides. The H-frames can be supported on heavy steel spreaders or specially cast concrete blocks. Ensure an adequate foundation under these frames and bases to prevent subsidence and over-turning during piling operations. This is particularly applicable during work in rivers, etc.
- If using concrete blocks ensure that they are suitably reinforced to withstand loads from lifting and shock loading. Vertical steel columns should have a good bottom fixing. Vertical timber should not be cast into the block but should be wedged and bolted. Where doubt exists over stability use guy lines or raking steel props.
- All horizontal gates with platforms over 2 m high, or over any potentially dangerous areas, must be provided with adequate guardrails, toeboards and correct ladder access.
- Ladders must be secured and extend at least 1 m above staging.
- If using a cantilever system, use a tie-back where possible, as well as kentledge to provide safe anchorage.
- When piling is progressing and temporary piles are used to support the gate system, use purpose-made brackets and bolt them to the piles. Any welding necessary should be carried out by competent welders.

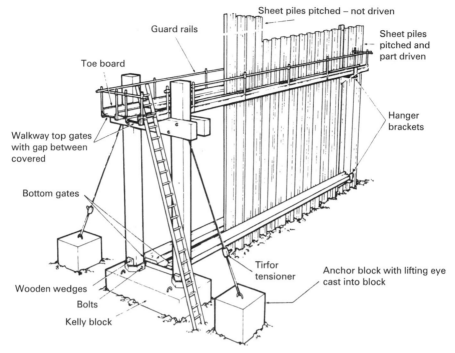

Figure 7.3　Gate system using concrete bases (showing piling, the use of hanger brackets and the provision of a safety walkway).

7.11.4　Pitching piles

- If shackle holes have to be burned in the pile, remove sharp burrs to prevent damage to shackle pins.
- Use quick-release shackles wherever possible:
  - the sheet pile must not be lifted vertically without first checking that the pin is properly engaged through the sheet
  - do not use a pull rope less than 5 mm diameter
  - the length of rope used must be less than the length of the pile, to prevent the extra rope snagging and pulling the release for the shackle
  - secure the rope around the sheet pile to prevent snagging
  - if a special lifting eye is to be welded to the pile for angled pitching, the weld should have a factor of safety of at least 2.
- Pitch long sheet piles with a pile threader, following the manufacturer's guidance for use. Where this is not possible, use a pile pitching cage. The cage is normally hung from an adjacent pile, the operatives wearing safety harnesses hooked to the adjacent pile before the crane hook is removed from the cage.
- When feeding sheet piles through the top and bottom gates, use wood blocks or a bent bar. Never use a straight pinch bar, as fingers can easily be trapped. Use a spacer block between the guides to keep the leading *free* clutch in its correct alignment.
- Where access and work is carried out from ladders:

- *Clutching*: the ladder must be placed in the valley of a previously placed pile; the ladder must be footed and when at the top of the ladder both hands are required for clutching, a safety belt must be worn and secured to the pile using a manlock.
- *Wedging*: the ladder must be placed against the H-beam and footed wedges should be pre-placed on the beam. A 4 lb. lump hammer must be used as this can be swung with one hand but if two hands are required, a safety belt must be used with the lanyard wrapped around the H-beam or used with a manlock.

Note: At all times safety helmets and footware must be worn. When working at height the helmet should be secured with retaining strap.

- Changes in work method:
  - The work method must not be changed without consultation of the senior site representative responsible for the piling operation.
  - If windy conditions (e.g. over 30 mph gusts) make the handling of the sheet piles difficult, stop work until the senior site representative responsible for the piling operation has been consulted and a safe method of continuing the work has been devised.

Further reading

Standards and Codes of Practice

BS1377: 1990 *Method of Test for Soils for Civil Engineering Purposes*.
Note: This BS is in several parts dealing with such topics as general requirements; sampling;classification; chemical, electro-chemical compaction-related compressibility, permeability, and durability tests; consolidation and permeability tests in hydraulic cells and with pore pressure measurement; shear tests (total and effective stress) and *in situ* tests.
BS5607: 1998 *Code of Practice for the Safe Use of Explosives in the Construction Industry*.
BS5930: 1999 *Code of Practice for Site Investigations*.
BS6031: 1981 *Code of Practice for Earthworks*.
BS8000–1: 1989 *Workmanship on Building Sites*: Code of Practice for excavations and filling.
BS8004: 1986 *Code of Practice for Foundations*.
CP3010: 1972 *Code of Practice for Safe Use of Cranes (Mobile Cranes, Tower Cranes and Derrick Cranes)*.
Note: See also Chapter 3.
BS7121: 1989–1999 *Code of Practice for Safe Use of Cranes*.
Note: This Standard is produced in several parts. See also Chapter 3.

Other related texts

Note: For HSE documents see Chapter 19.
Deep Excavations: M. M. Puller, London, Telford, 2nd ed., 2003.
Underpinning: S. Thorburn and J. Hutchison (eds), Glasgow UK, Surrey University Press, 1985.
Foundation Design & Construction: 5th ed., M. J. Tomlinson, Harlow UK, Longman, 1986.
Construction Materials Reference Book: D. K. Doran ed., Oxford, Butterworth-Heinemann, 1991.

Construction Materials Pocket Book: D. K. Doran, Oxford, Newnes, 1994.

Site Investigation: I. K. Nixon *et al.*, Oxford, Butterworth Civil Engineer's Reference Book 4th ed., (L. S. Blake ed.), 1989.

The Effect of Wet Weather on the Construction of Earthworks: CIRIA, London, CIRIA Report RR3, 1995.

Proprietary Trench Support Systems: CIRIA, London, CIRIA Report TN95, 1986.

Trenching Practice: CIRIA, London, CIRIA Report 97, 1992.

Specification for Highway Works: HA, London.

The Party Wall etc. Act 1996: Explanatory Booklet. ODPM, London, 2002.

Chapter 8 FOUNDATIONS

Michael Tomlinson

Basic requirements for best practice

- Clear drawings and specifications
- Good site investigation
- Adequate testing procedures
- Good safety practice
- Adequate protection to opened-up work

8.1 General

Some background design information is included in this section so that the reasons for necessary site actions are understood. Foundation design should have taken into account the local ground conditions, the possible types of ground movement and the response of the structure to the effect of this movement on its foundations. These movements can arise from the application of loads on the ground, or their removal from it, or from natural ground movements, e.g. frost, earthquakes, mining subsidence.

This movement can affect the function of the structure, external connections and relationships to adjacent structures. Relative or differential movement within structures, unless limited, can lead to distortion with damage to structural members, claddings and finishes. New construction can lead to additional settlement of adjacent structures.

Ground conditions affecting the choice of foundation type are:

- low bearing capacity soils: i.e. less than 100 kN/m²
- filled site
- trees and shrinkable clays
- mining (past, present and future)
- high water table or alteration of water table
- hill creep
- swallow holes or solution cavities
- high sulphate, acid or other aggressive conditions in the ground.

Note: If it is suspected that there are conditions likely to lead to a foundation problem developing then contact the designer, architect or engineer responsible and if necessary the quality manager.

8.2   Site investigations

It is false economy to omit site investigations and soil tests before designing foundations. Be fully aware of all information and requirements in the site investigation report before work starts on site. Every excavation on site is in effect a trial pit and every opportunity should be taken to observe the soils. This information can be useful to gain a general picture of the soils underlying the site.

Cohesive soils are liable to show volume changes and require careful consideration. Take care if present or past trees and hedges are located within or close to the site, and consideration must be given in foundation design to types of trees if future planting is envisaged.

8.3   Treatment of bad ground

There are a number of established methods to aid foundation construction in difficult ground conditions, e.g. groundwater lowering, injection of grouts, compaction and freezing (see also Chapter 7).

If consideration of these techniques is left until difficulties are encountered during construction, appreciable delays will result while the necessary geotechnical process is carried out.

8.4   Foundation construction

8.4.1   Site exploration and preparation

The mechanisation of site operations demands an efficient system of temporary roads and good site drainage to maintain a high rate of work in all weathers. Site preparation must include tracing and clear marking of all underground services, power, telephones, gas, water and sewers. The accidental cutting of electric cables and gas mains can kill people, as well as resulting in claims for damages if supply is cut off from, for example, a factory. Repairing damaged telephone trunk lines or television cables can cost tens of thousands of pounds. Burst water mains can cause the flooding and collapse of incomplete excavations with disastrous results. Collaborate with the supply authorities on safety measures whenever the site is approached closely or crossed by overhead power cables or underground services.

8.4.2   Dilapidation and condition survey of adjacent properties

Jointly with the owners, occupiers and their professional advisers if appointed, record and agree the condition of their properties before our operations start. This record

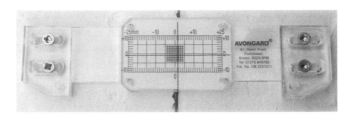

Figure 8.1 Tell Tale Plus (courtesy Avongard Ltd., Bristol).

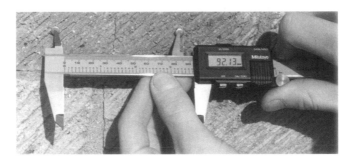

Figure 8.2 Digital caliper (courtesy Avongard Ltd., Bristol).

will be invaluable in the event of any subsequent damage claims and can provide information to protect your company. This information assists in agreeing the causes of movement.

Record in detail the decorative and structural condition. Photograph all significant points. Fix tell-tales across cracks and arrange for both parties to record readings at appropriate intervals during the progress of the works. Once fixed they will enable all movements to be monitored.

Proprietary graduated perspex tell-tales are available. (Fig. 8.1 shows a standard flat tell-tale). Corner, floor and displacement types are available. They can be read to 0.25 mm by eye. For more sensitive readings or less noticeable tell-tales, small discs set in each side of a crack can be monitored using vernier gauges (see Fig. 8.2 for digital version). Glass tell-tales are vulnerable to damage and should not be used.

Extensive and continuous monitoring can use electronic transducers, recording meters and printers providing automatic records. This system is useful when regular access to locations requiring monitoring is difficult or expensive.

Where necessary, establish and agree levelling points and record in a similar manner. Levels should be referred to a *stable* datum point, located clear of operations by mechanical plant.

Surveys should also include condition of roads, paving etc. This is particularly important where deep excavations, blasting, or dynamic compaction or piling works are involved.

8.4.3 Foundation depths

The drawings should show the site investigation report reference and state the allowable net ground bearing pressure used for the design of foundations, the depth at which foundations are to be placed and the soils expected at this level. Concrete mix details will take into account any aggressive chemical conditions discovered by the site investigation.

No variation must be made to the specified depths of foundations unless first agreed with the engineer or architect as appropriate.

Note: In case of doubt, or if ground conditions differ from those expected, notify the engineer or architect and obtain his instructions in writing. If there is any likelihood of adjacent foundations being at all affected, site management must obtain written instructions from the engineer or architect before proceeding with the work.

8.4.4 Safety in excavations

It is necessary to support the sides of all excavations properly to ensure stability and safety in all weather and groundwater conditions. (See also Chapter 7 and your company safety instructions and HSE requirements relating to confined spaces and toxic atmospheres.) Any underground space is a potential death trap due to possible poisonous, asphyxiating and explosive atmospheres. Before entering such spaces test the air, and assistance and breathing apparatus must be immediately to hand.

Note: *By law*, all excavations deeper than 1.2 m must have side support or be excavated to a stable slope.

8.4.5 Excavation for foundations and preparation of formations

Any soft spots encountered at formation level are normally replaced with approved and properly compacted granular material, or lean mix concrete. See the specified requirements. Clays and silts are highly susceptible to softening when in contact with water. Subject to the approvals referred to in Section 8.4.8, clay formations should be protected with blinding concrete or with the foundation concrete as soon as possible after completion of the excavation.

Formations in granular soils will usually have been loosened by the excavation process and should be compacted using suitable vibratory plant prior to placing blinding concrete.

Surfaces of rock are required to be sound, completely exposed, and generally normal to the direction of load and of a bearing capacity required in the design. Soft rock surfaces (e.g. chalk) are swept clean of loose debris then blinded with 50 mm of 10 N/mm^2 concrete to prevent softening by rain.

Hard rock formations are preferably cleaned with water or a water jet, followed by an air jet to remove excess water. Remove any standing pools of water. Keying of footings into the rock may be required. Study thoroughly the specification and drawings. If blasting is used, this must be carried out by specialists who must control charges to avoid damage to other foundations or other adjacent property.

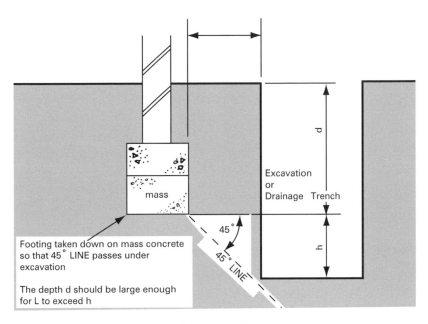

Figure 8.3 Avoidance of surcharge on excavation.

8.4.6 Avoidance of surcharge on adjacent excavations

It is essential that foundations do not surcharge existing drains, trenches, retaining walls and other excavations. To achieve this, take under-sides of new foundations down so that a line, at 45° from the nearest bottom edge of the new foundation, passes under the drainage or trenches (see Fig. 8.3). Conversely to prevent new excavations from undermining adjacent foundations, excavations must not be taken down below the 45° line. In these circumstances the Party Wall etc. Act, 1996 may apply. A useful *free* booklet explaining this act may be obtained from the Office of The Deputy Prime Minister.

8.4.7 Records of obstructions

Ensure that all obstructions are recorded precisely. Records must show nature, location, depth and dimensions of obstructions, and be similar to but separate from the record of foundations. The engineer must be promptly told about such obstructions so that any significance these may have on design may be investigated.

8.4.8 Bottoms of excavations (formation level)

Formation levels of excavations should be inspected and details passed by the architect, statutory authority, the engineer or client's representative as required by the contract. Immediately after approval, all formation levels other than hard rock are to be protected (normally by blinding with concrete) for protection against the weather and to provide a firm working surface for subsequent reinforcement fixing operations if needed.

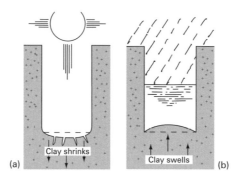

Figure 8.4   Trenches in clay need protection.

8.4.9   Foundations in shrinkable clay

The specified formation level will normally be below the zone of cyclical shrinkage and swelling due to seasonal rainfall, and taking into account the significant effects of trees, shrubs and hedges.

Site survey drawings normally indicate all past and existing trees and hedgerows. Account of these should have been taken in the design of foundations. Trees, stumps and root systems of dead trees can occur outside the site boundary, but close enough to affect new foundations. If old stumps or root systems are discovered or if there are any doubts about the possible effects of trees and hedgerows over the site boundary, the engineer should be consulted before proceeding.

The required depth of foundations as a safeguard against damaging movement caused by swelling or shrinking of clay soils in open ground or where influenced by growing or cut-down trees depends on three factors:

- the water demand of the clay
- the long-term rainfall conditions (climatic zone)
- the type and mature height of the tree or shrub.

The NHBC Standards, Chapter 4.2: 'Building near trees', provides detailed guidance on foundation depth requirements taking the above factors into account.

Formation surfaces in clay should be protected from the softening effects of rain and drying shrinkage until inspected, then immediately blinded with 50 mm of 10 N/mm² concrete, or otherwise cover in accordance with the specified requirements. Remove any soil inadvertently softened before blinding. See Fig. 8.4a and b.

Delay in concreting a strip foundation may cause problems, particularly in clay. Where compressible material or void formers are provided to the sides of trenches to absorb horizontal clay swelling they must be set truly vertical and secured against floating in the wet concrete.

8.4.10   Concreting foundations

Before placing concrete ensure that formation surfaces are clean and trimmed and that any side formwork is securely strutted with all gaps at joints between panels packed to prevent grout leakage. If this occurs and the concrete contains reinforcement, make good the honeycombed concrete before backfilling to prevent ingress of groundwater to the reinforcement.

Struts bearing against the earth sides of excavations must bear on adequate spreader plates or timbers.

The materials, the mix, the water content and the methods of mixing, transporting, placing and curing the concrete are to conform fully with the specification. Sulphate-bearing ground or groundwater will probably call for sulphate resisting cement to be used in the concrete mix. Surfaces against which concrete is to be placed must be firm and free from loose material. If concrete is to be placed upon or against, and is required to bond with, old concrete surfaces, clean the surface of the previously cast concrete of oil, grease, or other foreign matter and laitance, preferably by wet sand-blasting. Ultra-high pressure water blasting is also an appropriate method. Surfaces sometimes require roughening. Roughening for its own sake is not necessary to obtain bond if a thoroughly clean surface, comparable in cleanliness to a fresh break, is obtained. Such a clean joint surface approaching dryness without free water is best for bond strength.

Avoid damaging the formation if sand or water blasting methods are used. Give all reinforcement the specified amount of concrete cover. Spacers and chairs to BS 7973–1: 2001 should be used in accordance with BS 7973–2: 2001 to support reinforcement. Reinforcement should be tied in accordance with the requirements of BS 7973–2: 2001, clause 5 (see also Chapters 12 and 13).

When concreting foundations, arrange for competent and *continuous* supervision of this operation *until completion*, then check and record the finished concrete levels. Comply with the procedures in your company standing instructions or procedures.

- Sun and wind will drive the moisture out and shrink the exposed clay. Once placed, the foundation may suffer heave movement later when the clay takes up moisture and swells.
- Similarly, a foundation placed on a wet, swollen clay base will settle later as the clay compresses under the wall load.

8.5 Spread foundations

8.5.1 Strip footings

These are used to spread the linear load from walls of brick, blockwork or concrete directly onto the ground at the specified level. The significant design feature will be the width of the footing and it is necessary to ensure that this is maintained as specified. If the trench is excavated by machine, the formation level is normally cleaned off by hand for inspection and approval. If the strip footing is to contain reinforcement, blinding concrete is first laid to the correct level.

Ensure that no soil spills into the trench after blinding, during the fixing of any reinforcement and before concreting. Provide formwork to sides in vertical-sided trenches in bad ground, or splay trench sides back if space allows. If the footing is to support a reinforced concrete wall, fix the starter bars to their correct line with sufficient lacer bars, supplied and fixed to retain the starter bars in this position

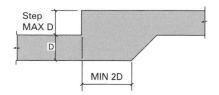

Figure 8.5   Stepped footing step detail.

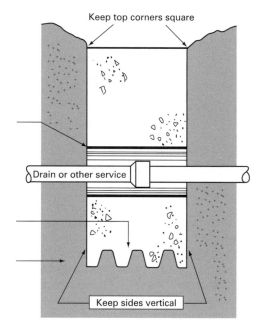

Figure 8.6   Deep strip or trench fill footing.

during concreting. Narrow or trench fill footings are a particular type of strip foundation for low-rise buildings foundation where trench sides will stand vertically without use of support.

When strip footings are required to be stepped, keep within the limits shown in the following step detail (Fig. 8.5). It is important that trenches for narrow strip foundations in clay soils should be set out and excavated accurately. Foundations in stiff clays may be no more than 375 mm wide and if they are incorrectly positioned the subsequently constructed load-bearing walls could be eccentric to the centre line of the foundation. There is an increased risk of such an occurrence if the concrete is allowed to spill out at ground surface thus obscuring the actual location of the trench.

Keep the trench sides vertical and the corners square at the top of the trench fill foundation (Fig. 8.6). Do not form a *mushroom head* by concreting higher against the rounded trench edges. Cutting away excess concrete is expensive.

When providing services through deep strips or trench fill footings, a pipe, duct or box is fixed into the trench before concrete is placed. The duct should be sensibly larger than the drain or service to be installed, and positioned accurately.

Concrete should be placed immediately after approval. In clay soil, an excavator bucket often leaves tooth marks in the bottom of the trench. In the case of deep strip or trench fill footings only, this does not matter if the clay between them is firm. Any loose soil must be removed.

8.5.2 Pad foundations

These spread the load from one or more columns, or brickwork piers. Except in hard ground, formwork is almost always needed to confine the concrete to the required shape and ensure that the side surfaces of the concrete foundations are sound. This is essential in sulphate-bearing ground even when the foundations are un-reinforced and in all cases when reinforcement is included. Formwork to sides of foundations must be grout tight to avoid honeycombed pockets of concrete being left. Provide the required reinforcement cover to bottom sides and tops of bases by following the requirements of BS 7973.

Position column starter bars accurately and provide adequate lacer bars to maintain this position during concreting. If the foundations are to receive steel columns then position the appropriate holding down bolts and assemblies and retain using appropriately substantial templates (see also Chapter 14).

Where the pad foundations are providing pockets for precast columns, position the formers for the pockets accurately and secure against flotation in the wet concrete. Do not place precast columns in the pockets until the concrete of the base has sufficient strength to withstand the high local stresses around the edges of the pockets, caused by the wedges commonly used in temporarily positioning these columns. In cold weather, prevent freezing water in pockets from damaging concrete by draining or filling pockets with rags, hessian etc.

Checklist for strip and pad footings

- Check the setting out of the foundation.
- Check that the foundation excavations conform to the setting out lines.
- Ensure that a level monitoring system for adjacent structures (if any) is in operation.
- Inspect the completed excavations to ensure that they have been taken below any fill or weak soil and below the root zone of any vegetation in clay.
- Ensure that obstructions and existing services have been recorded and properly dealt with.
- Inspect excavation supports (where necessary) for safety and stability.
- Inspect the bottom of excavations to ensure the removal of loose debris and possible softening due to recent rain.
- Check the alignment of compressible material or void formers (where required).
- Check that formwork (where required) has been fixed within the specified tolerance and has been properly supported.

- Check bar dimensions, fabric sizes, assembly and tying of any reinforcement.
- Check the spaces, chairs and cover to the reinforcement and the positioning of any inserted fixings (e.g. drainage pipes).
- Make a final inspection of the excavation bottom or blinding concrete.
- Check the proportioning of site- or ready-mixed concrete. Take test cubes as required.
- Check the surface level of concrete after placement.
- Protect green concrete from rainfall or surface water.
- In hot weather, protect green concrete from premature drying.

8.5.3 Ground floor slabs (small buildings)

For reasons of economy whenever possible these are designed as *floating*, i.e. carried directly by the ground. The depth of fill beneath a floating ground floor slab, including any hardcore or other sub-base, should not normally exceed 600 mm. Check with the designer if a greater thickness is shown on the construction drawings or results from the replacement of soft pockets of soil.

It is essential to follow specified requirements. Any sub-base laid must not contain degradable material. Check the compatibility of any fill material with the existing and foreseeable chemical ground conditions. Hardcore layers are blinded with sand or concrete to receive the specified membrane, which is usually lapped with the horizontal DPC in the walls. Do not puncture this membrane. All DPC and membrane details should be shown on architects' or engineers' drawings or specification. The concrete ground floor slab, with or without reinforcement, is then cast over the membrane to the required thickness and finished as specified. Use soft substrate 'A' spacers to BS 7973: 2001 to prevent DPC membrane from being punctured.

In shrinkable clay conditions a suspended slab must generally be provided, either using proprietary precast concrete units or adequately designed and reinforced *in situ* concrete casts on a void forming or compressible material.

8.5.4 Raft foundations

These are used on low-bearing capacity soils; in areas of past or future mining, or on ground of varying compressibility, as the stiff slab and beam construction is utilised to bridge over the more compressible soil. They are sometimes used where a high water table precludes digging foundation trenches.

8.5.5 Bad ground foundation slab (housing)

The full area of this is not required to carry the load from the structure to the ground and so it does not function as a raft. It is generally used on low-bearing capacity soil.

8.6 Piled foundations

8.6.1 General

Piles derive their carrying capacity from a combination of friction along their sides and end bearing at the pile point or base. The former is likely to predominate for piles in clays and silts and the latter for piles terminating in a stratum such as compact gravel, hard clay or rock.

If piles are installed through significant thickness of fill or soft materials *drag down* or negative skin friction forces can act on the piles. It is important to ensure that the piles have been designed with these forces in mind. The engineer should be advised if thicknesses of fill in excess of those anticipated from the site investigation report are encountered in the work. Systems for installing piles to the prescribed depth or set per blow of the driving hammer should not be changed without the agreement of the engineer and the piling contractor. Precast concrete piles should be at least 28 days old before driving. Check the subcontractor's casting records, and that the spacers for the cover to the reinforcement comply with the requirements of BS 7973.

Piles are normally designed to take axial load only. Inform the engineer where any pile positions have been changed to miss obstructions or where tolerances of position have been exceeded so that the pile design can be modified. Additional piles may be required to balance out eccentric loadings from columns above. Permitted tolerances on the position of the pile are shown in the relevant pile specification and working drawings.

The two main types of piling in general use are:

- Driven piles: preformed units, usually in timber, concrete or steel, driven into the ground by the blows of a hammer or by jacking.
- Bored and cast-in-place piles: piles formed by boring a hole with or without casing into the ground and filling it with concrete.

With driven piles, the ground is displaced as the pile is driven or jacked into the ground, and ground heave can occur especially in clay soils and silts. Allow for this before starting a reduced level dig. Check the elevation of pile heads after driving adjacent piles. Re-drive any piles which have lifted due to ground heave.

In all forms of bored piles, arrangements must be made to remove all bored spoil promptly and continuously. When concrete is placed in cast *in situ* piles by tremie pipe or through the stem of a continuous flight auger (CFA pile), the concrete emerging at the ground surface is likely to be contaminated with soil and may be segregated. Placing should be continued until properly mixed, uncontaminated concrete emerges at the ground surface or at the bonding level with the pile cap, whichever is appropriate.

A proposal to use driven piles on a site is normally preceded by an appraisal of the effects of the shock waves and vibration inherent with this method. These can affect nearby underground services and buildings. Such appraisals include preliminary surveys as outlined in Section 8.4.2 and where necessary agreeing with adjoining owners such proposals for monitoring the effect on their structures as may be necessary during the piling operations. Consultation with local authority environmental health departments is often necessary to agree acceptable limits of noise and vibration. The possible provision of such monitoring is best considered in the development stage of a project. Consult the engineer in good time. *Do not wait until after damage has occurred.*

Piling in water-bearing ground sometimes utilises bentonite in suspension to retain the bore profile. This is a specialised process. The retention of its material properties if recycled requires special monitoring, and the operational use of this material in both the hole boring and pile concreting requires particular supervision.

The engineer will no doubt wish to examine the subcontractor's proposals. Arrangements should be made with the local authority for disposal of used bentonite. It should not be discharged into water courses where it could cause pollution.

8.6.2 Pile caps

Protect the bars projecting from a pile into the pile cap from damage, displacement or cutting during the excavation for the pile cap and the cutting out of the surplus concrete at the pile head, which is usually concreted to 300–400 mm above cut-off level. The pile cut-off level is normally required to be 75 mm above the under-side of the pile cap. It is important that the pile projects into the pile cap, therefore take care to cut off the piles at the correct level. Any low cut piles should have the blinding around them dished down to achieve this. See the engineer's drawings for details.

8.6.3 Supervision, checking and testing

In any form of piling, the vital parts are not visible. It is, therefore, essential that all operations are properly supervised and recorded. Keep the engineer informed as piling work proceeds.

Particular attention is drawn to the design and maintenance of the piling (working) platform. Inattention to this has led to a number of accidents in which piling plant has collapsed due to insufficient support. Some accidents relate to poorly backfilled trenches that have been dug across piling platforms before or during the commencement of piling.

It is recommended that the provisions of *Notes on the Design, Installation of Working Platforms for Plant for Specialist Foundation and Geotechnical Works* is followed. This documentation is available from the website of The Federation of Piling Specialists (FPS).

It is the main contractor's responsibility to ensure that supervision of the piling subcontractor is by an engineer with previous piling experience. Where large numbers of piles are involved, this should be his sole task in view of the detailed work it entails and the time and money that may be saved by avoiding trouble during and subsequent to installation.

If you are the supervising engineer, you must be in possession of the specification and drawings included in the subcontract so that you know the type, length and size of each pile, together with its setting out, cut-off level, working load, concrete quality, reinforcement, cover and spacer details. You should also have a copy of the piling tender and the enquiry documents. These will state the terms of responsibility for setting out. Particular attention is drawn to the number of cube tests on the concrete that have been specified.

Maintain a record of the installation of every pile using the documentation procedures defined in your company standing instructions or procedures. In the event of a pile being suspect, these records are likely to be the only information available to your company on which future action may be based. Inform the engineer immediately if there are appreciable differences between the lengths or sizes of piles driven compared with those specified. Also make the engineer aware should the driven lengths of adjacent piles vary significantly.

Observe piling rigs and concreting equipment. Many delays have been experienced from subcontractors with badly maintained equipment; these are the ones who fail to have spares to hand for quick repairs. In addition to any preliminary test piles, it is sometimes required to test working piles after installation. The selection is best made in conjunction with the engineer and is usually based on:

- a location on the site which will least inhibit progress on substructure construction
- knowledge of any piles having had problems during installation: do not give piling subcontractor prior knowledge of which working piles will be tested.

In the case of cast *in situ* concrete piles it is now becoming usual to have the construction of these checked by one of a number of methods of 'integrity' testing. This is done with electronic equipment, sometimes installed in a vehicle that will require access to the pile positions. On a recent contract, ten piles (600–1800 mm diameter) were checked each day.

Where large-diameter bored piling is used it is essential that all necessary safety precautions are followed (refer to BS 8008).

Steel pile casings need to be true to line and diameter, and clean before each use; check their diameter as it has been known for piles to be put down with the wrong diameter. Measure and record the volume of concrete discharged into each cast-in-place pile and compare with the volume calculated from length and nominal diameter.

Errors of setting out require expensive remedial measures in the form of modification to caps and ground beams. On soils which heave and displace when driven piles are installed, check the setting out points prior to pitching each pile otherwise an accurately set out row of marker pins may be successively displaced by the piling to result in a very large final cumulative error.

The following list shows some of the main piling faults together with their cause and the supervision required to prevent their occurrence.

Formation of pile

Fault	Supervision	Correct practice
Concrete or steel permanent casing defective initially or damaged by • over-driving on obstruction or • driving mandrel moves out of plumb and lifts casing on withdrawal.	Visual—during piling and prior to concreting.	Casings should be withdrawn and pile re-driven.
Collapse of sides when casing not used—caused by soft, loose or fissured soils.	Visual—during piling.	Use casing.

Fault	Supervision	Correct practice
Excessive water in pile—caused by inflow of ground water.	Visual—during piling.	Use bentonite or steel casing. (These methods are normally agreed with the engineer before work starts.)
Collapse of bell for augured pile with enlarged base—caused by soft or fissured clay, or base left unsupported too long before concreting.	Visual—aided by measurement of depth.	Concrete should be placed as quickly as possible after boring has been completed.
Softening of clay at bottom of enlarged base. Effect is aggravated if water is present.	Probing bottom of bore with open-ended tube.	To be referred to the engineer.
Lateral movement or uplift of adjacent piles during driving of pile casings—caused by heave and displacement around piling.	Maintain checks on plan position and levels of adjacent piles.	To be referred to the engineer.
Settlement of adjacent ground surface or buildings—caused settlement of loose sandy soils due to vibrations in pile driving, or loss of ground in soft clays or loose water-bearing sands when sinking holes for bored piles.	Maintain checks on levels of ground or adjacent buildings before and during piling.	Loss of ground in sandy soils can be minimised by keeping a head of water in the casing while boring. These points should normally be decided by the engineer.
Displacement of setting out marker pins—caused by heave and displacement of soil with driven pre-cast or cast-in-place piles.	Visual—by check measurements.	Establish basic setting out grid well clear of piled area. Check each marker pin immediately prior to pitching pile.
Out of plumb—caused by insufficient strength and rigidity of rig, tilting of base of rig during driving.	Visual—by measurement with plumb line.	Correctly designed rig, properly packed up with means of jacking to re-level if on soft ground.

Concrete

Fault	Supervision	Correct practice
Insufficient strength—caused by bad mix design or concreting practice.	Cube tests—refer to the specified requirements—normally not less than three cubes per day for each mix. (The frequency could be reduced if satisfactory results are being obtained consistently.)	Mix to be properly designed to give required strength with suitable workability and adequate control applied to concreting.

Fault	Supervision	Correct practice
Water/cement ratio too high for bored or auguered in clay.	Check mix proportions on site and measure workability by slump test. Note particularly if there is excessive laitance on top of concrete pile shaft.	To be clearly specified by the engineer and correctly carried out by the contractor.
Voids when no casing is used or casing used and not withdrawn.	Visual—ensure sufficient compaction of concrete.	Compaction by falling weight or vibration.
Voids when casing is used and withdrawn—caused by concrete being drawn up by casing.	Check if volume of concrete used is appreciably less than pile volume. Check cleanliness of casing.	As above. Concrete should not be filled too high in casing. (Lifting is more likely to occur the smaller the diameter of the pile.)
Necking or waisting in cast-in-place piles, when casing is withdrawn—caused by insufficient head of concrete above the bottom end of the casing during withdrawl.	Ensure that there is sufficient head of concrete in casing (but not so high that casing will lift the concrete).	Establish and maintain correct techniques.

Reinforcement

Fault	Supervision	Correct practice
Displacement during compaction either laterally reducing cover or by binding being dragged down—caused by no or wrong spacers, careless technique, defective anchorage of steel in some pile types.	Inspection of cages during fabrication to ensure adequate tying at intersections. Visual supervision during installation.	Cages rigidly supported from ground level during initial concreting. Use spacers complying with the requirements of BS 7973. Consult BSI if in doubt.

Working load

Fault	Supervision	Correct practice
Failure under test load or permanent working load caused by		
Driven piles • No or wrong spacers used.	Check what spacers pile manufacture proposes to use before piles are made.	Use spacers to BS 7973. Consult BSI if in doubt.

Fault	Supervision	Correct practice
• Insufficient driving resistance. • Insufficient concrete in bulb for bulb ended, bored and driven piles. • Careless placing technique for concrete in core of driven shaft. • Ground heave causing lifting of piles.	Check hammer drop and blows per metre. Check volume and compaction of concrete used.	Before work starts ensure that the piling subcontractor has: • procedures that meet the specified requirements • adequate technical supervision on site. Ensure that these procedures are thoroughly understood by the Main Contractor's supervising engineer.
Bored piles • Insufficient depth in relation to character of soils or rock encountered. • Ground surrounding pile weakened by incorrect boring technique. • Misplaced reinforcement cage resulting in inadequate cover. • Careless technique in placing concrete.	Check consistency of spoil. Ensure correct boring and concreting techniques are used. Check spacers comply with BS 7973.	With test piles, the criteria for failure and the procedures to be followed in the event of a failure should be included in the engineer's piling specification. Use only BS 7973 compliant spacers. Check with BSI if in doubt. The supervising engineer should thoroughly appraise himself of the specified requirements before work starts. Any deviation from them or variation between similar adjacent piles to be referred immediately to the engineer.
Mechanically augured piles • Insufficient depth in relation to character of soils encountered. • Insufficient diameter to under-reamed base or bell. • Ground surrounding shaft or beneath base weakened by incorrect boring technique. • Misplaced reinforcement cage resulting in inadequate cover. • Careless technique in placing concrete.	Constantly monitor consistency of spoil during boring. Observing that the main drilling and under-ream tools are operated correctly in accordance with pre-check at surface. Check spacers comply with BS 7973. Ensure that contractor uses correct concreting techniques.	Comments as above. Comments as above. Use only BS 7973 compliant spacers. Check with BSI for compliant products. Comments as above.

8.7 Basements

8.7.1 General

Before constructing basement walls and floors, carefully study the specification and detail drawings to plan the sequence of any waterproofing operations. Construction planning must take account of this.

A careful examination of the soil conditions will determine whether or not the excavation will be self draining or will act as a sump collecting and containing surrounding surface and groundwater. Completely enclosed basements can be subject to flotation out of the ground in some circumstances. The engineer's design will normally allow for this and specify precautionary measures.

Excavation support measures must relate to the methods of excavation and construction sequence. They must fully account for the safety of adjoining buildings and underground services. Temporary works design must be by persons competent in this type of design.

8.7.2 Effect on surrounding ground

Excavation for deep basements can cause settlement of the surrounding ground surface, sufficient to cause structural damage to buildings near the excavation and disturbance of existing services. This settlement can be the result of:

- lateral movement of the face of the excavation due to the cumulative effects of the yielding of sheeting members, walings and struts, or anchors which support the face
- deflection of the basement retaining wall after placing backfilling
- lowering of groundwater level
- ground slips, heave of base or erosion as result of bad construction methods or poor execution
- vibration of the ground under adjacent buildings or streets caused by site operations, particularly when large-diameter bored piles are used where liners are sunk or extracted by heavy vibration equipment.

8.7.3 Adjacent structures

No matter how carefully work is executed it is not possible to prevent totally the yielding of excavation supports. Generally it is assumed that the settlement of the ground surface will not occur to any appreciable extent beyond a line drawn at a slope of 1 (horizontal) and 1 (vertical) from the base of the excavation. Structures inside this line are usually underpinned. Again, a small degree of risk can be taken on shallow basements after considering the cost of underpinning and the cost of repair to the structures and the effects on any activities within the premises. It must however be stated that each case must be assessed individually with the engineer responsible.

In all cases involving the stability of slopes where buildings could be affected, liaise with the owners.

By adopting rigid forms of excavation supports it is possible to avoid underpinning adjoining buildings. The supports incorporate jacking to counteract inward yielding. Such methods must be designed or supervised by specialists in this particular field.

Note: This work may impinge on the requirements of the Party Wall etc. Act, 1996.

8.7.4 Watertightness

The provision of a watertight basement is complicated and demands strict compliance with the specification and drawings and the provision of appropriate supervision. Ensure that adequate protection is specified and provided to any waterproof membranes on the outsides of retaining walls.

The problem of watertightness during construction will normally have been considered by the engineer and can be overcome by use of diaphragm walls, sheet piling, secant piling or contiguous bored pile walls, if no dewatering system is adopted. The gaps occurring in contiguous bored piles may need grouting to achieve watertightness.

The short upstand or *kicker* for a structural concrete basement wall should always be cast integrally with the slab on which it sits. Make the height of this section sufficient to vibrate the fresh concrete with a poker vibrator and accommodate any water bar that may be specified. The height required usually varies from 75 mm upwards.

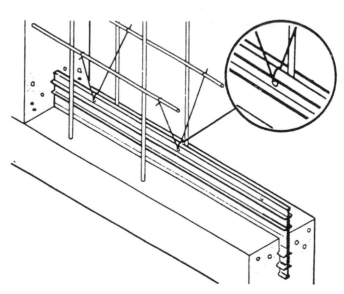

Figure 8.7 Typical eyeleted waterbar showing reinforced eyeleted flanges wired to steel rebar.

Where proprietary water bars are specified, the supplier normally prepares detail drawings, based on the engineer's general arrangement drawings.

With junction pieces fabricated by the manufacturer it is normal only to require straight butt splices on site. For this, purpose-made clamps and sealing irons are available and are to be used in accordance with the manufacturer's instructions.

Unless installed with care and diligence, waterbars can leave more tracks for water entry than if none were used in the first place. Waterbars that are fixed directly to formwork or laid on blindings must be securely fixed to prevent disturbance during concreting.

Waterbars for positioning between layers of reinforcement are supplied with eyelet holes at close centres along both edges. It is imperative that every eyelet hole is used for tying with tying wire to the adjacent reinforcement to maintain the water-bar in its required position during the placing and vibrating of the concrete. Use the vibrator to ensure sound concrete along both sides of the waterbar, especially the back of water-bars at the foot of retaining walls and the under-side of horizontal waterbars.

8.8 Backfilling

Check the compatibility of backfilling material with existing chemical ground conditions. Limestone quarry waste reacts with acidic rain and groundwater to produce carbon dioxide gas. Sea-dredged ballast must not be placed against concrete without first washing the aggregate to reduce the salt content to acceptable limits.

Road planings (bituminous material recovered from old pavements during reconstruction or resurfacing works) should not be used as a sub-base to any construction required to support static or intermittent stationary loads unless it is recycled by reheating, adjusting the bitumen content as necessary and laying hot in the normal manner for bitumen-bound materials. If laid and compacted cold, the bitumen content will almost certainly inhibit full compaction of the mineral aggregate. Under static loads, the material will subsequently undergo creep compression resulting in settlement, the magnitude of which will depend on the initial corn action achieved. This cannot be judged on the visual appearance of the compacted material, which may seem very dense even when the mineral aggregate is still in a relatively loose state.

When backfilling foundation excavations internal to a building, consolidate in layers not exceeding 150 mm thick. This procedure will safeguard against the cracking of a floating ground floor slab, and enable the full design load to be carried by the slab. Backfill to a uniform depth over the whole area of the excavation, equally on both sides of any internal walls, before proceeding to the next layer. This prevents damaging the wall by horizontal loading on one side only, when not allowed for in the design.

When backfilling behind retaining walls don't form earth ramps or bunds against the final exposed face without checking with the designer that the design will allow for this condition of reversed loading. Don't drive plant or stack materials on the ground at the high side of the wall. If not allowed for in the design, this type of loading will overload the wall causing excessive deflection of the vertical wall and the possible rotation of the base slab.

8.9 Shoring and underpinning

This is required:

- to support a structure which is sinking or tilting due to ground subsidence or superstructure instability
- as a safeguard against the possible settlement of a structure when excavating close to and below its foundation level.

The underpinning of a structure is required for the reasons given above and, in addition, to enable the foundations to be deepened for structural reasons, e.g. to construct a basement beneath a building. It is also used:

- to increase the width of a foundation to permit heavier loads to be carried, e.g. when increasing the storey height of a building
- to enable a building to be moved bodily to a new site.

Shoring and underpinning are highly skilled operations and must always be undertaken under the direction of the engineer responsible for designing the underpinning scheme. Sometimes specialist subcontractors are engaged. The engineer must always produce details and sequencing of underpinning work and these must not be varied without his prior agreement.

8.10 Cofferdams

These are essentially temporary structures designed to support the surrounding ground and to exclude water from an excavation. However, any inflow is removed using reasonable pumping facilities in such a way that soil particles are not also removed. Remember that changing existing groundwater levels can affect surrounding land and any buildings supported by it.

Cofferdams must be designed by an engineer with adequate experience in temporary works of this kind. They are constructed in a wide variety of types and material and the choice and design will have been based on a thorough knowledge of the soil and full details of the groundwater conditions.

Large cofferdams for civil engineering works require special consideration at the design stage.

8.11 Caissons

A caisson is a box or shell which is sunk through ground or water for the purpose of excavating within and placing a foundation at the prescribed depth and which subsequently becomes an integral part of the permanent work. The design and operational use of a caisson is governed by the method proposed for sinking, by the skin friction that may be expected during sinking and by the nature and dimensions of the permanent structure required. It may be constructed of framed steel plate, cast iron or reinforced concrete segments, *in situ* reinforced concrete or mass concrete. Sometimes working within the caisson utilises compressed air to combat water pressure.

8.12 Diaphragm walls

Water can be prevented from entering excavations by providing an impermeable surrounding concrete wall constructed *in situ* by one of several methods. Unless such a wall is carried down to an impervious stratum, water can enter through the bottom of the excavation.

The various construction methods are:

- contiguous bored piles (the gaps between which will require plugging with concrete or other approved material)
- mix-in-place walls
- thin cast-in-place diaphragm walls
- thick cast-in-place diaphragm walls
- secant piled wall.

Where diaphragm walls are constructed in made ground, old workings, and generally where drains, ducts and voids may be present, precautions have to be taken against a loss of mud which could lead to a collapse of the excavation.

Cofferdams, caissons and diaphragm walls are usually designed and constructed by specialist subcontractors, and their proposals and work on site should receive appropriate main contractor appraisal and supervision.

Further reading

Standards

BS 8008: 1996 *Safety Precautions and Procedures for the Construction and Descent of Machine-bored Shafts for Piling and other Purposes.*
BS 5837: 2005 *Trees in Relation to Construction.*
BS 6031: 1981 *Code of Practice for Earthworks.*
BS 8004: 1986 *Code of Practice for Foundations.*
BS 8102: 1990 *Code of Practice for Protection of Structures against Water from the Ground.*
BS 8103–1: 1995 *Structural Design of Low Rise Buildings.* Code of Practice for stability, site investigation, foundations and ground floor slabs for housing.
BS 8215: 1991 *Code of Practice for Design and Installation of Damp-proof Courses in Masonry* Construction.
BS 5973: 2001 *Spacers and chairs for steel reinforcement and their specification.*

Other related texts

Simplified Rules for the Inspection of Second Hand Timber for Load-bearing Use: TRADA, High Wycombe, Buckinghamshire, UK.
Timber in Temporary Works: Excavations: TRADA, High Wycombe, Buckinghamshire, UK.
Proprietary Trench Support Systems 3rd ed.: London, CIRIA, CIRIA Report TN95, 1986.
Trenching Practice: CIRIA Report 97, London, CIRIA, 1994.
Fill—Classification and Load Bearing Characteristics: BRE, Garston UK, BRE Digest 274 Pt 1, 1987.
Hardcore: BRE, Garston UK, BRE Digest 276, 1983.
Why do Buildings Crack?: BRE, Garston UK, BRE Digest 361.
Specification for Highway Works: London, HA.

Standards: NHBC, Amersham UK, NHBC.

Note: These standards are continuously updated and are available to members on CD or in loose-leaf hard copy, 2003.

Deep Basements: IStructE, London, IStructE, 2003.

Underpinning: S. Thorburn and J. Hutchison (eds), Glasgow UK, Surrey University Press, 1985.

Foundation Design & Construction 5th ed.: M. J. Tomlinson *et al.*, Harlow UK, Longman, 1987.

Foundation Design: M. J. Tomlinson, Oxford UK, Butterworth Civil Engineer's Reference Book 4th ed. (L. S. Blake ed.), 1989.

The Party Wall etc. Act 1996: Explanatory Booklet: ODPM, London, ODPM, 2002.

Specifiers Guide to Steel Piling: A. R. Biddle *et al.*, Ascot UK, SCI Publication P308, 2002.

Steel Bearing Pile Guide: A. R. Biddle, Ascot UK, SCI Publication P156, 1997.

Foundations in Chalk: J. A. Lord *et al.*, London, CIRIA Report No. 11, 1994.

Piling Handbook: BS, Scunthorpe UK, BS Report SP&CS, 1997.

Installation of Steel Sheet Piles: TESPA, Luxembourg, TESPA Report No. L–2930, 1993.

Durability and Protection of Steel Piling in Temperate Climes: CORUS, Scunthorpe UK, CORUS Report No. 202, 2001.

Notes on the Design, Installation of Working Platforms for Plant for Specialist Foundation and Geotechnical Works: FPS, Beckenham UK, FPS, 2001.

The Thaumasite Form of Sulfate Attack. Thaumasite Export Group, DETR, London, 1999.

Chapter 9   MASTIC ASPHALT

Robert Langridge

Basic requirements for best practice

- Independent mastic asphalt specialist advice

- Clear specifications and accurate details supplied promptly to all concerned

- Supply and use inspection record sheets

- Plan construction to allow continuity of laying of mastic asphalt

- Gully outlets of the correct type, already installed correctly and clean

- Good access to working area

- Competent site supervision, with sufficient time to oversee mastic asphalt work

- Trial area of mastic asphalt work

- Independent visiting on-site quality control from the outset

- Weekly inspections jointly with the mastic asphalt subcontractor

- Temporary protection of finished work

9.1   Introduction

Mastic asphalt is a long-established material for waterproofing flat roofs and the basements of buildings, as a finished floor or as an underlay for another flooring finish, and for paving footpaths and footbridges and vehicular trafficked areas, for example loading bays and the decks of multi-storey car parks. Mastic asphalt is usually laid by hand and does not require compaction.

When mastic asphalt has been correctly manufactured, well specified and detailed, and the workmanship on site is to a high standard, it should give many years of trouble-free service. When developing the design of a building, the potential for long service life of mastic asphalt is very important for environmental protection, as it should help to minimise any need for repairs, replacement materials and the depletion of natural resources. Furthermore, the expectancy for long life of good quality mastic asphalt work is an important monetary consideration for the building owner.

For new mastic asphalt work and for maintenance, there are key points requiring attention by the specifying authority. These are highlighted above under the heading of 'Basic requirements for best practice'.

149

Table 9.1   British Standards relating to mastic asphalt.

British Standard Specification	Grade
BS 6925 type R 988	Roofing
BS 6925 type T 1097	Tanking
BS 6925 type F 1076	Flooring
BS 6925 type F 1451	Flooring
BS 1447 Grade S and Grade H	Paving

9.2   Mastic asphalt material

Mastic asphalt consists of a mixture of graded aggregate, for example, limestone, and a binder. The binder may be bitumen, a blend of bitumen and Lake Asphalt, or a polymer modified bitumen. The mixing of the ingredients takes place hot at a manufacturing plant under carefully controlled conditions. At the time of manufacture, samples of the mastic asphalt are routinely taken and tested for composition by analysis and for hardness number in the plant laboratory. The freshly made mastic asphalt may be transferred to a lorry-mounted mobile mixer and delivered hot to the site where the material is to be used. Alternatively, the molten mastic asphalt may be poured into moulds and allowed to cool to form blocks. The block material may be stored and delivered to sites as required. On site, block material is made ready for spreading by melting in a mixer or a cauldron.

Mastic asphalt is manufactured to comply with British Standard specifications or with proprietary specifications. Polymer-modified mastic asphalts are manufactured to proprietary specifications. However, for most polymer-modified mastic asphalts many of the requirements of the British Standards and Codes of Practice for mastic asphalt are applied during manufacture, and during the laying process.

For British Standard materials, details of the requirements for composition by analysis, hardness numbers and grades of mastic asphalt may be found in the specifications listed in Table 9.1.

Mastic asphalts containing a blend of bitumen and lake asphalt may give better spreading properties and a superior finish. Polymer-modified mastic asphalts may be more flexible in cold conditions, when compared to products complying with British Standards.

Block mastic asphalt delivered to site may be identified by coloured labels fixed to each block:

• White: roofing
• Red: flooring
• Green: tanking
• Blue: paving.

The labels provide the name of the manufacturer, sometimes the batch number of the material and, if to a British Standard, details of the British Standard number and grade.

9.3   Some design considerations

9.3.1   Roofing

Mastic asphalt requires good support during and after laying. Where the roof is to be subjected to foot traffic and the deck is concrete, the deck and its supporting structure should, when possible, be designed to allow for the inverted warm deck roof, also known as the 'upside-down roof' or 'protected-membrane roof', to be installed. For an inverted, warm deck roof, the horizontal mastic asphalt is laid on a separating membrane of sheathing felt, on the roof deck or cement-bound screed. The separating membrane should have 50 mm wide lapped joints. The finished mastic asphalt is covered with a non-woven polyester fleece, followed by an extruded polystyrene thermal insulation board. The insulation board is covered with paving slabs on spacers, or 20 mm to 40 mm size rounded pebbles. The thermal insulation board protects the mastic asphalt from extremes of temperature and from mechanical damage.

On lightweight decks and some concrete decks, the thermal insulation except, for example, cellular glass, is usually laid on a vapour control layer. The thermal insulation is then overlaid with one or more layers of separating membrane before the mastic asphalt is applied. The vapour control layer should have fully bonded lapped joints and be fully bonded with hot bitumen. The mastic asphalt may be finished with solar reflective paint or washed light coloured stone chippings 10 mm to 14 mm nominal size bedded in gritting solution. This type of roof is known as a 'warm roof', and is widely used.

Any point loads, for example the legs of chairs or timber screens, placed directly on roofing grade mastic asphalt will cause indentation. Whenever possible, roofs to accept chairs should be designed as an inverted warm deck roof, or the mastic asphalt should be protected by suitable promenade tiles. Support timber screens, for example, on concrete plinths placed directly on the deck. Mastic asphalt should then be dressed up the sides of the plinths and into a chase or rebate. Weatherproof the tops of the plinths with suitable metal flashings.

On brickwork and blockwork upstands, the horizontal and vertical mortar joints should have a hollow key or bucket handle finish (see Fig. 9.1). Between upstands and plinths, allow sufficient space in the design to enable the mastic asphalt skirting, horizontal work and fillets to be correctly applied and finished.

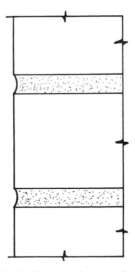

Figure 9.1   Jointing brickwork to receive mastic asphalt.

At all internal angles, form two-coat mastic asphalt fillets. The fillets should have a face width of 40 mm minimum. The face of the fillet should be at an angle of approximately 45° to the horizontal.

Lay flat roofs with falls to drain water. A minimum fall of 1:80 is often specified. However, to allow for deflection and building tolerances, a fall of 1:40 may be required. As a compromise, a finished fall of 1:60 is often used. This may avoid having an unacceptable increase in the thickness of the screed. On the other hand, many large flat roofs using mastic asphalt and the inverted warm deck roof assembly have been laid without falls. For a roof designed without falls, it would be important to have a reasonably plane and even deck to avoid excessive pools of water, and possibly local flotation of the thermal insulation board.

Avoid the use of internal gutters if possible. If internal gutters are required, install them with adequate falls to prevent pools of water forming.

Rainwater outlets should be located at the low points of the falls. The outlets should be firmly held in the deck. The top of the flange of the outlet should be flush with the top of the adjacent screed, concrete or timber deck. However, for syphonic drainage systems the outlet should be lowered to form a sump. Gravel guards should be securely fixed to roof outlets.

Movement joints are required in mastic asphalt only when there is a movement joint in the underlying structure. When possible, horizontal movement joints should be of the twin kerb type and covered with a suitable flashing. In car parks, for example, when flush movement joints are required, locate these at the high points of falls.

The upstands to receive a mastic asphalt skirting should be rigid. Any freestanding kerbs should be rigid and fixed securely to the deck (see Fig. 9.2). Except perhaps at thresholds, the height of the upstand should be sufficient to allow the mastic asphalt skirting to be finished not less than 150 mm above the adjacent finished horizontal surface.

The top of mastic asphalt skirting, except on timber or metal upstands, should be tucked into a 25 mm × 25 mm chase, and pointed with sand and cement mortar containing an adhesion agent.

On timber upstands, sheathing felt with 50 mm wide lapped joints is overlaid with expanded metal lathing (eml) and fixed with extra large head galvanised clout nails. Eml should have butt joints. The butt joints should be located on a plane surface and not at internal or external angles. On metal upstands, the eml is usually fixed by spot welding. On concrete upstands suitable drilled and plugged fixings should be used (see Fig. 9.3).

The top of mastic asphalt skirting should be covered with a metal flashing. The flashing should be fixed beneath the damp-proof course and in the same joint. It should vertically overhang the mastic asphalt skirting for a distance of not less than 75 mm. The damp-proof course should be located one course above the chase into which the mastic asphalt is to be tucked.

When possible, on a protected-membrane roof, protect the mastic asphalt skirting from solar heat with extruded polystyrene thermal insulation boards with weatherproof toppings, laid on edge. Chamfer the top edge of the board to an angle of approximately 45° and cover it with a suitable flashing to shed water. The metal flashing should hold the vertical thermal insulation board in place. To reduce the risk of indentation of the horizontal mastic asphalt, support the bottom edge of the

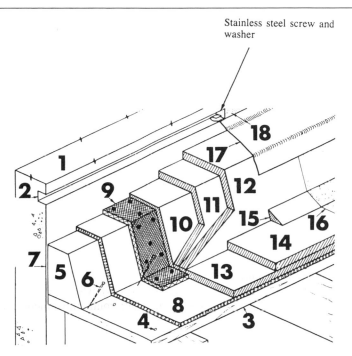

Stainless steel screw and washer

Order of installation

1. Reinforced concrete or brick upstand.
2. 25 mm × 25 mm chase for flashing. The flashing should be fixed at approximately 450 mm centres with stainless steel screws and washers. For a flashing fixed with lead wedges at not more than 450 mm centres a chase measuring 10 mm high × 25 mm deep should be provided.
3. Plywood or tongued and grooved timber deck.
4. Nail heads punched below surface of deck.
5. Plywood or timber free-standing kerb.
6. Nails for fixing free-standing kerb to deck and joists.
7. 13 mm to 20 mm gap between back of free-standing kerb and reinforced concrete or brick upstand.
8. Sheathing felt on kerb and deck.
9. Bitumen-coated or galvanised expanded metal lathing fixed overall at 150 mm centres with extra large head galvanised felt nails to BS 1202 Part 1. Extend lathing from the kerb onto the deck a distance of 75 mm and nail at 150 mm centres. Ensure lathing is close fitting in the internal angle formed between the deck and the upstand. Use butt joints.
10. First coat of mastic asphalt.
11. Second coat of mastic asphalt.
12. Third coat of mastic asphalt.
13. First coat of horizontal mastic asphalt.
14. Second coat of horizontal mastic asphalt well rubbed with sand.
15. Two-coat mastic asphalt fillet minimum 40 mm wide face at approximately 45° to horizontal.
16. Solar reflective paint.
17. Non-ferrous metal flashing fixed into chase and extending down over mastic asphalt for 75 mm minimum (see note).
18. For 25 mm × 25 mm chase, polysulphide gun applied sealant and for a 10 mm × 25 mm chase, sand and cement pointing containing and adhesion agent.

Note: To reduce the transfer of solar heat from the flashing to the mastic asphalt, 5 mm thick extruded polystyrene sheet may be used as a separator between the two materials.

Figure 9.2 Free-standing kerb detail and a cold 'non-insulated' deck.

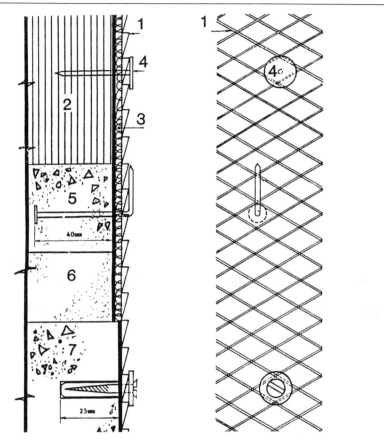

Methods of fixing expanded metal lathing

1. Bitumen-coated or galvanised plain expanded metal lathing to BS 1369, measuring between 6 mm and 10 mm short way of mesh and not less than 0.5 mm thick. Long side of diamond horizontal, and the pitch of the horizontal metal strands inclined upward and outward from the surface, providing a key for the mastic asphalt.
2. Plywood or timber. Lathing fixed through an underlay of sheathing felt (3) with extra large head galvanised felt nails (4).
5. *In situ* concrete. When it will not be possible to drill the concrete and fix the lathing with plugs and screws, nails can be partly driven into the timber shuttering at 150 mm spacing all over the area required.

   When the timber shuttering is removed the nails left projecting can be bent up after the lathing is positioned.

   Only use this method of fixing in confined spaces where the recommended method of drilling and plugging would be difficult.
6. Lightweight concrete blockwork. Fix lathing on an underlay of sheathing felt, using plugs and screws as described in (7) below, or use a suitable sand cement facing.
7. Concrete, already cast. Lathing fixed direct. 20 mm diameter metal washer and 32 mm × no. 8 countersunk screws fastened to drilled and plugged concrete at 150 mm centres all over with minimum embedment of 25 mm. This is the best method of fixing expanded metal lathing to lightweight and dense concrete already cast, and concrete blockwork.

Figure 9.3   Methods of fixing expanded metal lathing.

thermal insulation board on the horizontal thermal insulation board. The lower part of the vertical thermal insulation board should be kept in place by the paving slabs or pebbles.

Guidelines for the thickness and number of coats to be used are shown in Table 9.2. Structures located on, or passing through, a flat roof should usually by mounted on suitable plinths. Examples of details for this work are given in Figs. 9.4–9.6.

9.3.2 Tanking

Tanking is the protection of a building from water below ground level. On horizontal surfaces up to and including 10° pitch three coats of mastic asphalt to a total thickness of 30 mm are used. On sloping and vertical surfaces over 10° pitch, three coats of mastic asphalt to a total thickness of 20 mm are used. Lightweight concrete should be rendered with a suitable sand and cement mix, or coated with a proprietary slurry containing a plasticizer in accordance with the manufacturer's instructions, before the mastic asphalt is applied. At all internal angles, two-coat mastic asphalt fillets are formed. The fillets should have a face width of 50 mm minimum. The face of the fillet should be at an angle of approximately 45° to the horizontal.

The use of a separating membrane of glass fibre tissue, beneath horizontal tanking mastic asphalt should be considered. The separating membrane should have 50 mm wide lapped joints. The glass fibre tissue is not required to allow for movement because the mastic asphalt would be sandwiched. However, it may help to reduce any tendency to blowing by the mastic asphalt, and with the release of any trapped air or moisture during laying. Any *blow* holes not repaired in the mastic asphalt may allow water to penetrate the material.

No movement joints should be allowed in tanking work. The penetration by services of tanking mastic asphalt should be avoided as far as possible. Where

Table 9.2 Guidelines for the thickness and number of coats.

	Total thickness (mm)	Number of coats
Horizontal substrates		
Where the mastic asphalt will be fairly readily accessible	20	2
Where mastic asphalt will not be readily accessible, under roof gardens for example	30	3
Vertical substrates		
Skirtings up to 300 mm high, excluding timber and lightweight concrete blockwork	13	2
Skirtings on timber and lightweight concrete blockwork or metal	20	3
Skirtings which after completion will not be readily accessible, in roof gardens for example	20	3
Vertical work over 300 mm high	13 or 20	2 or 3

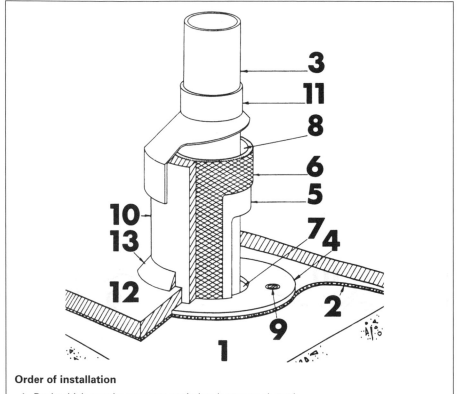

Order of installation

1. Deck which may be concrete or timber (concrete shown).
2. Sheathing felt or glass fibre tissue.
3. Pipe which may be of metal or plastic.
4. Metal flange of sleeve.
5. Metal sleeve. (**Note:** Top of sleeve must be minimum 150 mm above finished surrounding surface.)
6. Expanded metal lathing spot welded or wired to sleeve.
7. Clearance (specified by pipework designer) between deck and pipe.
8. Clearance (specified by pipework designer) between metal sleeve and pipe.
9. Screws for securing metal flange to deck.
10. Three-coat mastic asphalt skirting.
11. Metal collar fixed and sealed to pipe (maintain clearances in items 7 and 8).
12. Two-coat horizontal mastic asphalt.
13. Mastic asphalt fillet minimum 40 mm wide face approximately 35–45° to horizontal.

Figure 9.4   Pipe which may contain hot liquids or gases passing through a clearance hole in the roof.

required, service pipes at ambient temperature may be sleeved with mastic asphalt, and specialist advice should be obtained.

Mastic asphalt is attacked and softened by oil. Where necessary, suitable drip trays should be provided beneath plant. The drip trays should be fitted with suitable drain pipes connecting to a collection tank.

Any hot water pipes penetrating mastic asphalt tanking should be insulated to avoid transfer of heat to the mastic asphalt. Obtain specialist advice.

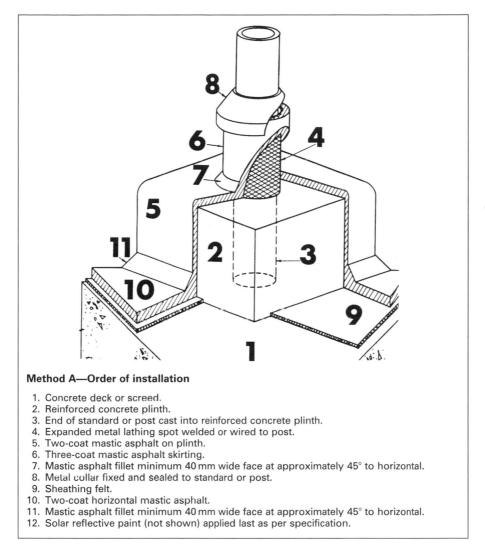

Method A—Order of installation

1. Concrete deck or screed.
2. Reinforced concrete plinth.
3. End of standard or post cast into reinforced concrete plinth.
4. Expanded metal lathing spot welded or wired to post.
5. Two-coat mastic asphalt on plinth.
6. Three-coat mastic asphalt skirting.
7. Mastic asphalt fillet minimum 40 mm wide face at approximately 45° to horizontal.
8. Metal collar fixed and sealed to standard or post.
9. Sheathing felt.
10. Two-coat horizontal mastic asphalt.
11. Mastic asphalt fillet minimum 40 mm wide face at approximately 45° to horizontal.
12. Solar reflective paint (not shown) applied last as per specification.

Figure 9.5 Method A—cast in. Fixing posts and handrail standards on mastic asphalt roofs.

There are two methods of tanking:

- external—the mastic asphalt is applied to the outside of the structure
- internal—the mastic asphalt is applied to the inside of the structure.

External tanking is preferred. The groundwater pressure forces the mastic asphalt against the structure. However, the method of tanking selected will depend on site constraints.

For external tanking, sufficient space between the structure and the side of the excavation is required for scaffolding; for an asphalter to work properly, and for a wall of protective brick or blockwork to be built. A 40 mm wide gap should be left between the face of the mastic asphalt and the wall. The gap should be filled with mortar.

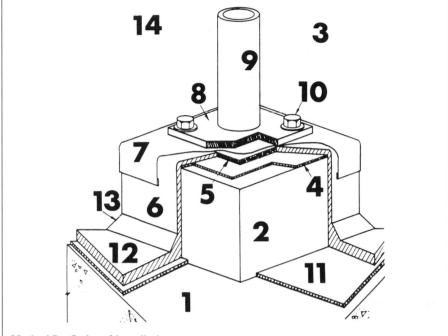

Method B—Order of installation

1. Concrete deck or screed.
2. Reinforced concrete plinth.
3. Holes previously drilled in reinforced concrete for anchors to receive holding down bolts. Baseplate can be used as a template for this.
4. Sheathing felt.
5. Rigid packing.
6. Two-coat mastic asphalt skirting and horizontal on plinth.
7. Metal flashing.
8. Baseplate to which standard or post 9 has been welded bedded in non-setting mastic.
9. Standard or post.
10. Holding down bolts, sealed with non-setting mastic.
11. Sheathing felt.
12. Two-coat horizontal mastic asphalt.
13. Mastic asphalt fillet minimum 40 mm wide face at approximately 45° to horizontal.
14. Solar reflective paint (not shown) applied last. Check the specification.

Figure 9.6 Method B—bolted on. Fixing posts and handrail standards on mastic asphalt roofs.

For external tanking the concrete base must extend for a distance of not less than 150 mm outside the basement wall. A 50 mm minimum thickness of sand and cement screed should be applied over the horizontal mastic asphalt. Before this is done, however, where the horizontal mastic asphalt and the vertical mastic asphalt is to be connected, a 150 mm wide strip of building paper should be placed over the horizontal mastic asphalt. The strip of building paper should allow the screed above it to be removed when the loading coat of concrete or the structural slab has been cast and the vertical wall has been constructed.

For internal tanking, the mastic asphalt work is normally undertaken after the structural walls and the concrete base have been completed. The horizontal mastic asphalt should be protected with a 50 mm minimum thickness of sand and cement screed. For the vertical mastic asphalt, a protective wall of brick or blockwork, with a 40 mm wide gap between the face of the mastic asphalt and the wall, is required. A structural concrete slab is laid on top of the screed.

9.3.3 Underlays and flooring

Flooring grade mastic asphalt may be used as an underlay for a floor covering material or as a finished surface.

Flooring mastic asphalt, to British Standard specifications, is available in four grades and a range of thicknesses. The grade and thickness selected will depend on the use to which the floor is to be put and the minimum and maximum ambient temperature in the building.

Special mastic asphalt, for example spark resistant and electrically conductive mastic asphalt, may be available to order.

Mastic asphalt becomes softer when heated and harder when cooled. In a warm hospital ward or in a domestic dwelling, for example, the mastic asphalt needs to be sufficiently hard to resist indentation by furniture, but not so hard that it might crack if the building is left unheated for a period of time in very cold weather.

Mastic asphalt skirting usually comprise two coats, to a total thickness of 13 mm. Where a superior finish is required, or the skirting is over 300 mm high, three coats of mastic asphalt to a total thickness of 20 mm should be applied. At internal angles, two-coat mastic asphalt fillets are formed. The fillets should have a face width of not less than 40 mm. The face of the fillet should be at an angle of approximately 45° to the horizontal. Alternatively, the face of mastic asphalt fillets may be coved, that is, it will have a concave finish.

It is essential that the flooring mastic asphalt is well supported. Any underlying deck or screed should be capable of withstanding loads imposed on the floor and that deflection criteria are met. The surface of the underlying concrete deck or cement-bound screed should have a wood float or power float finish.

A separating membrane of glass fibre tissue should generally be used beneath the flooring mastic asphalt except, for example, on timber boarding when black sheeting felt should be used. The separating membrane should have 50 mm wide lapped joints. The mastic asphalt should be laid to uniform thickness.

The finish on horizontal mastic asphalt will depend on the use for the floor. A sand-rubbed finish or a natural float finish may be required on mastic asphalt used as an underlay. The manufacturer of the floor covering should be consulted. If required, specialist advice should be sought.

Where outlets are required, for example in shower rooms, these should be of a type suitable for use with mastic asphalt.

9.3.4 Paving

Paving grade mastic asphalt to British Standards is available in two types:

- S: For general use
- H: For heavy-duty areas. Loading bays and bus stops for example.

Polymer-modified paving mastic asphalt to proprietary specifications is available. The finished thickness, and the percentage and nominal size of any added coarse aggregate, will be determined by the site requirements. Advice is available in BS 1447, or specialist advice may be obtained.

On the top decks of multi-storey car parks, paving mastic asphalt is usually laid on a waterproofing layer of two coats of roofing grade mastic asphalt, or a proprietary sheet waterproofing material. On intermediate decks, paving mastic asphalt alone may be used. A separating membrane of glass fibre tissue with 50 mm wide lapped joints should be used between the mastic asphalt and the deck.

To allow for drainage of surface water, a minimum finished fall of 1:80 is recommended.

Where possible, the fall should be incorporated in the deck. Under no circumstances should the falls be formed in mastic asphalt, which should be laid to even thickness. Avoid the use of internal gutters. Locate the outlets at the low points of the falls.

Any movement joints should be located at the high points of the falls. Where possible, water should not be allowed to flow over movement joints.

On footpaths, a separating membrane of glass fibre tissue placed on the substrate may reduce blowing of the mastic asphalt at the time of laying. During the service life, the glass fibre tissue should reduce any tendency for the mastic asphalt to blister. Paving mastic asphalt may have a sand rubbed finish, or a sand rubbed and crimped finish.

For skirting, roofing grade mastic asphalt is generally used. The mastic asphalt should be in two coats, to a total thickness of 13 mm. On expanded metal lathing or where a superior finish is required, three coats of mastic asphalt to a total thickness of 20 mm should be applied.

At all internal angles, two-coat, roofing-grade mastic asphalt fillets should be formed. The fillets should have a face width of 40 mm minimum. The face of the fillet should be at an angle of approximately 45° to the horizontal.

On access ramps having a gradient of not more than 1:10 and unheated, the concrete should have a cross-tamped finish to provide a key for the mastic asphalt. The mastic asphalt should be laid directly on the concrete. A 15 mm thickness of roofing grade mastic asphalt laid in one coat should be covered with 25 mm thickness of paving grade mastic asphalt. The paving mastic asphalt should have a sanded and crimped finish. For heated ramps, seek specialist advice.

9.4 Requirements for good quality work

- Prepare a schedule of checking, inspection and testing requirements.
- Ensure that the specification is entirely suitable for the purpose.
- Ensure that the drawings are complete and show correct details.
- The falls should be formed in the substrate.
- Where the thermal insulation is laid beneath roofing mastic asphalt, it is important to ensure adequate falls to avoid pools of water.

- On mastic asphalt paving, ensure that there are adequate falls to avoid pools of water.
- Surfaces (substrates) should be suitable to receive the mastic asphalt. Cement-bound screeds and horizontal concrete surfaces should have a wood float finish. Any footprints or other holes in the surface of the substrate should be made good with suitable mortar before the separating membrane and mastic asphalt are laid. Upstands should be free from visible surface voids (unless expanded metal lathing is to be used as a key for the mastic asphalt), and loose material. All surfaces to receive mastic asphalt must be dry. Any primer used must be applied in accordance with the manufacturer's instructions and be allowed to dry properly.
- All gullies should be of a design suitable for use with mastic asphalt and firmly fixed in place. Their flanges must be clean and any with threaded holes protected from contamination.
- At drainage channels, the mastic asphalt waterproofing should be continued beneath the channels. A clamping cone type of outlet is usually installed in the mastic asphalt beneath the channel outlet. The spigot from the channel may be sleeved into the clamping cone outlet.
- Good access.
- On site, to minimise heat loss and the need for carrying the material further than necessary, locate the supply of hot mastic asphalt (mixer or cauldron) as close as possible to the point of laying.
- For melting block mastic asphalt, use mechanically stirred mixers. Use cauldrons for melting mastic asphalt only when required by site conditions, for example, when the throughput of mastic asphalt is small, or site conditions are impractical for a mixer. (Note: controlling the temperature of mastic asphalt in cauldrons is difficult and can lead to local overheating and rapid hardening of the material.
- Careful programming to ensure that the laying of the hot mastic asphalt is a continuous operation.
- Lay the mastic asphalt to a constant thickness.
- All lap joints and mastic asphalt fillets should be made after first warming and cleaning the *in situ* mastic asphalt with temporary applications of hot mastic asphalt.
- The mastic asphalt contractor should supply a Project Specific Test and Inspection Plan detailing how they will ensure that the materials and workmanship on site comply with the specification. Details of supervisory staff, including their experience, job description, qualifications and frequency of proposed site visits should also be supplied.
- At the commencement of the mastic asphalt work, lay a trial area, which includes all type conditions. Include the requirement for a trial area of mastic asphalt work in the subcontract agreement. At the pre-order meeting, agree the location of the trial area with the mastic asphalt subcontractor and mark its location on a site plan. If the trial area is acceptable, it might be included in the works.
- For mastic asphalt work, each day, whenever possible, apply the mastic asphalt to upstands before laying the horizontal mastic asphalt. Install the mastic asphalt fillets each day as the spreading of the mastic asphalt proceeds.

- On partly finished and completed mastic asphalt work, provide immediately adequate temporary protection against fouling and damage by other trades.
- As the mastic asphalt work proceeds, check each day's work for compliance with the specification using a checklist. Make any repairs immediately. An example of a checklist for use when inspecting mastic asphalt work is given in Fig. 9.7.

9.5 Quality control

At the outset, the specifying authority should require an independent mastic asphalt specialist to be employed on a part-time basis. The mastic asphalt specialist should check drawing details and the specification, attend the pre-start meeting, be present on site at the start of the mastic asphalt and associated work, and make a sufficient number of visits to site to ensure the work is satisfactory in every respect.

To ensure, as far as possible, a long and trouble-free life for the construction, good site quality control of mastic asphalt and associated work is essential. On site, quality control should be the responsibility of everyone. Establish as soon as possible close cooperation on every aspect of the work, between the designers and the main contractor, the main contractor and mastic asphalt subcontractor, independent mastic asphalt specialist and, for example the screeders, bricklayers and plumbers.

On site, during the melting of the block material the mastic asphalt may be damaged, for example by overheating or prolonged heating. At no time should the temperature of the molten mastic asphalt be allowed to exceed 230 °C. The temperature of the hot mastic asphalt from the mixer should be checked at least once every hour during the working day by the mastic asphalt subcontractor and recorded on a temperature record sheet. An example of a temperature record sheet is given in Fig. 9.8.

When melting block mastic asphalt, particularly in cauldrons, some segregation of the aggregate in the mix may occur.

Each working day take samples of the mastic asphalt from the mixer or cauldron for testing for compliance with the specification. Safety during sampling is of paramount importance. Provide training in the taking of samples of mastic asphalt. The greatest care is essential to avoid accidents, particularly falls and burns. Use safety equipment as necessary, including heat-resistant gloves with a close fitting wristband. BS 5284 provides information on taking bulk samples of mastic asphalt from site mixers or cauldrons for submission to a laboratory. At the laboratory, the samples may be tested for composition by analysis and for hardness number. However, tests for composition by analysis are expensive and may only be required from time to time. For example, if a visual examination of the samples or hardness test results suggest something may be wrong with the material.

Mastic asphalt hardens during heating and after laying. To provide an indication of the hardness of the mastic asphalt and nothing more, each morning and afternoon take a bulk sample of the molten mastic asphalt. Cast representative portions of the bulk sample into British Standard size moulds and float finish them on site. Provide the hardness number specimens with an identification number and the date of sampling and dispatch for testing. The remainder of the bulk sample should be retained on site until the results of the hardness number tests are known.

Check that gully outlets are at the correct levels, firmly secured in the deck, with the tops of their flanges clean and flush with the adjacent screed or deck surface. Ensure that in the body of each outlet the threaded holes for the securing studs are clean and temporarily protected, for example by pieces of waterproof adhesive tape.

Check that all chases in upstands are of the specified size and are in the correct positions.

Where a sand-rubbed finish is required on mastic asphalt, the sand should be applied as the laying of a bay of mastic asphalt proceeds, and be well rubbed with a wood float into the surface of the mastic asphalt while still at working temperature. The purpose of the sand rubbing is to remove the film of bitumen from the surface of the mastic asphalt.

Check that all lap joints in bays of mastic asphalt are properly warmed and cleaned using temporary applications of hot mastic asphalt. It is important that all traces of rubbing sand and any slivers of timber from timber gauges or rust from steel gauges are removed and that the new mastic asphalt is completely fused to the *in situ* mastic asphalt. The surfaces of adjacent bays of mastic asphalt should be flush.

Check for blows and sinkers in the surface of the coats of horizontal mastic asphalt. Blows and sinkers represent holes in the mastic asphalt. Some holes may be almost full depth. Blows and sinkers are due to air or moisture trapped at the time of laying of the mastic asphalt. They should be removed by the asphalter while the mastic asphalt is at working temperature. Blows and sinkers may be repaired afterwards using temporary applications of hot mastic asphalt to warm and soften the in situ mastic asphalt. However, where there is excessive blowing of the mastic asphalt, this can be a very lengthy process, and may not be completed before the roofing is to be covered, for example by roof-mounted plant. In these circumstances, it might be preferable to prevent the occurrence of blows and sinkers in the mastic asphalt beforehand.

Check the mastic asphalt for finished thickness. Take samples if required. Ensure that any sample holes are repaired using properly made lapped joints.

When making mastic asphalt fillets warm and clean the horizontal and vertical surfaces of the in situ mastic asphalt using temporary applications of hot mastic asphalt. It is important that all of the temporary application of mastic asphalt is removed and a 1–2 mm thickness of the *in situ* horizontal and vertical surfaces is also removed. Apply the first coat fillet to the warm and clean surface immediately. Apply the second coat fillet as soon as possible.

At gully outlets, the clamping ring should be fitted and secured by the asphalter while the mastic asphalt is at working temperature. It is important that the clamping ring be in completely close contact with the mastic asphalt.

Completed mastic asphalt roofing may be checked for water leaks using an electronic leak detection method, or a flood test. Before flood testing, it is generally recommended that all outlets should be covered and sealed. However, sometimes reportedly leaking roofs are subsequently found to be leaking at the outlets, particularly if the outlet is blocked and local pools of water form above the outlet concerned. To minimise the risk of this, test each outlet by plugging and locally flooding with water for a period of 48 hours. Before flood testing, it would be important to check that the roof and building structure will withstand the imposed load of water, and that should rainfall occur none of the upstands could be over-topped by water.

9.6 Examples of quality control record sheets

INSPECTION OF MASTIC ASPHALT AND ASSOCIATED WORK

Contract:..

Location:.. Grid reference........................

Date:.. Inspection made by:.....................................

1. Concrete deck/screed finishes satisfactory — Yes/No/NA
2. Falls on concrete deck/screed satisfactory — Yes/No/NA
3. Upstand satisfactory (including the position of the DPC) — Yes/No/NA
4. Dimensions of chase or rabbet satisfactory — Yes/No/NA
5. Rainwater outlets correctly fixed — Yes/No/NA
6. Rainwater outlets at correct levels — Yes/No/NA
7. Rainwater outlet bodies and flanges free from cement grout and rust — Yes/No/NA
8. Blows, sinkers or blisters in mastic asphalt skirting — Yes/No/NA
9. Measured thickness of mastic asphalt skirting — _____mm/NA
10. Reinforced bitumen membrane (RBM) separating membrane butted to skirting — Yes/No/NA
11. Measured widths of side and head laps in RBM separating membrane — _____mm/NA
12. Edges of bays of mastic asphalt free from contamination — Yes/No/NA
13. Rubbing sand well rubbed into the surface of the final coat of horizontal mastic asphalt — Yes/No/NA
14. Blows or sinkers in horizontal mastic asphalt — Yes/No/NA
15. Measured thickness of horizontal mastic asphalt — _____mm/NA
16. Measured width of lap in horizontal mastic asphaltc — _____mm/NA
17. At edges of bays. Wet RBM separating membrane removed or allowed to dry — Yes/No/NA
18. Lap joints properly made between bays of mastic asphalt — Yes/No/NA
19. At rainwater outlets. Clamping rings, slotted collars and gratings correctly installed and fixed — Yes/No/NA
20. Mastic asphalt fillets correctly made — Yes/No/NA
21. Finished mastic asphalt temporarily protected from damage — Yes/No/NA
22. Horizontal thermal insulation correctly laid — Yes/No/NA
23. Vertical thermal insulation satisfactory — Yes/No/NA

Signed:.. Date:............................

On behalf of subcontractor:...

The subcontractor should complete this form and hand it to the main contractor's package manager at the end of each days work.

Figure 9.7 Checklist for inspection of mastic asphalt when using the inverted warm deck roof design.

Record of temperature readings taken in mastic asphalt ex mixer

Contract:_____ Date:_____

Temperatures recorded by:_____

	Time	Reading °C*
8 a.m. – 9 a.m.		
9 a.m. – 10 a.m.		
10 a.m. – 11 a.m.		
11 a.m. – 12 noon		
12 noon – 1 p.m.		
1 p.m. – 2 p.m.		
2 p.m. – 3 p.m.		
3 p.m. – 4 p.m.		
4 p.m. – 5 p.m.		
5 p.m. – 6 p.m.		

*Readings to be recorded once each hour during the working day.

Signed:_____ Date:_____

On behalf of subcontractor:_____

The completed form should be handed by the subcontractor to the site agent at the end of each day's work.

Figure 9.8   Sheet for recording the temperature readings taken in hot mastic asphalt from the mixer.

Further reading

Standards and Codes of Practice

BS 1202–1: 2002 *Specification for nails. Steel nails.*
BS 1447: 1988 *Specification for Mastic Asphalt (Limestone Aggregate) for Roads, Footways and Pavings in Building.* (Note: BS EN 13108–6:2006 supersedes BS 1447:1988; however, it is understood from the Mastic Asphalt Council that it is envisaged that reference to BS 1447 will continue in the UK for some time to come.)

BS 5284: 1993 *Methods of Sampling and Testing Mastic Asphalt used in Building and Civil Engineering.*

BS 6925: 1988 *Specification for Mastic Asphalt for Building and Civil Engineering (limestone aggregate).*

BS 8000–4: 1989 *Workmanship on Building Sites.* Code of Practice for waterproofing.

BS 8204–5: 1994 *Workmanship on Building Sites.* Code of Practice for screeds, bases and *in situ* floorings for mastic asphalt underlays and wearing surfaces.

BS 8218: 1998 *Code of Practice for Mastic Asphalt Roofing.*

Other related texts

Note: For HSE documentation see Chapter 19.

Roofs and Roofing. Design and Specification Handbook: D. T. Coates, Whittles Publishing, Scotland, 1993.

The Blue Book: Flat Roofing: a Guide to Good Practice: Ruberoid, Rochester, Ruberoid, 1982 (but continuously updated to reflect revisions in Building Regulations and British and European Standards).

Design and Construction of Deep Basements Including Cut-and-cover Structures. I Struct E, London, 2004.

Note: In addition to the above, the Mastic Asphalt Council produces guides on roofing, tanking, flooring and paving which may be accessed through: *www.masticasphaltcouncil.co.uk*

Chapter 10   MASONRY

Brian Barnes

Basic requirements for best practice

- **Correct selection of brick or block**

- **Adequate storage to prevent damage before use**

- **Keep cavities clear—see them as a drain**

- **Check that DPCs are continuous and that they direct water outwards**

- **Take extra care with shelf angles, cavity ties, special supports and restraints**

- **Mix facing bricks from as many packs as possible**

10.1   Introduction

10.1.1   General

This chapter applies mainly to the use of clay bricks. The principles of clay bricks and brickwork relate closely to other forms of masonry including concrete bricks, concrete blocks and to natural or reconstituted stone.

It is important to obtain and to retain skilled bricklayers. This increases the pressure on site managers. Good site management with constant and knowledgeable supervision is essential for defect-free construction.

Hundreds of pages can and have been written concerning brickwork and other forms of masonry. The selected advice below is the result of analysing costly mistakes witnessed by the compiler and his working colleagues.

10.1.2   Responsibilities

All the site engineer's actions and decisions must be in accordance with the conditions of the contract. Before any brickwork starts, ascertain what the site engineer

167

and his company are responsible for, by establishing answers to the following questions:

- What are the required properties and functions of the completed brick wall? For example:
  - aesthetics
  - durability
  - structural
  - weathertightness
  - thermal
  - acoustic
  - fire
  - movement accommodation
  - other properties particular to that contract.
- What level of quality is required?
  This is the most difficult question to answer, but, to start the process, all perform-ance requirements should be quantified and all other requirements should be fully described. Quality is difficult to describe in words. It may have to be expressed in terms of a nearby, completed building but it is recommended that a sample panel is erected and the results agreed with the supervising officer. This is also the stage at which discussions on accuracy and tolerances should be started.
- Who is responsible for each of the following?
  - Main materials
  - Ancillary materials
  - Assemblies
  - Testing, pre-construction
  - Testing, during construction
  - Design
  - CDM regulations
  - Specification
  - Working details
  - Interfaces with the structural frame
  - Interfaces with other construction elements
  - Site management
  - Procurement delivery periods
  - Supply chain management
  - Delivery, distribution, protection and wastage
  - Quality control
  - Accuracy and tolerances
  - Workmanship
  - Supervision
  - Inspections
  - Maintenance requirements
  - Maintenance manual.

10.1.3 Components in wall construction

In order to establish the complexity or otherwise of the wall construction, extract the list of components that need to be incorporated in each type of wall, for example:

- bricks
- brick specials
- brick bond and layout
- copings or cappings
- cills and other features
- mortar strength
- mortar colour
- mortar additives
- mortar profile i.e. jointing or pointing
- inner skin
- cavity (clear or partial-fill)
- insulation (partial- or full-fill)
- insulation clips
- cavity wall ties
- lintels
- supports and brackets
- special hangers
- sliding anchors
- windposts
- cavity insulation and clips
- openings (e.g. for windows and doors)
- damp-proof courses (generally insulated at window reveals)
- damp-proof trays
- supports to damp-proof trays
- damp-proof cloaks or specials
- cavity, low level fill
- weepholes
- movement joints, horizontal and vertical
- joint fillers
- sealants
- interface with structural frame
- interfaces with DPM (damp-proof membrane)
- interfaces with roof membrane
- interfaces with other elements
- winter working conditions
- other specific requirements.

With many of the above, it is important to obtain the manufacturers' recommendations for installation.

10.1.4 Aspects of conformity

The notes that follow assume that the site engineer's main responsibility is for 'good *site* practice'. This would include site management, supervision, workmanship and inspections. However, a high and increasing proportion of contracts now include design responsibilities. The latter cannot be covered adequately in this manual (note there is an extensive section in BS 5628 Part 3 that deals with masonry design). On the other hand, it is impossible to ignore design completely and some references to design need to be made.

On many sites there is great emphasis on the aesthetics of the brickwork. The importance of looks should be recognised, but it is arguably as important to ensure that the finished wall will perform its other specified requirements satisfactorily.

The site engineer should check that all activities conform to all existing health and safety requirements and that 'others' have established any requirements by the local planning authorities.

Starting with the contract specification, list the technical compliance requirements, for example:

- Contract Specification
- Building Regulations
- Construction Products Directive
- British Standards
- European Standards
- ISO Standards
- Agrément Certificates
- National Federation requirements
- Manufacturers' requirements
- Other contract-specific standards.

Obtain copies of all relevant requirements and then extract the guidance that pertains to the relevant site activities and, in particular, to recommendations that cover workmanship, supervision and inspections. Before any work starts, establish how these recommendations will reach the site operatives and what site check lists should be drawn up.

The construction industry is widely promoting the use of *best practice*. The difficulty with *best practice* in technical matters is relating this to British Standards that claim to be a level of *good practice*. Any organisation promoting a higher level than *good* will have to establish its own technical guidance. This manual assumes a level of *good practice*.

Do check the wording in the foreword to each British Standard which warns that *compliance with this standard does not of itself confer immunity for legal obligations.*

10.2   Planning and specification

The main requirements of good masonry are strength and stability, durability, resistance to frost and rain, thermal performance and a good appearance. Recent and forthcoming changes in the building regulations require greater air tightness and acoustic performance.

All the technical guidance needed is available to the designer for meeting these requirements. Extensive design information can be found in Section 3 of BS 5628–3. This can be updated by using specialist publications from the Brick Development Association. Useful information on clay bricks can be found in BS EN 771–7. The main standard for workmanship is BS 8000 Part 3, but the latter, by its own foreword, acknowledges that it is *not necessarily complete.*

The contract specification, drawings and working details should be studied to extract all relevant information for material ordering and for construction. Discuss any gaps or ambiguities with the designer before ordering any materials. Company standing instructions or procedures and technical guidelines should also be followed.

Establish delivery periods before placing any order. Some such periods may be unacceptable and alternatives may need to be considered. A good example of this is the twelve weeks that are often needed for the delivery of 'special' specials in clay brickwork. Alternatively special bricks formed by 'cutting and bonding' (using special adhesives) can be obtained considerably sooner. It is best to avoid the use of freshly cast concrete blocks and bricks or the use of recently fired clay bricks.

Before any brickwork starts the designer may require sample panels of brickwork. Even if these are not specified it may still be prudent to provide such samples for approval. After the brick has been agreed, but before any brickwork starts, a reference panel should be constructed in accordance with PAS 70, annex to BS EN 771-1. It is essential that the reference panel be built to standards that are achievable in normal daily working. This applies to the materials, which should meet the specified requirements without being especially selected, and the workmanship should correspond to current good practice. After its acceptance, this establishes the standard for the bricks, the bond, the mortar colour and profile and for the workmanship.

In the case of calculated load-bearing masonry the designer may also require evidence of the mechanical and physical properties of the masonry units.

10.3 Detailing

10.3.1 General

Inadequate or impractical detailing and specification of interfaces, openings, features, damp-proof courses or trays, flashings, supports or restraints, movement joints and so on can influence the performance of the finished wall. This applies particularly to the weathertightness, stability and durability of the completed wall.

If, upon examination, information is missing or the details shown are considered to be suspect or impractical the matter should be drawn to the attention of the designer to take the necessary action (as explained in Section 10.3.8 Responsibility).

10.3.2 Setting out dimensions

An experienced designer should plan the dimensions of masonry walls, returns, pier widths and other details to suit the sizes of whole masonry units and mortar joint widths. The Brick Development Association publish their Brickwork Dimensions Tables particularly for clay bricks (see BDA Design Note No. 3).

The same designer should be familiar with both inherent and induced deviations and should take both into consideration. Particularly useful guidance is provided in Section 3 of BS 5606 (particularly page 18).

Discussions with the designers will establish those locations where agreed adjusted dimensions will effect economies, having due regard to the bearings of beams and lintels, minimum returns and pier widths and services. A practical aspect of this is to establish any location where a clay brick *special* special can be replaced by a *standard* special.

10.3.3 Damp-proof courses and cavity trays, methane and radon barriers, flashings and interfaces

Examine the drawings and details before work starts to check that the DPC (damp-proof course) and cavity tray detailing is complete, continuous and practical. If the building has a damp-proof membrane or a gas-resistant membrane, check that it is continuous with any adjacent DPCs or trays. Check whether insulated cavity closers have been specified and if so how they will affect the position of the DPCs.

Practical and comprehensive advice on DPCs and trays can be found in BS 8215. The advice that follows is to minimise repetitive problems that are known to occur. DPCs should be sandwiched between layers of fresh mortar and, with normal face work, should project 5 mm beyond the face of the wall. The common practice of laying the DPC dry onto the brickwork beneath creates a slip-plane, which reduces the stability of the finished wall. It is essential that this should be looked at at parapet level where dislodgement could create a safety hazard. All DPCs and trays should be lapped by the manufacturer's required minimum dimensions. All DPCs preventing horizontal or downward movement of water should be both lapped and sealed.

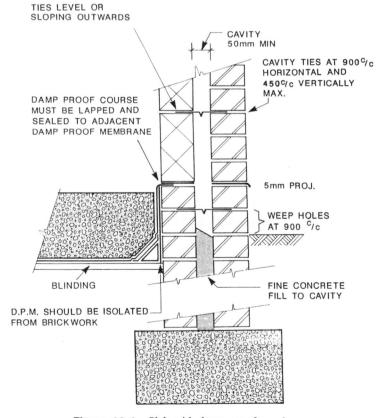

Figure 10.1   Slab with damp-proof membrane.

DPCs at window and door reveals require correct detailing and careful installation. BS 8215 offers clear guidance. As the opening obstructs the cavity a damp-proof tray is required at the head. It should extend at least 25 mm beyond any vertical DPC and the tray should be provided with stop-ends. If the cill is jointed the DPC beneath the cill should be turned up at the back and at the ends. As far as it is practical, the vertical DPCs should be lapped under the tray at the head of the opening and lapped over the cill DPC or tray. The principle to be maintained is that all laps should permit any water to move downwards and outwards. The vertical (reveal) DPC should extend into the opening to make its designed contact with the frame. The same DPC should extend, at its other edge, into the cavity (a minimum of 25 mm, but preferably 50 mm). Rain frequently penetrates at window and door positions and great care is needed to prevent this problem.

Where cavity trays, lintels or cavity (weak concrete) fill are installed, weep holes should be provided. These should be formed to permit the free escape of any water and should be provided at a maximum of 1 m spacings. Extra weepholes should be provided where short lengths of wall are involved and where walls are subjected to greater exposure to weather.

Cleanliness of the cavity has a major effect on the working of the cavity as a drained air space. Mortar droppings should be kept to a minimum and any droppings should be removed before they have time to harden. Steel rods should not be used for this purpose, as they will damage cavity trays (see also Section 10.6.6).

Cavity trays should be inspected for correct laps and seals, for damage and for cleanliness as work proceeds.

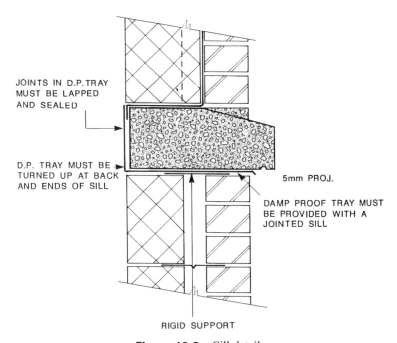

JOINTS IN D.P.TRAY
MUST BE LAPPED
AND SEALED

D.P. TRAY MUST BE
TURNED UP AT BACK
AND ENDS OF SILL

5mm PROJ.

DAMP PROOF TRAY MUST
BE PROVIDED WITH A
JOINTED SILL

RIGID SUPPORT

Figure 10.2   Cill detail.

10.3.4 Movement joints and sealants

Masonry walls that clad buildings are generally attached to a steel or concrete frame. A pattern of vertical and horizontal movement joints should be provided to cope with any movement of the masonry (mainly moisture and thermal movements), with any relative movement of the frame (mainly caused by dead and live loadings, by shrinkage or elastic deformation and by differential settlements). The general principle is to *accommodate* movement where it is difficult to prevent or minimise such movement.

In addition to the general pattern, movement joints should also be provided at particular wall features that are likely to allow a crack to occur. These are:

• movement joints in the structural frame
• changes of wall thickness
• changes of wall height
• changes of direction
• locations of large chases
• interfaces with other elements
• parapet walls.

The location of movement joints should be included in the specification or the designer's details. They can sometimes be dictated by the modular dimensions of the building. If such joints are not shown this should be brought to the designer's attention before any work proceeds.

Some of the functions of a movement joint are:

• to permit the calculated movement to take place
• to maintain the designed performance of that element
• to prevent stresses or damage occurring during movement
• to accommodate both induced and inherent deviations
• to make what would be a crack or a gap visually acceptable.

The designer should liaise with the material producer and the structural engineer. The manufacturers of the bricks or blocks can provide guidance regarding the movement of their materials. The structural engineer should provide details of the movement characteristics of the frame.

BS 5628 for clay brick walls stresses that the spacing of movement joints should never exceed 15 m. Clay brick manufacturers may recommend 12 m, 10 m or occasionally 9 m. Concrete blocks generally require movement joints at 6 m spacings and the same dimension is frequently recommended for natural or reconstituted stone. Darker coloured materials, insulation in cavities and strong mortars can also affect the movement characteristics of the wall.

The designer should calculate the width of each movement joint. This is particularly important with lighter structural frames that result from more efficient designs. The filler material should then be chosen appropriately depending on whether the predominant movement is shrinking away from the filler or compressing it. In clay brick walls BS 5628 recommends a highly compressible, closed-cell filler board. It also states that fibreboards should not be used.

The shape of the joint should then be considered. In most cases this will be a flush or a butt joint in which the sealant is very exposed to the weather including ultraviolet degradation. The sealant should then be chosen not only for its colour or cost but rather for its:

- durability
- resistance to weathering
- resistance to water penetration
- resistance to local pollution
- movement accommodation factor
- adhesion to the substrate (or use of a primer?)
- ability not to leach into, or stain, the substrate
- fire rating (in a fire-rated wall).

The costs of premature failure including the costs of access to remove and replace a failed sealant means that the sealant manufacturer should be consulted as early as possible in case any verifying tests are needed.

10.3.5 Substructure

For work near or below ground level and where ground conditions are likely to be aggressive, the specification of the materials requires greater care. For example, the brickwork between ground level and DPC should be built with bricks of adequate frost resistance. The strength of the mortar also needs extra attention and, in some cases, sulphate-resisting cement should be incorporated in the mortar. One reason for this is that some clay bricks are high in soluble sulphates. Concrete blocks or bricks will also have to be carefully chosen.

The designer should check with material suppliers for evidence that the materials can withstand these particular conditions for the design life of the wall.

10.3.6 Parapet walls, free-standing and retaining walls

This section also applies to masonry facings to concrete walls.

The degree of exposure means that such walls are frequently subjected to permanent or long periods of dampness, combined with extremes of temperature (saturation combined with frost is generally the most severe condition). The choice of masonry units, mortars and ancillary materials should therefore be specified accordingly. It is essential to check the adequacy of the specified materials with prospective suppliers. Copings, cappings, DPCs, supports and restraints, movement joints and so on will all need attention and then should all be installed in accordance with the specification and the manufacturer's recommendations. Design advice for exposed conditions can be found in BS 5628–3.

Check that the material retained behind a masonry wall is adequately drained to remove the build up of water pressure. Weep holes may be needed. In addition, the back face of the wall is normally treated or covered with an impermeable membrane to protect the wall from damp conditions and to minimise staining which results from water penetration.

10.3.7 Cavity wall insulation

The use of full cavity insulation increases the risk of rain penetration. Check with Table 12 in BS 5628–3 for this and other design guidance on such risks. Recent and anticipated increases in the thermal requirements of walls together with maximizing

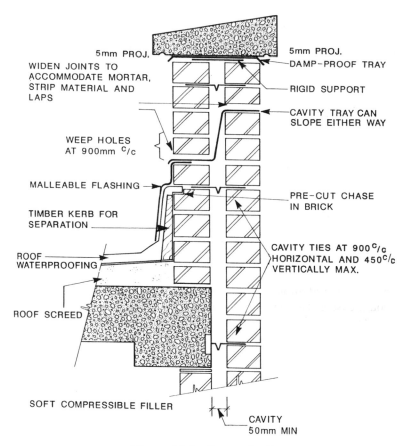

Figure 10.3 Parapet detail.

the available floor area mean that the demand for full-fill cavity insulation will probably increase. Wherever full cavity insulation is specified, the design and construction of the outer leaf probably needs to be higher than the minimum required by BS 5628.

The installation of board, batt or quilt cavity insulations is still a relatively (or reluctantly) new activity for bricklayers who often do not appreciate the importance of care at all joints, of continuity at slab edges, openings, corners etc. This should also be fully appreciated by site managers.

The designer should also appreciate that a batt or board partially filling the cavity will not result in a constant width drained air space. Normal permissible deviations, particularly reductions to the nominal width, make it more difficult to keep the cavity clean. The designer should be aware of the warnings in BS 5628–3 regarding the exclusion of moisture. Building Regulations require a minimum of 50 mm air space in partially filled cavities. Experience has shown that a wider air space may be needed where there is an increased risk of rain penetration.

So, before commencing work:

- ensure that the insulation can be incorporated as detailed and can comply with the manufacturer's installation requirements

- check that the correct ties and insulation clips will be used to hold the insulation satisfactorily against the chosen leaf (normally the inner one)
- check that the pattern of the ties is consistent with the size of the boards or batts.

10.3.8 Responsibility

The foregoing sections indicate that a number of important design matters need to be fully taken into consideration before materials can be ordered, let alone before construction starts. In order to maintain the momentum of work, queries, alternative design proposals and the management of risk have to be sorted out beforehand. In a traditional form of contract it is important that any alternative proposal is submitted for the designer's approval and will only be executed upon receipt of an instruction, in writing, from the designer. Failure to do this has frequently led to disputes regarding responsibility and payments, which have also proved to be disruptive and expensive to *traditional* contractors.

Where important design queries are raised without proposals, work should only proceed on the written instructions of the designer responsible.

When the design of all items is agreed and considered to be practical, site managers should ensure that the work is carried out in accordance with the specified requirements and good site practice.

10.4 Materials

10.4.1 Clay bricks

Clay bricks create one of the most attractive forms of cladding in construction but bricks have their limitations. Knowing the properties, and hence the limitations, of the proposed brick will help in managing the brickwork. High-density clay bricks may be used as engineering bricks for use in manholes and as DPCs.

BS EN 771–1 has procedures for the following:

- measurement of dimensions
- determination of soluble salt content (check also for soluble sulphates)
- determination of compressive strength (this can be related to the mortar strength)
- determination of water absorption
- determination of frost resistance
- Manufacturers declare the performance of their products as a result of the test procedures.

It is also worth considering:

- frost resistance
- efflorescence
- movement characteristics
- vulnerability to damage (and hence wastage).

The more exposed the brickwork the greater the need for careful selection. Some such locations are copings, cappings, parapets, cills, paving, free-standing walls or retaining walls. BS 5628 recommends that a fully frost-resistant brick should be used in combination with one of the two higher grades of mortar.

With some bricks, because of firing practices, the colour of the header face can differ from the stretcher face and this should be checked with the designer. If special bricks are to be used the designer should check that the colour of the specials is close enough to that of the bricks to be visually acceptable.

All brick deliveries should be inspected in accordance with the established quality management system. This is particularly important when bricks arrive from a different firing. Any queries should be raised immediately with the supplier. Typical faults are poor shapes, excessive fire cracks, damaged edges and corners, lime nodules or black hearts due to incomplete firing. If sand-faced bricks are used, the facing should cover the exposed face and the return of the brick.

10.4.2   Other bricks

Calcium silicate (CS) (also known as sandlime or flintlime)

These contain lime, silica sand and water. CS bricks are a uniform product of regular shape. As they are susceptible to damage, particular care in storage and handling is essential. When tested to BS 3921 their compressive strength is usually in the range 7 to 35 N/mm^2.

Concrete

These are made, to fairly close tolerances, from carefully controlled concrete mixes, often incorporating additives as colouring agents.

10.4.3   Concrete blocks

There are three basic concrete block types: dense, lightweight aggregate and aerated. They are all used extensively and can replace bricks for a number of applications. Concrete blocks are normally specified to comply with the requirements of BS 6073.

All concrete blocks, particularly lightweight blocks are subject to drying shrinkage and moisture movement. Precautions are therefore necessary to minimise problems of cracking. For example, blocks which have not been autoclaved should not be used when they are *green*. Four weeks or more should elapse between manufacture and use. Autoclaved blocks do not require the same curing period but should be allowed to cool before being used.

10.4.4   Natural stone

The procurement period for natural stone is likely to be considerably longer than that for other forms of masonry. It requires considerable cooperation with and from the stone supplier and the stone masonry subcontractor. Because of variations in sizes of stone blocks, particular care must be taken in bedding this material when used in load-bearing applications.

As it is a naturally occurring material, availability in colour, sizes needed, strength and consistency should all be considered. An early visit to the intended quarry is most important. The Building Research Establishment has published helpful technical directories on British stone quarries. The Stone Federation also publishes technical guidance on use, design and handling of stone. In addition BS 8298 give guidance on design and workmanship.

10.4.5 Brickwork mortars

Mortar is normally required to comply with BS 5628–3. It is important for the mortar to be consistent in both strength and colour. With delivered mixes, that onus is on the mortar supplier. If mortar is being mixed on site, gauge boxes are the only practical solution. Any attempt to mix mortar using shovelfuls as a measure, should be stopped.

The sand should be from an approved supplier. The sand grading and colour should be consistent when used in face brickwork since its colour frequently affects that of the mortar mix.

The consistency of coloured mortars is more difficult to achieve on site and may need to be supplied by a ready-mix mortar company. The latter source should be constantly checked for consistency. The proportion of the brickwork face that is mortar is roughly 20%, which means that variations in the set mortar can show up as visually unacceptable 'banding'.

Colour variations can also arise from different weather conditions as the mortar sets; the dampness of the bricks being laid, or drying suction from previously laid bricks, and also from the way the each bricklayer finishes his jointing or pointing. Colour variations may be minimised by:

* ensuring that all the mix ingredients are consistent
* controlling the moisture content of the bricks and the recently laid brickwork
* protecting the work during intermittent wet or dry weather
* allowing, wherever possible, the minimum number of men with the necessary skill to finish the joints in the face brickwork to achieve a uniform appearance.

Plasticisers or retarders are commonly used in building mortars, particularly in delivered ready-mixed mortars. The latter should comply with BS 5628–3. The retarders act as air-entraining agents and their use should be strictly controlled. Their use should also be discussed with the designers before work starts. Areas for concern include high-rise cladding, brick cladding to timber frames, areas of high wind exposure or areas where movement could cause damage to low-strength or immature mortars. For load-bearing brickwork the use of any additives should have the designer's approval in writing. The retarded mixes will also have an effect on the construction height limits for brickwork before further height is added.

The designed retardation of these mortars is a minimum of 36 hours but the mortars can remain unset in their containers for a much longer period. The average will be of the order of 50 hours and, if there is a lowering of the ambient temperature after delivery, the mortar may not stiffen for some 90 hours. Furthermore the mortar can stay soft for several days in joints between bricks of low absorption. Although such mortars will usually stiffen in joints between absorbent bricks and will appear to harden, this can be misleading so far as its true strength is concerned. Use retarded mortar mixes strictly in accordance with the manufacturer's instructions and do not allow re-tempering the material once hardening has begun.

Reservations on the use of retarded mixes are due to these uncertain setting times and to variations in early strength in adjacent areas of brickwork. In load-bearing brickwork there are difficulties in forecasting when load can be carried safely. The stipulated maximum height to be built in a day should not be exceeded.

There is no evidence, to date, that delayed setting has any adverse long-term effect. In the short term, these mixes are more prone to frost attack and are more likely to stain the brickwork during wet weather.

Before brickwork starts, establish the rules for taking mortar cubes for testing. This may be a set for each specified volume of mortar but it is more logical to take a set for a specified area of brickwork. Do remember how much brickwork will have been added before the earliest cube tests are available. Quality control by cube testing at seven days is not always possible because the retarder can make it impractical to de-mould cubes early enough for testing at this age. For this reason, do not use retarded mortars without obtaining the designer's or structural engineer's approval ensuring that all concerned fully understand the issues referred to above.

The higher the cement content of the mix, the more likely it is that drying shrinkage cracking will be visually unacceptable. Many mixes are too strong. Check that the minimum strength has been specified. Ensure that higher strength mortars are not inadvertently used due to lack of control in mortar batching.

Mortar mixes that are too high in cement content can result in the cracking of concrete block walls due to the inability of the stronger mortar mix to accommodate movement. Calcium silicate bricks are particularly prone to this type of defect. The mortar strength for all masonry should be checked with the brick or block manufacturer.

10.4.6 Metal supports and restraints

Masonry walls generally require stabilization using metal ties in cavities and restraints tying walls into floors, columns and frame. Useful information can be found in BS 5628–3 and BS 1243. The approach in the latter standard for prescribed ties can be replaced by the principles of DD 140, which leads to a performance tie.

Following the lead of DD 140, the tie and restraint manufacturers have researched for and developed ranges of supports, ties and restraints for most forms of masonry construction. Their supply catalogues indicate what is available, but several manufacturers have published design guidance as well.

The use of supports, ties and restraints is not an exact science but the current recommendations from the manufacturers are frequently linked to earlier problems and failures. Their technical assistance should be sought and it would be unwise not to follow their recommendations.

Cavity wall ties are required to enable two thin skins to act together structurally. The technical reason for this is because, although masonry is good in compression, it is very weak in tension. Before ordering or installing ties check that:

- the ties and straps are as specified regarding type, material, length, spacing and corrosion protection
- there are no backward-sloping ties which, if mortar droppings were present, would convey moisture towards the inner leaf
- embedment is correct despite variations in cavity widths
- the types and lengths of ties and retaining clips where cavity insulation is installed is correct
- the designers will not normally permit ties to be built into one leaf prior to bringing up the remaining leaf—if it is necessary to use this practice check carefully

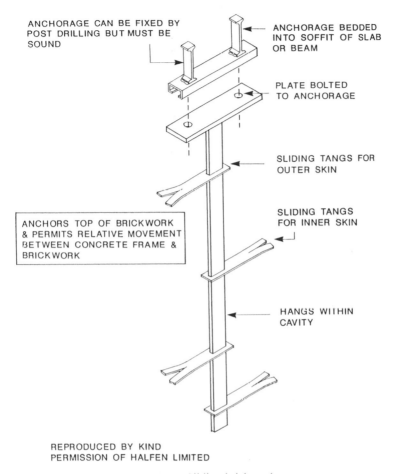

ANCHORAGE CAN BE FIXED BY
POST DRILLING BUT MUST BE
SOUND

ANCHORAGE BEDDED
INTO SOFFIT OF SLAB
OR BEAM

PLATE BOLTED
TO ANCHORAGE

SLIDING TANGS FOR
OUTER SKIN

SLIDING TANGS
FOR INNER SKIN

ANCHORS TOP OF BRICKWORK
& PERMITS RELATIVE MOVEMENT
BETWEEN CONCRETE FRAME &
BRICKWORK

HANGS WITHIN
CAVITY

REPRODUCED BY KIND
PERMISSION OF HALFEN LIMITED

Figure 10.4 Sliding brick anchor.

the protruding length and ensure that the ties are level or marginally sloping
outwards
- dovetail slots in concrete are to be fitted with ties to restrain a masonry wall
- the ties do not have to be bent into position as this may affect the tie, its embedment or both
- additional ties are provided at all window or door openings and either side of any
movement joints
- sleeves will be used where needed.

Careful consideration should be given wherever such ties or restraints are to be
anchored to the frame. Inadequate anchoring could reduce the effectiveness of the
tie or restraint. This applies particularly to shelf angles that are extensively used to
support brickwork at upper floor levels. The added difficulty is that the brickwork
above requires a minimum of two-thirds of its width as a bearing on that angle or
any support. In some circumstances floor restraints to brickwork may be provided
by the use of dovetail anchors engaging with dovetail slots.

Other structural components will include wind posts and hanging anchors. These require careful consideration in design and careful installation. An early decision must be taken as to who should install these items. For convenience, it is frequently put into the bricklayer's package but it would be appropriate to check if the bricklaying subcontractor has operatives with the right skills to install these important items.

There are also specially shaped anchors that are designed more to hang bricks from the concrete frame and to it. In some of these, each brick is separately supported or tied. In others, the threading of stainless steel rods through the hangers, and through perforations in the bricks, provides the means by which the bricks hang in place. Again considerable care is needed both in design and installation. Such work should not be entrusted to bricklayers with average skills.

10.4.7   Reinforcement

The use of bed-joint reinforcement in brickwork or blockwork is sometimes specified for special and limited applications, e.g. gable walls to two-storey houses in the area adjacent to a stairwell at first floor; over openings to supplement other means of support; to extend the design spacing between vertical movement joints, and to restrict cracking.

External work would normally require the use of stainless steel or non-ferrous reinforcement. Check with the designer if this is not specified and ask for written instructions before work proceeds. When the work is carried out it should be in accordance with the designer's requirements.

With other special applications such as reinforced brickwork columns, reinforced brickwork retaining walls or reinforced brickwork lintels, agree all details with the engineer before work proceeds.

10.4.8   Rendering

If an external wall is to be rendered, the bricks should be of low sulphate content. If not, then sulphate-resisting cement should be incorporated into the render mix and used for both the backing coat and surface mortar to minimise the risk of problems of sulphate attack. Draw the designer's attention to the need for such precautions. Where Hanson flettons are used, Hanson's technical department usually recommend this specification. Normally the key afforded by most clay bricks with recessed joints or keyed varieties is adequate, but where there is any doubt, it is advisable to provide a key by a splatterdash treatment. Apply the mixes for rendering in accordance with BS EN 13914–1.

Rendering can be successfully applied to external skins of concrete blockwork with techniques based upon experience of materials and skill. One vital factor is the use of a block with relatively low movement characteristics to avoid cracking. Technical guidance should be sought from the block manufacturers.

10.5   Storage and handling

Bricks and blocks need careful handling to minimise damage. Stack off the ground on a level and clean area capable of withstanding their weight and keep covered during storage. As many clay bricks are prone to colour variation, bricks should be intermixed from as many packs as possible but not less than three packs to avoid *banding* between lifts. If all the bricks for one contract can be obtained from one firing this reduces the risk of banding.

All masonry materials need to be safely stored as close as possible to the place where they are to be used to avoid unnecessary handling and damage. Concrete blocks that arrive wet on delivery should be protected from further wetting and allowed to dry by adequate ventilation.

10.6 Supervision, workmanship and inspections

10.6.1 General

The clay brick manufacturers strongly recommend that all bricks, and also all fresh brickwork, are kept as dry as possible for as long as possible. Work should be protected at night and over the weekends and, if it starts to rain, during the day. This limits efflorescence and lime staining and enhances rain resistance. In hot weather it may be necessary to dampen the bricks before laying to ensure that water is retained in the mortar long enough for setting to take place satisfactorily. The degree of dampening should be the result of experiments, as overwetting can result in bricks slipping, or of the mortar squeezing out because of the slow rate of setting.

This section will look at recommendations for good site practice in a number of important aspects:

- Accuracy and tolerances
- Appearance
- Durability
- Structural performance
- Weathertightness
- Thermal performance
- Movement accommodation
- General aspects.

There are a number of recommendations that could be placed in more than one category. As an example, fully filling all cross-joints with mortar enhances the durability, structural performance, weathertightness and acoustic performance of the brick wall.

10.6.2 Accuracy and tolerances

Establish the accuracy requirements for the finished brickwork in relation to the structural frame, incorporated openings, adjacent elements and any visual requirements. Then set out all work correctly for line and level prior to commencing the brickwork. The use of storey gauge rods will assist considerably. Where good quality brickwork is needed, set out two courses of bricks dry, particularly to check the bond, to minimise cutting and to compare the bond with proposed openings further up the wall. Pay special attention to any shelf angles or other brickwork supports (in line, level and bearing) which will be part of the structural frame or attached to it.

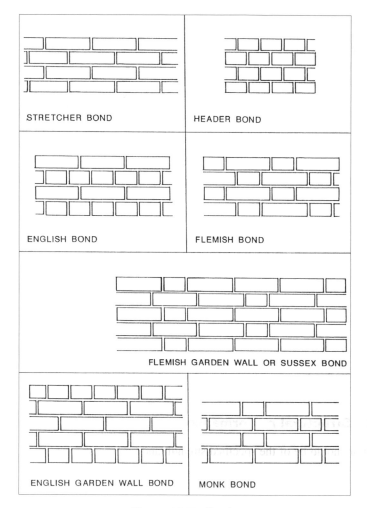

Figure 10.5 Bonds.

10.6.3 Appearance

Clay bricks vary in size and hence brickwork is a craft skill rather than a process of assembling accurate components. Any specification for accuracy must reflect practical considerations. It is unreasonable to insist that all perpends (the visible part of the cross joint between bricks is commonly referred to the perp or the perpend) are plumb and that every bed joint is 10 mm wide. If such visual matters are critical then the specifier should call for concrete bricks.

The positioning of the perps must be pre planned. Good visual brickwork is achieved if every fifth perpend is plumbed with pencil marks at appropriate points and that this process is carried right up the wall. More accurate bricks are needed to plumb up the fourth or even the third perpend. Between the plumb marks, the bricks and cross-joints are evened out as far as possible.

Clean down the brickwork as work proceeds and protect the fresh brickwork against rain, frost and adjacent activities, which could affect the brickwork visually or physically. Even light rain falling onto fresh mortar on a scaffold board adjacent to finished brickwork can cause unsightly mortar splashes. At the end of any working day, boards should be moved or turned to prevent this happening.

Acid cleaning of brickwork should only be considered as a last resort. This treatment can alter the appearance of the brickwork and can have a detrimental effect on the brickwork and any other materials that are in close proximity. This procedure is not necessary if the work has been carried out to a good standard and adequately protected. If acid cleaning is absolutely unavoidable, proceed with care, wet the work before treatment and thoroughly wash the area with clean water immediately afterwards.

10.6.4 Durability

The brick should have been chosen for its durability taking into account the known degree of exposure. Many of today's cement-based mortars can be too strong. The brick manufacturers would prefer the mortar to be related to the strength of the brick. If the mortar strength were nominally less than the strength of the brick then any small, movement cracks would occur in the mortar rather than the bricks.

The mortar should be given every opportunity to cure properly. This applies particularly to fresh brickwork when frost damage could severely attack the mortar and, for that reason, insulating materials should be incorporated underneath the waterproof protection until the mortar is capable of withstanding freezing conditions.

10.6.5 Structural performance

The mortar should be of the specified strength and be consistent. All bed- and cross-joints should be filled with mortar. The mortar in the cross-joints can consist of two small dabs as a time- and material-saving short cut. This practice is often referred to as 'tipping and tailing' the bricks. It is not recommended as it considerably weakens the structural properties of the brickwork and permits greater ingress of water. Solid cross-joints are achieved by trowelling a solid dab of mortar onto the end of each brick before laying it. Bearings on shelf angles and the like should not be less than two-thirds of the brickwork width. Unless otherwise agreed, the frog of any brick should be laid uppermost. In general bricks for load-bearing work will be laid frog-up with the frogs filled with mortar. BDA have, however, noted that a frogged brick could be laid frog-down *if the completed wall still met all its performance requirements.*

The structural performance of the wall will rely heavily on all wall supports and restraints. These can include footings, lintels, shelf angles, wind posts, special hangers or anchors and cavity wall ties. All of these should be correctly installed and this generally means in accordance with the manufacturer's recommendations.

The rate of laying should not normally exceed 1.5 m in height per day.

Cavity wall ties, because of their size may seem to be relatively insignificant, but the two thin skins of masonry can only act together if such ties have been correctly designed and properly installed. Cavity wall ties, regrettably, are often missed or poorly installed. Inspections for wall ties can sometimes be a cursory

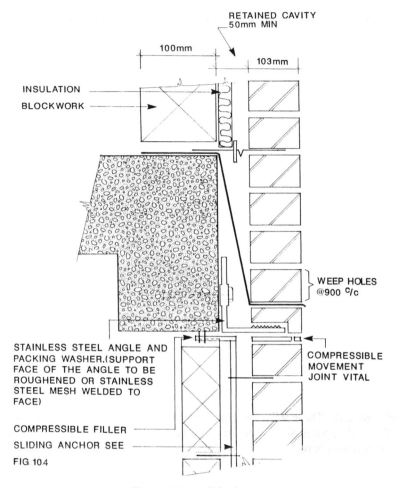

Figure 10.6   Edge beam.

glance followed by a tick or two on a quality checklist. Consider what a thorough check might consist of in the light of the following list:

Cavity tie checklist

- Compliance with contract specification for cavity ties
- Compliance with DD 140
- Type—to comply with Building Regulations. Generally stainless steel
- Type—shape
- No sharp edges or ends
- Not damaged or bent
- Clean before installation (especially oil)
- Special ties (e.g. to steel frame)
- Correct length (including cavity width deviations)

- Minimum embedment lengths (usually 50 mm min., but check)
- Embedment (mortar both above and below i.e. no dry bedding)
- No disturbed embedment e.g. by knocking or bending
- Correct insulation clips
- Free of mortar droppings
- Slope—preferably level, but if sloping, slightly outwards
- Drip—right way down
- Drip—centre of cavity or of residual cavity
- Pattern (staggered or rectangular)
- Frequency/general spacing (no./m²)
- Extra ties at all openings
- Extra ties at either side of movement joints
- Square—horizontally (as a structural member)
- Square—vertically (as a structural member)
- Manufacturer's technical guidance
- Test data

From the above, a site checklist could be drawn up. Consider similar lists for other ancillary items.

10.6.6 Weathertightness

The success of external cavity walls depends on the crucial principle of the cavity acting as a barrier i.e. a drained air space. The cleanliness of the cavity is critical. Mortar protrusions into the cavity should be removed and collected with the trowel as work proceeds. There will, however, be mortar droppings that accumulate at the base. These droppings should be removed regularly, after the mortar has set but before it becomes hard.

BS 5628–3 recommends two methods for keeping the cavity clean, firstly the use of cavity battens or laths and secondly the use of coring holes. The cavity battens are cut to fit the width of the cavity. They collect a high proportion of the droppings and as they are drawn upwards with cords before each set of cavity ties are installed, this enables the collected mortar to be removed.

Coring holes are left immediately above the base of the cavity at approximately 1 m centres. The removal of the loose mortar is then greatly assisted by the introduction of a continuous length of 150 mm wide hessian, at this level, as indicated in Fig. 10.7. The onus of keeping the cavities clear of mortar falls on the bricklaying specialist. His practical recommendations should be considered and the agreed method adhered to.

Because any external wall thermal requirements are now statutory there is greater use of partial- or full-fill cavity insulation. These alternatives both need additional care if the wall is also to keep out rain penetration. In the case of full-fill cavity insulation, where there is no residual cavity, the advice of the specialist insulation manufacturer must be sought to check how the wall is to be constructed. Consider asking for evidence that the manufacturer's advice has worked. With full-fill cavity insulation, any rainwater penetrating the outer brick skin seeks any penetration or joint in the insulation and tracks across, by capillary action, to the inner skin.

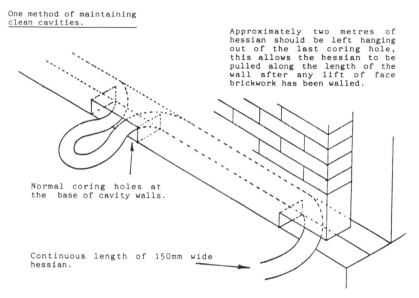

One method of maintaining
clean cavities.

Approximately two metres of
hessian should be left hanging
out of the last coring hole,
this allows the hessian to be
pulled along the length of the
wall after any lift of face
brickwork has been walled.

Normal coring holes at
the base of cavity walls.

Continuous length of 150mm wide
hessian.

Figure 10.7   Removal of mortar from cavities.

Mortar droppings at board joints or on ties increase the amount of water reaching the inner skin.

In the case of partial-fill insulation the residual cavity still becomes the principal barrier to rain penetration and its cleanliness is crucial. Again, before commencing work, ensure that the insulation can be incorporated in accordance with the manufacturer's recommendations. In particular check that the insulation clips are compatible with the cavity wall ties and that the drip of the cavity tie is centred on the residual cavity. Any cutting around wall ties, openings and angles should be carried out accurately and all small pieces of insulation must be securely held in position. All horizontal and vertical joints between insulation boards or batts should be tightly butted. Very small amounts of mortar between batts will permit water to be conveyed to the inner skin. Insulation built in should not be allowed to get wet and the tops of cavities containing insulation should be protected against rain during construction.

The importance of fully filling all bed and cross-joints has been stressed elsewhere. This has a major effect on the weathertightness of the wall. So also does the shape and tooling of the profile of the jointing or pointing. The two best profiles are weatherstruck and bucket handle because the finish is produced by a tooling action and they are both well shaped for deflecting rainwater. The flush finish has a good shape but the surface is generally produced by a scraping action and hence is relatively porous. The square recessed profile has the disadvantage of a ledge on which rainwater can rest together with a surface that is scraped rather than tooled. The use of the last mentioned profile has been responsible for water getting through the outer skin in many exposed situations.

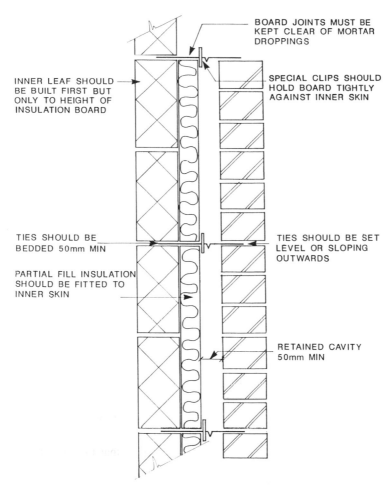

INNER LEAF SHOULD
BE BUILT FIRST BUT
ONLY TO HEIGHT OF
INSULATION BOARD

BOARD JOINTS MUST BE
KEPT CLEAR OF MORTAR
DROPPINGS

SPECIAL CLIPS SHOULD
HOLD BOARD TIGHTLY
AGAINST INNER SKIN

TIES SHOULD BE
BEDDED 50mm MIN

PARTIAL FILL INSULATION
SHOULD BE FITTED TO
INNER SKIN

TIES SHOULD BE SET
LEVEL OR SLOPING
OUTWARDS

RETAINED CAVITY
50mm MIN

Figure 10.8   Cavity insulation partial fill.

10.6.7   Thermal performance

Ensure that there are no gaps in the insulating layer and that all boards, batts or
pieces are properly clipped back. Keep all insulation as dry as possible before and
after fitting.

The need to comply with ever-increasing thermal requirements often leads to
the insulation fully filling the cavity. In such a case it is essential to check with the
insulation manufacturer how his material is to be installed without permitting water
penetration. The greatest practical difficulty will be to prevent the capillary movement of water through joints and perforations for cavity wall ties.

10.6.8   Movement accommodation

Movement joints should be correctly formed and be at the appropriate spacings. The
designer should have established the movement joint pattern in conjunction with the
brick manufacturer and the structural engineer. As a brick can grow by 1 mm/m

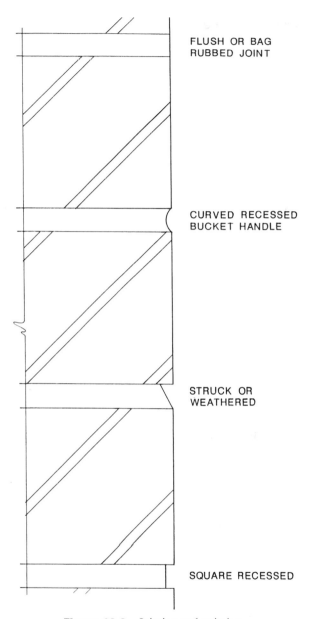

Figure 10.9   Jointing and pointing.

(over its lifetime) it is important to use a closed-cell and highly compressible filler material. The spacing of movement joints is best established with the help of the manufacturer. The formula for calculating the width of the joint can be found in BS 5628–3. Good workmanship is essential in forming these joints.

Care should be taken to ensure that sleeved ties used across a movement joint have the sleeves on the same side of the joint. Establish a rule with the bricklayers before work commences.

10.6.9 Holes and chases

Holes in the brickwork/blockwork walls for services can cause major problems for the brickwork subcontractor. Services engineers may dimension holes for multiple services (which need to pass through the wall) without a consideration or understanding of how these holes are to be formed.

All holes required in brickwork or blockwork walls should be shown in elevation, to scale, on the builders' work drawings. In addition, the services passing through them should be drawn in their correct location. The services from each aspect of the mechanical and electrical network should be gathered onto one set of drawings in order to combine service holes wherever possible. This set of drawings also serves to establish those changes that are instructed too late to be incorporated. From this information the brickwork subcontractor can assess the sizes of lintels required and how to incorporate them.

10.6.10 Lead flashings

Where it is necessary to incorporate a lead flashing underneath a damp-proof course or tray, it is important that a clean rebate be provided in the brickwork beneath or a wide bed joint incorporating a timber lath to line up with the front face of the brickwork. This will enable the lead flashing together with its supporting clips to be properly installed.

10.7 Protection of work

10.7.1 Curing and protection of freshly executed work

Brick manufacturers recommend that the brick and fresh brickwork be kept as dry as possible for as long as possible. This minimises efflorescence, staining and rain penetration. Waterproof covering also assists the mortar to cure properly in periods of very cold or very dry weather.

Note that polythene sheet that is hung so that it is in contact with newly built masonry can cause staining. Hang it on timbers or similar so that the sheet is kept 50–100 mm away from the face of the newly built wall.

All clay bricks contain salts. If these salts are soluble they can be carried by moisture to the outer surface of the brickwork. Sun and wind can evaporate the water, leaving the salts as a powdery deposit on the surface of the brickwork. This deposit is referred to as efflorescence. The liability for efflorescence should be established when the brick is chosen. If the brick is categorised as moderately or heavily efflorescent, the salt deposits cannot always be prevented, but they may be minimised by the following steps:

- Keep the bricks, and the freshly built brickwork, as dry as possible for as long as possible.
- Do not use water to remove the surface salt deposits.
- During dry weather, use a soft brush to remove as much surface salt as possible.
- Avoid the use of acids.

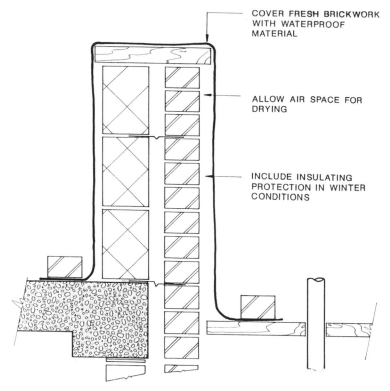

COVER FRESH BRICKWORK
WITH WATERPROOF
MATERIAL

ALLOW AIR SPACE FOR
DRYING

INCLUDE INSULATING
PROTECTION IN WINTER
CONDITIONS

Figure 10.10   Protecting fresh brickwork.

Further reading

Statutory documents

The Building Regulations (England and Wales), HMSO.
Approved Documents to the Building Regulations, HMSO.
Construction Products Directive.

Standards and Codes of Practice

BS 743: 1970 *Specification for Materials for Damp-Proof Courses.*
BS 1243: 1978 *Specification for Metal Ties for Cavity Wall Construction.*
BS 5606: 1990 *Guide to Accuracy in Building.*
BS EN 771–1: 2003 *Specification for Masonry Units: Clay Masonry Units.*
BS EN 934–3: 2003 *Admixtures for Concrete, Mortar and Grout. Admixtures for Masonry Mortar. Definitions, Requirements, Conformity, Marking and Labelling.*
BS 4729: 2005 *Clay and calcium silicate bricks of special shapes and sizes. Recommendations.*
BS 5628–3: 2005 *Code of Practice for Use of Masonry.* Materials and components, design and workmanship.
BS EN 13914–1: 2005 *Design, Preparation and Application of External Rendering and Internal Plastering. External rendering.*

BS 5642–1: 1978 Sills and Copings: Specification for Window Cills of Precast Concrete, Cast Stone, Clayware, Slate and Natural Stone.

BS 5642–2: 1983 *Sills and Copings: Specification for Copings of Precast Concrete, Cast Stone, Clayware, Slate and Natural Stone.*

BS 6073–1: 1981 *Precast Concrete Masonry Units: Specification for Precast Concrete Masonry Units.*

BS 6073–2: 1981 *Precast Concrete Masonry Units*. Method for specifying concrete masonry units.

BS 6398: 1983 *Specification for Bitumen Damp-proof Courses for Masonry.*

BS 8000–3: 1989 *Workmanship on Building Sites*. Code of Practice for masonry.

Note: See also Chapter 1.

BS 8200: 1985 *Code of Practice for Design of Non-loadbearing External Vertical Enclosures of Buildings.*

BS 8215: 1991 *Design and Installation of Damp-proof Courses in Masonry Construction.*

BS 8298: 1994 *Code of Practice for Design and Installation of Natural Stone Cladding and Lining. DD 140: 1986 to 1987 Wall ties.*

Note: This Draft for Development is produced in several parts.

Note: Incorporated materials, such as DPCs, often fall outside British Standards and information must be found in Agrément Certificates. The manufacturers' recommendations for installation should also be followed.

Other related texts

Construction Materials Reference Book, Chapters 11, 17: D. K. Doran ed., Oxford UK, Butterworth, 1991.

Construction Materials Pocket Book: D. K. Doran ed., Oxford UK, Butterworth, 1994.

Concrete Masonry Designers Handbook: J. J. Roberts *et al.*, Leatherhead UK, Viewpoint, 1983.

Note: The Brick Development Association (BDA) produces a number of useful documents which may be accessed through *www.brick.org.uk*

These include:

* *A guide to Successful Brickwork.*
* Building Note 1 *Bricks and Brickwork on Site.*
* Building Note 2 *Cleaning of brickwork.*
* Design Note 3 *Brickwork Dimension Tables.*
* Design Note 7 *Brickwork Durability.*
* Design Note 10 *Designing for Movement in Brickwork.*

Chapter 11 FORMWORK, SCAFFOLDING, FALSEWORK AND FAÇADE RETENTION

Peter Pallett

Basic requirements for best practice

- Clear accurate drawings and specifications
- Good safety procedures and practice
- New or guaranteed second-hand materials

11.1 General

11.1.1 Introduction

It is widely accepted that the temporary works account for over 39% of the cost of a concrete structure, and can be up to 55% of the cost in civil engineering structures. Making the best use of information and materials makes economic sense. This chapter is intended to give some practical guidance on the latest philosophies and economic use of materials. There is a further reading list at the end of the chapter in which all the references in the text are quoted in full.

Few contractors carry out their own temporary works. The operations are often subcontracted to smaller specialist organisations. On a large building contract it is often found that the entire concrete frame is contracted out to a specialist frame contractor. The relevant trade association is the Concrete Structures Group (CONSTRUCT). Thus, although the responsibility as *principal contractor* remains with the main contractor, the day-to-day management is left to others, often supervised by the site engineer with a *watching brief* on temporary works. Unfortunately, site engineers are rarely, if ever, given any significant training in the subject, thus, unless the site has experienced trades-based foremen or supervisors, the active control of the temporary works can be fraught with problems.

The three principal documents in the industry on the subjects are:

- *Formwork—A Guide to Good Practice* (2nd edition)
- TG20:08 *Scaffolding Guidance Launch: Guide to Good Practice for Scaffolding, with Tube and Fittings.*
- BS 5975: 2008 *Code of Practice for Temporary Works Procedures and the Permissable Stress Design of Falsework.*

11.1.2 Definitions

Backpropping

Propping installed at levels below the slab supporting the falsework to distribute the load applied to the uppermost slab to suitable restraints, such as lower slabs or to the foundations. They may be required at more than one level.

BCA

British Cement Association

CIRIA

Construction Industry Research and Information Association

CONSTRUCT

Concrete Structures Group

Contractor

Any organisation carrying out the construction operations on the site.

CPF

Controlled permeability formwork

ECBP

The European Concrete Building Project, a seven storey concrete flat slab building constructed in 1998 inside BRE Hangar No. 2 at Cardington, Bedfordshire, England.

Falsework

Any temporary structure used to support the permanent structure until it can support itself.

Formwork

A structure, usually temporary, used to contain poured concrete to mould it to the required dimensions and support it until it is able to support itself. It consists primarily of the face contact material and the bearers directly supporting the face contact material.

ggbs

Ground granulated blastfurnace slag

LOLER

Lifting Operations and Lifting Equipment Regulations

NASC

National Access and Scaffolding Confederation

PASMA

Prefabricated Aluminium Scaffolding Manufacturers Association

Permanent Works Designer (PWD)

Person or organisation appointed for the structural design of the permanent works under construction. Under certain contracts, such as design and build contracts, this responsibility can be the same organisation as the constructor.

pfa

Pulverised fuel ash

Postpropping

Temporary propping installed beneath permanent formwork prior to steelfixing and concreting, where the permanent formwork will carry its own weight and the construction loads, but the load from the *in situ* concrete is carried by the postpropping.

Repropropping

System used during construction in which the original supports to a slab being cast are removed and replaced in a sequence planned to avoid any damage to the partially cured concrete.

Scaffold

A temporarily provided structure which provides access, or on or from which persons work or which is used to support materials, plant or equipment.

Supporting slab

Suspended floor slab immediately below the floor slab under construction. The soffit formwork and falsework for the next slab will be erected and used on this slab.

Temporary Works (TW)

A structure used in the construction of the permanent structure. It is usually removed on completion.

Temporary Works Coordinator (TWC)

Person appointed to control and manage the technical and procedural aspects of the design, procurement, erection and use of the formwork, falsework and scaffolding.

Temporary Works Designer (TWD)

Person appointed to design the temporary works. The task may be split between different organisations, for example proprietary equipment suppliers and in-house designers.

Temporary Works Design Brief (TWDB)

The basic information necessary for the Temporary Works Designer to carry out the formwork, falsework and scaffolding design for a particular section of the project.

VERA

Vegetable oil-based release agent

11.1.3 Temporary works design brief (TWDB)

The TWDB is simply the basic information necessary for the TWD to carry out the design for a particular section of the project. In many cases it will be the site engineer who has the task of collating the relevant information and passing it through the TWC to the relevant designers.

Ideally, the TWDB should be prepared at the pre-construction planning stage. It will most likely be prepared by the appointed TWC for the project with assistance from the site engineers. On a small building contract this person might well be the agent or foreman.

BS 5975 introduces procedures for *all* temporary works which includes the appointment of a temporary works coordinator and the preparation of suitable design briefs. On larger sites a further appointment of a temporary works supervisor may be appropriate. In formwork, a similar procedure is recommended in the CONSTRUCT document *Guide to Flat Slab Formwork and Falsework*. The use of a TWDB will improve safety, ensure compliance with CDM and construction regulations and, in many cases, provide more economic formwork, falsework, safer handling and improved efficiency.

Preparation of the brief will involve collating information from a variety of sources: from site, the office, suppliers, consultants, experts etc. The brief can be intended solely for the contractors' temporary works designer, or for a proprietary supplier, and may actually be intended for a combination of both the TWD and the supplier. On soffit formwork there will also be an important input from the PWD as the brief will need to state the design loadings to allow the TWD to complete a realistic design.

The checklist shown below is only the very basic information needed and is not exhaustive. The items will not all be required on every job, but are intended to cover most of the likely information necessary for both building and civil engineering contracts. Each contract will be different. See also the detailed lists in the Concrete Society book *Formwork—A Guide to Good Practice* (Sections 2.9 and 3.6.1) and in CONSTRUCT's *Guide to Flat Slab Formwork and Falsework* (Annex B). See also CS 144 *Formwork Checklist*.

The TWD will be responsible for ensuring that he is in possession of all the relevant information; this is also a statutory requirement. There will be occasions when assumptions have to be made by the TWD: these should be clearly stated.

The following list covers the basic information relevant to all temporary works. More specific details may be needed for technical information to and from suppliers, and information needed when obtaining schemes from a supplier.

Checklist of basic information needed in the TWDB

- Contract name, company reference number. Name, address and contact details for the parties involved. These should include the contact telephone and facsimile numbers and email addresses. The parties will include the PWD, the principal contractor, the subcontractor, the TWD, the TWC (if appointed), the planning supervisor and others.
- Up-to-date (*not* superseded ones!) general arrangement and detailed drawings and, relevant, specification clauses for the formwork, which should highlight important notes and paragraphs.
- The increased use of electronic data transmission means that significant savings can be made in the TW design if the layout and arrangements do *not* have to be replotted by the TWD. Thus, whether or not the layout drawings are available electronically, and to what format, can be important. AutoCad 14 with a DXF format for transmission is often used.
- Copies of any existing relevant risk assessments, current health and safety plans will provide important additional information to the brief.
- Location of the site, the altitude (m), the surrounding topography (i.e. on a hill, cliff top etc.) and the time of year intended for the work. This will affect the choice of concrete temperature and the design wind speeds for the site.
- Meteorological data, including rainfall and water levels are helpful.
- Statutory restrictions on the site, such as local air rights for over-flying adjacent properties, pavement licences for debris fans and walkways.
- What is the mix design? Does it contain any admixtures that might affect the appearance or early strength of the finished concrete?
- What density of reinforced concrete is to be used in the falsework design? (Values for backpropping will generally be established for a density of 24 kN/m³.)
- The type of finish required for all formed surfaces: what are the details from the specification? Is it a performance or method (prescriptive) specification? For example, the proprietary formwork panel systems will generally give a DETR Class F2 finish, but are the joint lines acceptable to the PWD?
- Where form liners or enhanced durability materials (e.g. controlled permeability formwork (CPF) Zemdrain) are specified, the suppliers' requirements should be obtained and followed.
- Deflection limits for the formwork—normally 1/270th of the individual span of the formwork member, unless the PWD specifies otherwise.
- Are the soffits flat? Details of any residual cambers to be left in the structure will be required as these can seriously affect the selection of the formwork concept to be used. On flat slab buildings this requirement needs consideration at a very

early stage by the PWD. Typical values stated in specifications are to set the formwork to an upward camber of 1:600 on spans over 5 m. It should be pointed out that predicting the final soffit level after striking is impossible because of the number of variables that affect the final shape—thus it is unrealistic for a contractor to have to guarantee a final post camber after striking.

- Does the contract require independent checks or certification? BS 5975 introduces specific categories of temporary works design check depending on complexity. This can affect who is responsible for the calculations and their level of detail.
- What is the outline method of placing and compacting the concrete? Is it by skip or by pump, and are internal or external vibrators to be used?
- What is the intended rate of placing of the concrete in m³ per hour? The TWD will need this to estimate the vertical rate of rise (R) in metres per hour on the formwork.
- What is the type and position of construction joints? Do they have waterstops or water bars? Information on the type to be used is important as it affects the formwork solution. Are they surface mounted, internal or hydrophilic waterstops?
- The method of restraining the stop-ends on both slabs and walls can affect the design of the formwork.
- Striking times required by the contract documents will generally be conservative; site personnel will obtain faster times by using one of the methods of assessing the *in situ* gain of strength of the concrete. This will influence the number of sets of formwork required.
- What is the intended method of assessing the early age concrete strength for striking of formwork? Are cubes, LOK tests etc. required?
- Site restrictions—overhead cables, adjacent to railways, working over water, floodwater levels (mark on sections the anticipated maximum levels). Include any planning restraints, such as position of tower cranes etc.
- Any limitations on the size of formwork, (e.g. pour height, length limits).
- Capacity of plant to be used for handling the forms. Do not forget to include the height and weight of the lifting equipment when calculating the craneage necessary! LOLER requirements for lifts are very specific.
- Is this temporary works unique or can re-use be planned by studying other similar parts of the work and the intended overall programme?
- Anticipated material or equipment known to be available to complete the work—this might include the contractors' own items, or anticipated items becoming free from another part of the work, or a site preference for a particular system. Include any preferred sizes of members and stress levels normally used. Is there a type of plywood that is in common usage on the site or in the area?
- What method of curing the concrete is to be adopted? Note that certain spray-on chemical curing compounds are designed to degrade under ultraviolet light and thus cannot be used in underground and hidden work!
- What types of release agents are to be used? Are they compatible with any subsequent surface treatments to be applied to the concrete?
- Method of sealing the form faces. (Avoid the use of sealing tapes on quality work!)
- Can ties, if required, be fitted to the façade to restrain the scaffolding?
- Is the scaffolding to be fitted with protection nets or impervious sheeting?

- What activities will be carried out on the scaffold; how many platforms will be required?
- Is a walk-through pavement lift required at the bottom of the scaffold?
- Who is intended to carry out the statutory regular inspections of the scaffold?
- Safe working platforms, ladders and hoists for the operatives, including guard-rails, toe boards etc. Safety considerations should *not* be left to the contractor, the issues need to be considered as part of the pre-tender health and safety plan.
- Scaffolders working should comply with the recommendations in SG4:04 and may be required to wear harnesses. What provisions are to be made for access, egress, safety nets, harnesses?
- What procedures will be used to check that scaffolders are CITB registered?
- Dates for submitting draft and final schemes.
- Is there a QA procedure to be followed? How many copies are required and to whom should they be distributed?
- Is any training required for operatives in use of equipment, such as proprietary systems or the method of concrete strength assessment (e.g. LOK test).

11.2 Surface finishes

11.2.1 Specifications

The formwork surface will be specified either by method (prescriptive) or by performance. Most United Kingdom specifications are performance specifications, stating the standards and the tolerances required, but leaving the contractor to decide the way to achieve it. By contrast, the method specification will tell the contractor exactly what to use—and in some ways it removes the onus from the contractor to think about the method adopted.
In building the two most common specifications are:

- the *National Structural Concrete Specification* (NSC)
- the *National Building Specification* (NBS) *Formwork for in situ concrete* (Section E20).

Common performance specifications in civil engineering works are the:

- Water Services Association *Civil Engineering Specification for the Water Industry*
- BS 8110 *Structural Use of Concrete*, Section 6.2.7
- Highways Agency (HA) Specification Volume 1 Clause 1708 which specifies five classes:

  - Class F1—basic, e.g. for pile caps
  - Class F2—plain, e.g. rear, unseen faces of retaining walls (most UK panel systems will give Class F2)
  - Class F4—high class, e.g. quality surfaces to walls
  - Class F3—high class but *no* ties, e.g. on visible parapets (the hardest to achieve!)
  - Class F5—high class—embedment of metal parts allowed (intended for pre-cast work).

The NSCS uses reference panels for the surface finish and seven sample sets of panels have been erected in various UK locations. There are also other special finishes often specific to a contract.

The final concrete surface together with the surface zone of the concrete will be effected by the choice of material used to create the face. It will also create a mirror image result in the structure—for example, the mark on the plywood from a hammer head caused when 'over-hitting' a nail, will create a depression in the plywood, but leave a projection on the concrete. Different concretes, varying the plywood and the type of release agent, can all affect the finish. A strongly recommended reference is the Concrete Society Report TR 52 *Plain Formed Concrete Finishes* which illustrates in detail twenty-two different sites and how a plain concrete finish was achieved.

There can be blemishes from the use of formwork. A useful guide to the understanding of blemishes in concrete surfaces is the BCA booklet *The Control of Blemishes In concrete*, which illustrates their cause and possible solutions.

The secret to the production of high-quality concrete surfaces is attention to detail. Good results are obtainable on site with correct supervision and control. Correct use of the form face, correct application of release agent, together with care when vibrating and placing the concrete, are all major factors in the preparation of good surfaces. **Making-good costs money**, so getting it right first time has to be sensible. The achievement of satisfactory finishes, begins at the design stage, and in timely discussion with operatives and supervisors. It should not be left, and then agreed with the client, after the formwork has been struck—that is far too late.

The use of controlled permeability formwork (CPF) is covered in CIRIA Report C511. This allows a blowhole-free surface, with enhanced durability, which is produced by placing a liner on the formwork face (see also Chapter 13).

11.2.2   Tolerances

The aim is to construct a concrete structure which complies with the drawings and the specification. Absolute accuracy is impossible, and the level of inaccuracy that is acceptable should be considered. Tolerance is an absolute value, and the deviation is the measured value from a baseline, thus an item with permissible deviation +5 mm and −10 mm, has a tolerance of 15 mm.

The three sources of deviations in the finished structure are:

- *inherent deviations*, such as the elastic movement under load (e.g. deflections)
- *induced deviations*, such as the lipping tolerance on two sheets of plywood (possibly both sheets within their manufacturing tolerance)
- *errors*, such as inaccurate setting out (see also Chapter 4).

The magnitude of the deviations will depend on whether the work is normal, high quality or special, and also on the distance from which the final surface is viewed. Typical deviations for normal work for verticality of a 3 m high wall form will be ± 20 mm from a grid line. See, in particular, Table 3 in the *Formwork Guide*. The final position depends on the accuracy of the starting point, e.g. the kicker or base.

The deflection of the form is one of the inherent deviations. The rule of thumb in formwork is that the appearance and function is generally satisfied by limiting the deflection of individual formwork members to 1/270th of the span of the individual member. In certain cases, such as for direct decoration onto concrete walls in housing projects, tighter deflections may need to be specified. Note that it is the deflection of the individual members and not the final concrete shape that is the limit.

11.2.3 Release agents

Correct use of a release agent will ensure better finishes. The late 1990s saw the introduction of new products and there have been recent developments from Europe (see *European Developments on Concrete Release Agents*) to improve the safety and use of these products on site. The main categories are still classified by numbers, (see *Formwork Guide*), but some general information follows:

Category 1: Neat oils and surfactants

Rarely used and not recommended for production of high quality formwork

Category 2: Neat oils

Synthetic oils are now available which have good performance qualities, are totally non-harmful and non-flammable, even with high rates of application. They have been developed for precast and automatic machinery use.

Category 3: Mould cream emulsions

Rarely used

Category 4: Water-soluble emulsions

Quite successful in the precast industry, but can leave a dark porous skin on the concrete. Recent developments have improved the performance and usage rates, but rarely used on site.

Category 5: Chemical release agents

Remain the most popular. They act by having various blends of chemicals carried in light oil solvents. When sprayed onto formwork the light oil evaporates leaving the chemical to react with the cement to form the barrier. They are of a *drying* nature, and suitable for plywood, timber, steel and most forms. However, they do have the problem with over-application leading to build-up on forms, which in turn leads to dusting of the resultant concrete surface. As with chemicals dissolved in oils, there are inherent health and safety implications of these products if not used correctly.

Category 6: Paints, lacquers and waxes

Not strictly release agents, but the waxes are often very useful to provide the 'first' coat on very smooth impermeable surfaces, such as glass reinforced fibre moulds and brand new steel moulds. They can provide a surface for the normal release agent to stick.

Category 7: Other specialist release agents

When using elastomeric (rubber) liners a specially formulated product is often required.

Category 8: VERAs

Recent research (see *European Developments on Concrete Release Agents*) into use of non-toxic biodegradable vegetable oil release agents, generally based on rape seed oil, provides a new and safe type of release agent, suitable for use in confined spaces and for all formwork. They are supplied in plastic containers and diluted with water for use, but they do need more careful storage as they can freeze and separate out after a length of time. Use of VERAs on steel forms and moulds have produced improved finishes and rain resistance. Disposal is also improved, as the products are non-toxic.

Site engineers should ensure that they have the product information and safety leaflets from the supplier, prior to use of the products. Subsequent treatments on the concrete surface can be affected by your choice, and the supplier's recommendations should always be sought.

11.3 Materials

11.3.1 Timber and wood-based products

BS 5268–2 is for the structural properties and permanent use of timber. Timber is used in formwork as walings on wall formwork, and as bearers under the soffit in falsework. There are ten strength classes for softwood timber, so purchases can be related to strength. In formwork and falsework the *minimum* class used is C16 (originally called SC3). A separate standard, BS 2482–1, applies to 38 mm timber scaffold boards.

Timber

Several factors affect the strength of timber on site. It is exposed to the elements so wet exposure service Class 3 stresses are used (assumes moisture content greater than 20%). Also relevant are the duration of load, how the load is distributed and the depth of the timber section used. Timber is actually used in one of three ways on site, and documents such as '*Formwork—A Guide to Good Practice*' and BS 5975 give safe working properties for:

- Wall formwork i.e. walings (where the load is applied for a short time and there are other timbers to share the load).
- General wall and soffit members (where the duration of load is longer than on walls, and load-sharing can be applied).
- Primary members (where the duration of the loading is also longer than on walls, but the loads cannot be shared).

In falsework, there is also an in-built stability requirement, with a limiting allowable depth to breadth ratio of 2:1 when there is no lateral support, but when the

ends are held in position (i.e. wedged in forkheads) the allowable depth to breadth
ratio increases to 3:1.

Plywood

Perhaps the most common *sheet material* in formwork is plywood—a layered
material, made up in sheets (normally 1.22 × 2.44 m i.e. Imperial 4′ × 8′), with
properties related to the direction of the face grain of the outer ply. The stated shear
stresses are low in BS 5268, and users should refer to *Formwork—A Guide to Good
Practice* (Appendices G-W & G-S) which gives working properties for its use in
either wall or soffit applications.

Users of wood-based sheets should be aware that their properties can vary significantly depending on the orientation of the panel. In many cases deflections are
doubled when used the wrong way around. Generally panels are stronger when the
face grain of the material is in the same direction as the span of the panel, referred
to as *face grain parallel*. Certain plywood, e.g. *Good One Side*, is often quoted with
properties in one direction *only*, i.e. the strong way, so it is important to make certain that you are using the plywood the *strong way around*.

The number of uses from plywood can be an emotive subject, but as a general
guide for producing a quality finish on wall formwork, you will reasonably expect
about four uses from *Good One Side* 19 mm 'Canply', increasing to between eight
to ten uses for a better *PourForm* 17.5 mm Canply, and increasing again to over
forty uses for 19 mm Finnish *Wysaform* plywood.

Scaffold boards

Boards are usually cut from sawn and seasoned timber. They are available in three
thicknesses (38 mm, 50 mm and 63 mm). The most common timber board in use
is the 38 mm thickness board, 225 mm wide and 3.95 m (13′-0″) long. Boards
should be butt jointed. Each board should be identifiable by having the BS 2482
number, the suppliers name, a letter M or V (denoting Machine or Visual stress
graded) and the words *support at xx metres max*. BS 2482–1 specifies two classes
of 38 mm board: as 1.2 m span with visual and machine stress grading rules, and
a higher quality board as 1.5 m span only acceptable from machine stress grading.
Target spans are introduced to allow for inaccurate setting out of the bearers, with
maximum tolerance 100 mm. Many boards in use do *not* comply with BS 2482,
and these need supporting at 1.2 m centres maximum.

Normal 38 mm boards should not over-sail more than 150 mm, nor less than
50 mm. A scaffold board less than 2.13 m long requires the end to be held down
to avoid a tipping hazard. Proprietary scaffolds often have their own decking systems. It is important to use the correct boards and decking for the system and to
follow the suppliers' recommendations.

11.3.2 Proprietary formwork equipment

Many of the proprietary panel systems have plywood fitted within a protective
steel frame. This gives a greater re-use of the plywood, and reduces damage to the
sheets during handling. They are generally in small panels, (typically 1200 × 600)

Figure 11.1   Typical panel formwork (courtesy PERI UK Ltd).

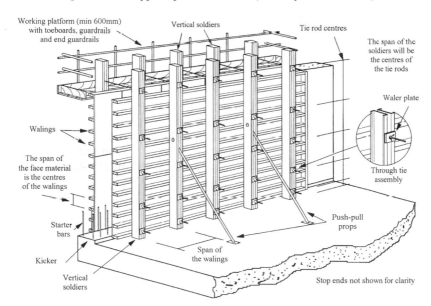

Figure 11.2   A typical arrangement of double-faced wall formwork.

designed for manual handling or, more recently introduced, larger panels, often 2700 × 2400, which require crane handling. A typical site is shown in Fig. 11.1. The significant advantage being the speed of initial use, with little *make-up* time. It should be noted that such panel systems may not give high quality concrete surfaces direct from the face.

The more traditional UK wall formwork will comprise face material, backed by horizontal members, called walings, with vertical soldiers (see Fig. 11.2). Walings may be of aluminium and there are a variety of such proprietary walings available for hire or purchase from several of the proprietary suppliers. These sections have

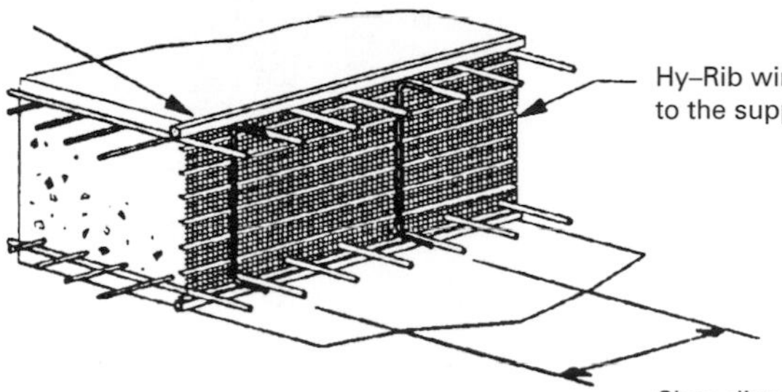

Timber rail, wire tied to top re-bar.
Hy–Rib generally given same
cover as the reinforcement.

Hy–Rib wire tied
to the supports.

Clear distance

Figure 11.3 Typical Hy-Rib stop end (source, Expamet Building Products).

high strength to weight ratios, allowing much stronger forms to be assembled for less weight than when using steel. A benefit is also the increase in span of a waling, which means fewer soldiers. Designers might need to check the deflection of the walings at these larger centres as aluminium is more flexible than steel or timber. Special arrangements for access platforms for operatives might also be needed if the scaffold boards cannot safely span between the platform brackets attached to the soldiers.

Recently introduced are proprietary timber beams, such as the Peri Vario beam. They are laminated and glued proprietary items, generally fitted vertically in the formwork with stiff horizontal steel channels as walings. The European product standard on prefabricated timber formwork beams is BS EN 13377.

Proprietary steel soldiers are the vertical stiff members associated with the wall formwork. They are available from different suppliers, and reference should be made to the supplier for the load characteristics of the soldiers. They should *not* be mixed on site as they can have different properties and stiffness which could affect site safety.

The use of expanded metal steel products, (e.g. Hy-Rib) such as in stop ends (see Fig. 11.3) has been shown by the BCA to give significant reductions to the concrete pressure. Trials on 5 m high walls using retarded concretes, poured at rates of rise in excess of 30 m/hr, showed that the maximum likely concrete pressure was limited to only 38 kN/m². See also the *Hy-Rib Designers Guide*. This low pressure limit obviously reduces the number of tie rods and supports needed and can give significant economic advantages. It should be noted that such expanded metal products are intended to be left in place and are not struck off the stop end. Trials of the shear strength have shown the resulting joint to be stronger than a scabbled joint, with excellent compaction at the Hy-Rib, even when using very fluid concrete with 60% ggbs cement replacement (see *Concrete* Vol. 34 No. 10).

11.3.3 Proprietary falsework equipment

There are many different types of systems on the market, many for hire as well as purchase. Always refer to the supplier's recommendations for load capacity etc. The

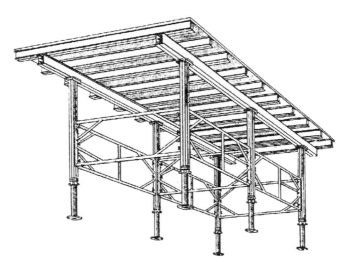

Figure 11.4   Typical aluminium table form (source, Ischebeck Titan brochure).

load capacity is not always obvious because although the vertical standards will be 48.3 mm outside diameter to suit the scaffold fittings, they can have different wall thicknesses and be of different grades of steel. They are often made of lighter material to reduce transportation weight and to make erection and dismantling easier.

Fast track aluminium systems (Fig. 11.4), comprising standards and bearers all of aluminium, are mostly used in building construction. A recommended reference being the CONSTRUCT's *Guide to Flat Slab Formwork and Falsework* which gives practical guidance to clients, designers and to site engineers on use of such systems. The method of handling such tables, with crane coordination, trained operatives and suitably designed buildings will give turn around times, floor to floor, of at best three to four days. The speed of operation is often dictated by the striking time, as discussed in Section 11.7.4.

11.3.4   Adjustable steel and aluminium props

The European product standard for adjustable steel props, BS EN 1065 classifies props into five strength classes, totalling thirty-two types of prop, by length. It gives the characteristic strength values, but for only one load case, with 10 mm eccentricity. The traditional props made in the UK, to the earlier BS 4074, have five types of prop (known as Nos 0, 1, 2, 3, & 4); their properties and two tables of safe working load are stated in BS 5975 for:

- axially loaded and maximum up to 1½ out of plumb
- allowing the load 25 mm eccentric and maximum 1½ out of plumb.

There is no equivalent standard for proprietary aluminium props, and the supplier's information must be sought. The properties of aluminium are such that the elastic shortening under load of aluminium propping will be greater than that of steel props, and needs to be considered, particularly in backpropping situations.

The elastic shortening (in kN/mm) for a typical prop length should be requested from the supplier.

Note: Many aluminium props are long, with extensions such that when erected the handles are out of reach for adjustment!

11.3.5   Scaffold tube and fittings

Steel scaffold tube—3 mm and 4 mm wall thickness

BS EN39 gives information on *two* wall thicknesses of scaffold tube, with 3.2 mm and 4 mm wall thickness for the same outside diameter of 48.3 mm. The UK practice has been to use the one referred to as 'Type 4' which means the thicker 4 mm tube. Until the early 1990s steel used for scaffold tube was about 5% weaker, and it was made to the earlier BS 1139: 1982. TG20:08, the guide to good practice for scaffolding, has precalculated basic scaffolds based on the strength of the 'Type 4' tube, but they are *not* relevant for the lighter 'Type 3' tube.

It should be noted that using the term *scaffold tube* in many European countries would imply use of the 3.2 mm tube which has a *lower* load capacity than *our* normal tube. 'Type 3' tube is being used by some scaffolding contractors in the UK—this lighter tube should *not* be mixed with the thicker walled tube.

Steel scaffold tubes are considered in design as either in the *as new* or the *used* condition. The *used* tube is allowed to be corroded by 15% and an allowance is made in both the bending and strut capacities. To overcome any controversy when erected if the quality of tube is not known, it is general practice to design to *used tube* values and make use of the inherent extra strength as an additional factor of security.

The safe load of any strut (member in compression) is limited by its buckling. The effective length and the slenderness ratio (l/r) should be considered. Table 11.1 gives the safe working load for scaffold tubes in compression. It is recommended that the slenderness ratio be not less than $l/r < 207$ for standards (i.e. 3.25 m), but increases to $l/r < 271$ for struts and braces intended to carry wind and lateral loads, i.e. the lacing and diagonal bracing should be less than 4.25 m long.

To increase the load capacity of a strut, it is necessary to provide adequate sideways restraint in *two* directions at right angles, known by designers as *creating effective node points*. The value of the restraint force is usually only 2.5% of the load in the strut.

The amount of fixity of the tube by the coupler is ignored in tube and fitting designs and the effective length of the tube is generally taken as the actual centre to centre distance of the couplers, i.e. the effective length factor is 1.0 L (see in particular Fig. 11.5). Be aware that the effective lengths of scaffold tube struts which have a free cantilever at the end, such as at the bottom of a freestanding scaffold, are regarded as having an effective length of the adjacent strut length *plus* twice the length of the free cantilever. See BS 5975 Appendix C and Figure 33 and Fig. 11.5.

Steel tube in tension

The safe axial tensile load is 79.1 kN but is usually limited by the connections. Normally the couplers and fittings have a safe slip load capacity of only 6.3 kN on *one* coupler, so this tends to be the practical limit in tension.

Table 11.1 Permissible safe axial load in steel scaffold tube struts.

| Effective length mm | Permissible AXIAL LOAD | | | | | | Comment |
| | Type '4' | | BS 1139 1882 | | Type '3' | | |
	As new kN	Used kN	As new kN	Used kN	As new kN	Used kN	
500	74.5	63.3	68.5	58.2			
1000	64.3	54.7	57.7	49.1	52.7	44.8	
1500	45.3	38.5	42.0	35.7	37.7	32.0	
2000	29.3	24.9	27.9	23.7	24.6	20.9	
2500	19.8	16.8	18.9	16.1	16.6	14.1	
3000	14.1	11.9	13.5	11.5	11.8	10.0	
3250	12.1	10.3	11.6	9.9	10.2	8.6	Limit of strut <<< as column
3500	10.5	8.9	10.1	8.6	8.8	7.5	
4000	8.1	6.9	7.9	6.7	6.8	5.7	
4250	7.2	6.1	6.9	5.9	6.0	5.1	Limit as a strut <<< or brace
4500	6.4	5.5	6.2	5.3	5.4	4.6	
5000	5.2	4.4	5.1	4.3	4.4	3.7	
5500	4.3	3.7	4.0	3.6	3.7	3.1	Limit of member <<<< as a tie
6000	3.6	3.1	3.5	3.0	3.1	2.6	
7000	2.5	2.1					Not
8000	1.4	1.1					recommended

Note: Values derived from BS 5973 and BS 1139 Part 1.1.

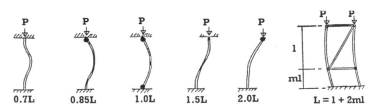

Figure 11.5 Effective length factors for struts (source, author).

Scaffold couplers

Also known as fittings, there is a great variety, manufactured from pressed spring steel, drop forged. Additionally there are different types such as doubles, swivels, putlogs, band and plates, and sleeves. The more common fittings are illustrated in Fig. 11.6. BS 1139–2–1 covers the specification and testing of many couplers for use with steel tubes. The safe working loads given in Table 11.2 are based on the characteristic strength from BS EN 12811–1 and using a 1.65 combined safety factor. All couplers should be tightened correctly using the appropriate spanner.

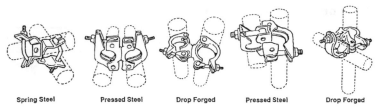

| Spring Steel | Pressed Steel | Drop Forged | Pressed Steel | Drop Forged |

Figure 11.6 Typical right angle and swivel couplers (source, PFP & CITB).

Table 11.2 Safe working loads for scaffold fittings.

Type of fitting	Class	Type of load	Safe load
Mills right angle coupler and SGB Mark 3A coupler	n/a	Slip along tube	12.20 kN
Right angle couplers	Class A	Slip along tube	6.10 kN
Mills swivel coupler	n/a	Slip along tube	6.25 kN
Swivel couplers	Class A	Slip along tube	6.10 kN
Parallel coupler	Class A	Tension	4.70 kN
Sleeve coupler	Class B	Tension	3.00 kN
Putlog coupler	–	Force to pull tube axially out of coupler	0.63 kN
Internal joint pin	–	Tension	nil
		Shear Strength	21.00 kN
Band and Plate coupler	–	Slip along tube	≈4.00 kN

The tightening torque for the test is preferably between 40 and 80 Nm, but use of the scaffolding spanner will generally provide the correct torque when done up tight.

Right angle couplers

Right Angle Couplers, known also as doubles, connect two tubes at right angles. They are a load-bearing coupler and will be used on all standards to connect together the ledgers, ties and main transoms. They should also be used on diagonal bracing to connect the brace to the horizontals. Safe working slip load on a Class A right angle coupler is 6.10 kN.

Swivel couplers

These connect tubes together through 360° and the slip load capacity has reverted to the same value as for the right angle coupler, following publication of BS EN 74–1. It is recommended that they should *not* be used in place of right angle couplers because of their greater flexibility and movement at the connection device. Safe working load on a Class A swivel coupler is 6.10 kN (slip load).

11.3.6 Proprietary scaffold systems

System scaffolds are relatively simple arrangements of scaffolds using patented connections to reduce, and sometimes eliminate, the need for scaffold couplers. Many different types of connections have been introduced. Essentially they comprise separate standards, ledgers and transoms in modular lengths and are designed for assembly by semi-skilled operatives. Training is still needed in the systems. BS 1139–5: 1990 is the specification for working scaffolds made of prefabricated elements.

Although the vertical standards are generally the same outside diameter as scaffold tube, they are *not* always scaffold tube, and often are a stronger grade of steel. This means that they can have a thinner walled tube, so reducing the weight and therefore making transporting and handling easier. The connections are usually very strong and when used in falsework will often be considered to contribute to the strength of the whole system. Under the HSW Act, Section 6, all manufacturers, importers or suppliers of such systems of scaffolding *must* not only test and verify the scaffold for safety, but must supply information about the use for which the scaffold can be put and mention any safety points. Some typical connection details used by proprietary systems are shown in Figure 11.7.

The boards of such systems are often of fixed length to suit the modules. Decking members can be in steel or wood, and often fit into special board transoms.
Note: Decking members are *rarely* interchangeable between systems, so make sure you have the correct staging.

11.3.7 Proprietary aluminium access towers

A common sight in building construction is the use of simple proprietary aluminium towers. They are easy to erect, but made with light components and have different rules for stability than the heavier, and more stable, steel towers (see Section 11.6.2). Figure 11.8 shows a typical aluminium stairway. The trade association, PASMA, produce a very helpful code of practice on the safe use of such towers.

Aluminium towers should not be exposed, and preferably should be tied to adjacent buildings. BS EN 1004: 2004 specifies the requirements for the prefabricated mobile and access working towers. Two service class loadings are stated and for each a horizontal load has to be allowed for in the design. It is the responsibility of the manufacturer to establish how these loads are to be restrained and the tower stabilised, e.g. with outriggers, stabilisers or weight boxes. Suppliers' instructions *must* be followed (see also Section 11.6.2).

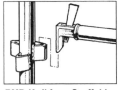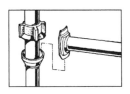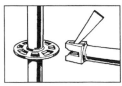

RMD-Kwikform Scaffold   SGB Cuplok   Layher All-Round

Figure 11.7 Typical system scaffold connections (source, CITB).

11.4   Loads

11.4.1   Vertical loads

Self weight

The weight of soffit formwork is usually taken as 0.50 kN/m², up to a maximum of about 0.85 kN/m² for a steel panel system. Wall and column forms vary in weight, from about 0.40 kN/m² for panel systems up to about 1.20 kN/m² for purpose made steel forms. Generally, for plywood and timber/aluminium waling formwork, a value of about 0.50 kN/m² is used.

The self-weight of scaffolding used to be quoted as a mass per bay per lift of scaffold. This did not allow for additional loads in the inside standards when inside boards are used. TG20:08 Table 15 gives the weight per standard of unboarded lifts, e.g. a 2 m bay × 2 m lift has a weight per standard of 0.42 kN.

The self-weight of falsework is very critical to the design. These structures are rarely tied to the ground and their self-weight is often the only load preventing them from overturning under maximum wind loads. Typical weights expressed in cubic terms for systems vary from 0.10 kN/m³ for aluminium systems to up to 0.15 kN/m³ for steel systems.

Imposed loads

Obviously the weight of the structure, i.e. the permanent work, is the imposed vertical load on the falsework and formwork. The usual value for the density of

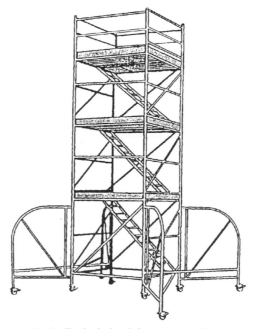

Figure 11.8   Typical aluminium tower with outriggers.

reinforced concrete is 24 kN/m³ for the permanent design, but in the construction stage there will be more free water in the mix that has not hydrated: thus falsework designers often use the value of 25 kN/m³. The weight can be just vertical (e.g. static and impact) or it may have some non-vertical components, such as from the pressure of concrete on other works.

Note: The structure will have been designed by the PWD for a service imposed load. This can include partitions, allowances for screeds, as well as the imposed floor loadings. Many of these will not be applicable during construction. Knowledge of the designer's imposed load—*vital* on multi-storey work—in building work is necessary so that the striking times and striking procedures can be safely established (see Section 11.7.5).

Construction operations loads (c.o.l.)

Imposed loads onto scaffolds and working platforms are classified in the standards by service class loads, recently introduced. They apply to *all* working platforms. They are:

- Service Class 1   0.75 kN/m²   Inspection and very light duty access
- Service Class 2   1.50 kN/m²   Light duty, e.g. painting
- Service Class 3   2.00 kN/m²   General building work, brickwork etc.
- Service Class 4   3.00 kN/m²   Masonry scaffolds—heavy duty.

To allow for the operatives placing the concrete/steelwork on falsework, including a permanent formwork, an allowance is normally made for a minimum load of 1.5 kN/m²—Service Class 2. When placing in situ, concrete thicker than 300 mm, this value increases up to a maximum of 2.50 kN/m² for slabs of 700 mm thickness or greater. (See also BS 5975: 2008, Clause 17.4.3.1.)

Note: This minimum of 1.50 kN/m² represents only 55 mm of excess concrete!

A reduced imposed load is allowed for inspection and access, such as beneath the soffit, at 0.75 kN/m²—Service Class 1 loading.

The imposed load from activities on scaffolding is often quoted as an additional weight per standard per bay lift of the scaffold. The self-weight of the basic scaffold has been allowed separately, so the additional load is the weight of the boards, toe-board and guard-rails, plus the weight of the operatives carrying out the activity on the platform. TG20:08 Table 16 gives the extra weight per standard of boarded lifts, e.g. for a 2 m bay × 2 m lift with a five board wide platform with two inside boards and Service Class 3 loading, to be 2.35 kN per standard. Adding this to the self-weight of 0.42 kN means that placing boards to one platform has increased the likely load per lift by a factor of 6.

Material storage and loading-out platforms

In construction, the procurement and handing of materials has been shown to be a critical and cost-sensitive item. The published research from ECBP in the *Guide to Flat Slabs* highlights this aspect in particular and recommends that site personnel consider very carefully the movement of equipment. Areas for material storage should be correctly signed, particularly above ground level. Handling material, including TW equipment should be considered. A common method is to insert

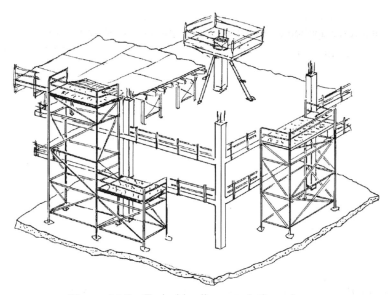

Figure 11.9 Typical loading-out platforms in use.

loading-out platforms, either as projecting beams or as free-standing towers as the work commences. Fig. 11.9 shows an example of the latter.

The likely imposed load from materials should be considered. A typical loading for general material storage is between 4.0 and 4.5 kN/m² (Service Class 5 loading) and would be generated by pallets of equipment weighing up to only 400 kg. The legs of pallets create very high localised loads and special platform surfaces are often required to cater for the impact and point loads applied.

Where brickwork is to be handled, the pallet loads will increase significantly—for example the footprint load from a single pallet of bricks can vary from 14 to 17.5 kN/m² depending on the weight of commonly used bricks. TG20:08, Clause 20.1 recommends that loading-out platforms intended for use by brick pallets be designed for a distributed imposed load of 10 kN/m² or the actual weight of the material to be stored—significantly larger than for simple material storage!

11.4.2 Horizontal loads

Erection tolerance

Falsework cannot be erected exactly vertical, and BS 5975 allows for an additional 1% of the vertical load as a horizontal load to allow for the structure being initially erected out of plumb. This erection tolerance load is regarded as a horizontal force to be transferred and restrained by suitable supports.

Specified loads

Investigations during the 1970s into the cause of falsework collapses, identified that one of the principal causes was the absence of overall structural stability. For this reason BS 5975 (Clause 19.2.9.1) recommends that *all* falsework be designed for a horizontal force applied at the top of the falsework, as the greater of:

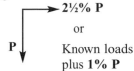

- either a specified 2½% of the vertically applied loads
- or all known horizontal loads plus 1% of the vertical or load as an erection tolerance load (Section 19.2.4) to cater for workmanship.

Note: This specified load could, in some cases, be restrained by connecting directly to the permanent works at the top of the falsework, such as at column tops—but only if the permanent work is strong enough to withstand the loads!

Environmental loads

The method of calculating the wind load stated in BS 5975 and TG20 is based on a simplified method using data from BS 6399–2 and uses a basic mean hourly wind speed. Note that you need the site altitude (in metres) and the local topography using this method. Generally on most falsework structures, placing concrete will not occur when the wind speed exceeds the operating limits set for the plant. This is usually at a Beaufort Scale of strength Force 6, and corresponds to a wind speed of 18 m/s. This is known as the working wind, and gives a wind pressure of 0.20 kN/m².

Whenever you place sheeting or debris netting onto scaffolding or falsework, you will alter the wind forces. Check with the designer what has been allowed—for example, all sheeted scaffolds and structures in exposed locations may need separate calculations because of the increased wind forces.

11.4.3 Pressure of concrete on formwork

The pressure of the concrete affects the choice of face contact material, the waling and soldier centres, and tie rod spacings. Correct calculation is very important in walls and columns where the pressures are high, typical values being 75 kN/m² on walls and 100 kN/m² on tall columns. Putting this into context, a pressure of 75 kN/m² represents *eight* cars parked on one square metre of formwork! On soffit formwork the pressures are generally much less.

The maximum pressure P_{max}(kN/m²) of a fluid at any position, is the vertical height h in metres, multiplied by the density of the fluid in kN/m³ such that:

$$P_{max} = h \times \text{density}.$$

Design pressures can be reduced below hydrostatic because concrete is not a fluid and stiffens with time. Work in the 1980s showed that there were many factors affecting concrete pressure: this resulted in the publication of CIRIA Report R108 which includes a simple table of the maximum design pressure.

Exceeding the design pressure will cause unplanned extra deflection of the face, extra deflections in the members, increase proportionally the load in the tie rods and, in the worst case, initiate a failure of the tie rods. Thus controlling the pressure is very important. The relevant factors are:

- Is it a *column* or a *wall* or a *base*?—A dimension greater than 2 m is considered a wall.

- What is the likely *concrete temperature?*—The tables are used for concrete temperatures of 5 °C (winter), 10 °C (spring and autumn) and 15 °C (summer).
- Increases in design pressure will occur when the temperature drops.
- What is the *height of the form* (m) being used?
- What is the *rate of rise* (m/hr) of the concrete uniformly vertically up the face of the forms measured in metres per hour—*not the placing rate of the concrete in m³/hour!*
- The use of *admixtures* and normal super plasticisers will not generally alter the pressure of the concrete because they only act on the concrete for about one hour, it then reverts to more normal concrete. But watch out for extended-life super plasticisers and certain water reducing admixtures! These can have significant retardation effects on the concrete, particularly at lower temperatures. The laboratory tests are often completed at 20 °C, whereas the concrete temperature on site might be down to 8 °C. This can increase the pressure significantly.
- The use of *additions* to the mix as cement replacement materials, such as pfa or ggbs, in blended mixes will always *increase* the concrete pressure. In parts of the UK blended cement concretes are becoming the norm. The *additions* cause the pressure increase by retarding the stiffening of the concrete, which results in a greater fluid head of concrete. Different types of mix are categorised into *groups*. The effects can be severe, so it is necessary to know the concrete mix to be used early in the contract.

Combining the effects of a blended cement with certain admixtures can give significant alterations to the stiffening time, particularly in cold weather. One admixture supplier states that the effect of a blended cement with a water reducing admixture on concrete at 8 °C is to give a staggering stiffening time of 14 hours!

A recent introduction into the industry is *self-compacting concrete* designed to eliminate the need for vibrators and the operative problem known as *white finger*. Although relatively new, and even after extensive research, little is known about its effect on the pressure of the concrete. Users should be aware that it might significantly retard the concrete, thus assuming a retarded mix might be prudent.

Formwork will be designed for a concrete pressure, and drawings should have a relevant note as *maximum concrete pressure not to exceed xx kN/m²*. CIRIA Report R108 gives design pressures for both columns and wall forms.

To assist site engineers, Tables 11.3 and 11.4 show the allowable *rate of rise* for different design pressures, related to the type of concrete mix—thus knowing the designer's pressure for the formwork, the engineer can establish on a particular day the permitted *rate of rise* to keep the pressure to that as designed.

11.5 Formwork

11.5.1 Double faced vertical formwork

Figure 11.2 shows an example of double-faced wall formwork using walings and soldiers. The principle of wall formwork design is straightforward: you follow the concrete pressure from the face, through the backing members to suitable restraints, knowing that all the forces will balance from one face to the other! Remember that the force acting on a face is the pressure multiplied by the area acted upon.

Having established the design pressure, the safe span of the face contact material is determined. This obviously then gives the maximum spacing of the next level of supports—usually horizontal walings. The span of the walings is then determined by the location of the next level of support, usually vertical soldiers. The soldiers have a central slot to allow tie rods to be passed through. These ties connect to the opposite face of the wall and provide the restraint in tension to avoid the walls 'bursting' apart under the pressure. As the pressure reduces near the top of the wall, fewer ties are required near the top, but increasing in numbers near the base.

In a design, the tie spacing and hence the soldiers might be pre-determined by the PWD. A particular surface finish with, for example, striations will dictate the acceptable locations of ties, i.e. in the recesses, thus establishing the tie spacing.

The design must take careful account of deflections and tolerances. Remember that the deflection limit of 1/270th on members is applied to the plywood, the walings and then the soldiers individually—it is not cumulative.

Best practice dictates that it is best to cast large walls by first casting a kicker to give a better horizontal alignment and allow fine vertical height adjustment. In certain situations kickerless construction can be adopted.

The principles when using panel formwork are very similar (see Fig. 11.1). One difference is that the essential design has already been carried out by the supplier, who has established the permitted maximum design pressure for the system. The safe span of the face contact material in the panels is controlled by this maximum pressure not being exceeded. The maximum spacing of the panels to the tie rods is determined by the strength of the panel frame and the tie locations. It is most important that the correct type of ties are used and the supplier's recommendations followed. The user of the panel system should know in advance the maximum design pressure, and can limit the rate of rise accordingly (see Tables 11.3 and 11.4).

HSE and CDM safety requirements need careful consideration by the TWD, with correctly designed lifting points for crane handled forms, with working platforms (minimum 600 mm wide), toe boards, double guard-rails etc. Procurement of column formwork needs to consider lifting points, possible permanent ladders, stability and, most importantly, the working platforms for concrete-placing operatives.

A check will need to be made on the stability of the formwork and propping designed for a minimum factor of safety on overturning of 1.2 when considering the most adverse conditions.

The three checks used for stability of wall formwork are:

1. Maximum wind plus overturning from a nominal Service Class 1 load on any working platforms attached to the forms.
2. Working wind on day of concreting, plus construction load as Service Class 2 load on all the platforms. The working wind represents the upper operating limit for the plant on site.
3. A minimum stability force representing 10% of the self-weight of both faces of formwork, considered to act at 3/4 up the form.

Often forgotten in formwork design is the attention to detail at the stop ends, which theoretically have the same pressure applied to them as that on the main part of the formwork.

Table 11.3 Rate of rise for Groups 1 and 2 concretes for walls and columns up to 6 m.

Concrete group	Conc. temp	Form design pressure kN/m²	Walls and bases — Height (H)					Columns — Height (H)				
			2m	3m	4m	5m	6m	2m	3m	4m	5m	6m
(1) CEM I (OPC), SRPC, CEM I/42,5R (RHPC), CEM I/52,5R (RHPC) without admixtures	5 °C	30	n/a	n/a	n/a	n/a	n/a	n/a	n/a	n/a	n/a	n/a
		35	0.2	n/a	n/a	n/a	n/a	n/a	n/a	n/a	n/a	n/a
		40	0.3	0.2	n/a	n/a	n/a	0.2	n/a	n/a	n/a	n/a
		45	0.8	0.2	n/a	n/a	n/a	0.5	0.2	n/a	n/a	n/a
		50	1.5	0.5	0.2	n/a	n/a	0.7	0.5	n/a	n/a	n/a
		55	–	0.9	0.4	0.2	n/a	–	0.7	0.2	n/a	n/a
		60	–	1.5	0.7	0.3	0.2	–	1.2	0.3	0.2	n/a
		65	–	2.4	1.2	0.7	0.3	–	1.6	0.6	0.3	0.2
		70	–	3.5	1.9	1.1	0.7	–	1.9	0.9	0.5	0.3
		75	–	4.9	2.7	1.7	1.1	–	–	1.3	0.8	0.6
		80	–	–	3.7	2.5	1.7	–	–	1.7	1.1	0.8
		85	–	–	5.0	3.4	2.4	–	–	2.3	1.6	1.1
		90	–	–	6.4	4.4	3.2	–	–	2.9	2.0	1.5
		95	–	–	8.1	5.6	4.2	–	–	3.8	2.6	2.0
		100	–	–	10.3	7.0	5.3	–	–	4.6	3.1	2.4
(2) CEM I (OPC), SRPC, CEM I/42,5R (RHPC), CEM I/52,5R (RHPC) with any admixture except a retarder	10 °C	30	0.3	0.2	n/a	n/a	n/a	0.2	n/a	n/a	n/a	n/a
		35	0.6	0.3	0.2	n/a	n/a	0.3	0.2	n/a	n/a	n/a
		40	1.1	0.6	0.3	0.2	n/a	0.5	0.3	0.2	n/a	n/a
		45	1.8	1.0	0.6	0.4	0.2	0.9	0.5	0.3	0.2	n/a
		50	2.8	1.6	1.0	0.7	0.5	1.3	0.7	0.5	0.3	0.2
		55	–	2.3	1.6	1.1	0.8	–	1.1	0.8	0.6	0.4
		60	–	3.1	2.2	1.7	1.3	–	1.4	1.0	0.8	0.6
		65	–	4.1	3.0	2.4	1.9	–	1.9	1.4	1.1	0.9
		70	–	5.4	3.9	3.1	2.6	–	2.4	1.8	1.4	1.2

75	1.5	1.9	2.3	2.9	–	3.3	4.0	5.0	7.1	–
80	1.9	2.3	2.8	–	–	4.2	5.0	6.2	–	–
85	2.4	2.8	3.5	–	–	5.3	6.2	7.6	–	–
90	2.9	3.3	4.1	–	–	6.4	7.5	9.2	–	–
95	3.5	4.0	5.1	–	–	7.6	8.9	11.0	–	–
100	4.0	4.7	6.0	–	–	9.0	10.5	13.4	–	–

15 °C

30	n/a	n/a	n/a	0.2	0.3	n/a	0.2	0.2	0.4	0.6
35	n/a	n/a	0.2	0.4	0.5	0.2	0.3	0.5	0.7	1.0
40	0.2	0.3	0.4	0.5	0.7	0.5	0.6	0.8	1.1	1.6
45	0.4	0.5	0.6	0.8	1.1	0.8	1.0	1.3	1.6	2.3
50	0.6	0.7	0.8	1.0	1.5	1.3	1.5	1.8	2.3	3.4
55	0.9	1.0	1.1	1.4	–	1.8	2.1	2.5	3.1	–
60	1.1	1.3	1.5	1.8	–	2.4	2.8	3.3	4.0	–
65	1.4	1.7	1.9	2.3	–	3.1	3.6	4.1	5.1	–
70	1.8	2.0	2.3	2.8	–	4.0	4.5	5.1	6.3	–
75	2.2	2.5	2.8	3.2	–	4.9	5.5	6.3	8.0	–
80	2.7	3.0	3.4	–	–	5.9	6.6	7.5	–	–
85	3.2	3.6	4.0	–	–	7.1	7.8	9.0	–	–
90	3.7	4.1	4.7	–	–	8.3	9.2	10.5	–	–
95	4.4	4.8	5.6	–	–	9.7	10.7	12.4	–	–
100	5.0	5.5	6.5	–	–	11.1	12.3	14.6	–	–

Notes:
1. The two tables state the rate of rise for five out of the seven main groups of cementitious materials.
2. The weight density of concrete is assumed to be 25 kN/m³.
3. The tables assume compaction by internal vibration.
4. In the lower heights of pour the pressure cannot exceed the fluid head of concrete and thus concreting at faster rates of rise will not cause the pressure to exceed the values stated.
5. A minimum rate of rise practically achievable of 0.2 m/hr has been considered and rates less than this value are not stated in the tables. They are shown as 'not applicable' as n/a. In such cases it is recommended that either the concrete is warmed to higher temperature, or the formwork is redesigned for a more suitable design value.
6. The tables are *not* applicable to underwater concreting.
7. Tabular information on Rates of Rise for WALLS up to 10 m high, and COLUMNS up to 15 m is given in Tables 17 and 18 of *Formwork—A Guide to Good Practice*.

Table 11.4   Rate of rise for Groups 3, 4 and 5 concretes for walls and columns up to 6 m.

Concrete group	Conc. temp	Form design pressure kN/m²	Walls and bases					Columns				
			Height (H)					Height (H)				
			2m	3m	4m	5m	6m	2m	3m	4m	5m	6m
(3) CEM I (OPC), SRPC, CEM I/42,5R, I/52,5R with a retarder	5 °C	30	n/a	n/a	n/a	n/a	n/a	n/a	n/a	n/a	n/a	n/a
		35	n/a	n/a	n/a	n/a	n/a	n/a	n/a	n/a	n/a	n/a
		40	n/a	n/a	n/a	n/a	n/a	n/a	n/a	n/a	n/a	n/a
		45	n/a	n/a	n/a	n/a	n/a	n/a	n/a	n/a	n/a	n/a
		50	0.2	n/a	n/a	n/a	n/a	n/a	n/a	n/a	n/a	n/a
(4) CEM III A (36%–65%), CEM II (6%–35%), CEM II A (6%–20%), CEM II B (21%–35%) or blends containing less than 65% ggbs or 35% pfa without admixtures		55	—	n/a	n/a	n/a	n/a	—	n/a	n/a	n/a	n/a
		60	—	0.2	n/a	n/a	n/a	—	n/a	n/a	n/a	n/a
		65	—	0.3	n/a	n/a	n/a	—	0.2	n/a	n/a	n/a
		70	—	0.8	0.2	n/a	n/a	—	0.4	n/a	n/a	n/a
		75	—	1.6	0.3	n/a	n/a	—	0.6	0.2	n/a	n/a
(5) CEM III A (36%–65%), CEM II (6%–35%), CEM II A (6%–20%), CEM II B (21%–35%) or containing less than 65% ggbs or 35% pfa with any admixture except a retarder		80	—	—	0.8	0.2	n/a	—	—	0.4	n/a	n/a
		85	—	—	1.4	0.4	0.2	—	—	0.8	0.2	n/a
		90	—	—	2.3	0.9	0.3	—	—	1.1	0.4	0.2
		95	—	—	3.5	1.5	0.6	—	—	1.7	0.8	0.4
		100	—	—	5.0	2.4	1.2	—	—	2.3	1.1	0.5
	10 °C	30	n/a	n/a	n/a	n/a	n/a	n/a	n/a	n/a	n/a	n/a
		35	0.2	n/a	n/a	n/a	n/a	n/a	n/a	n/a	n/a	n/a
		40	0.4	0.2	n/a	n/a	n/a	0.2	n/a	n/a	n/a	n/a
		45	0.8	0.5	n/a	n/a	n/a	0.4	0.2	n/a	n/a	n/a
		50	1.6	1.0	0.2	n/a	n/a	0.7	0.3	n/a	n/a	n/a
		55	—	1.6	0.4	0.2	n/a	—	0.5	0.2	n/a	n/a
		60	—	2.5	0.8	0.4	0.2	—	0.7	0.4	0.2	n/a
		65	—	3.6	1.3	0.7	0.4	—	1.1	0.6	0.4	n/a
		70	—	—	2.0	1.2	0.7	—	1.6	0.9	0.6	0.2

75	—	5.1	2.9	1.9	1.2	—	2.1	1.3	0.9	0.4
80	—	—	3.9	2.6	1.8	—	—	1.8	1.2	0.6
85	—	—	5.1	3.5	2.5	—	—	2.4	1.6	0.9
90	—	—	6.6	4.6	3.4	—	—	3.6	2.1	1.2
95	—	—	8.3	5.8	4.4	—	—	4.7	2.6	1.6
100	—	—	10.5	7.2	5.5	—	—	6.7	3.2	2.0

15 °C										
30	0.2	n/a	n/a	n/a	n/a	n/a	n/a	n/a	n/a	n/a
35	1.5	0.2	0.2	n/a	n/a	0.2	n/a	n/a	n/a	n/a
40	1.0	0.5	0.3	0.2	n/a	0.5	0.2	n/a	n/a	n/a
45	1.7	0.9	0.5	0.3	0.2	0.8	0.5	0.2	n/a	n/a
50	2.7	1.4	0.9	0.6	0.4	1.2	0.7	0.4	0.3	0.2
55	—	2.1	1.4	1.0	0.7	—	1.0	0.6	0.5	0.3
60	—	2.9	2.1	1.5	1.1	—	1.3	0.9	0.7	0.5
65	—	4.0	2.8	2.1	1.7	—	1.8	1.3	1.0	0.8
70	—	5.2	3.7	2.9	2.3	—	2.4	1.7	1.3	1.1
75	—	6.9	4.8	3.8	3.1	—	3.1	2.2	1.7	1.5
80	—	—	6.0	4.8	3.9	—	—	2.7	2.1	1.8
85	—	—	7.3	5.9	4.9	—	—	3.3	2.6	2.2
90	—	—	8.9	7.1	6.0	—	—	4.0	3.2	2.7
95	—	—	10.8	8.6	7.3	—	—	5.0	3.9	3.3
100	—	—	13.1	10.1	8.6	—	—	5.9	4.5	3.9

Notes:
1. The two tables state the rate of rise for five out of the seven main groups of cementitious materials.
2. The weight density of concrete is assumed to be 25 kN/m³.
3. The tables assume compaction by internal vibration.
4. In the lower heights of pour the pressure cannot exceed the fluid head of concrete and thus concreting at faster rates of rise will not cause the pressure to exceed the values stated.
5. A minimum rate of rise practically achievable of 0.2 m/hr has been considered and rates less than this value are not stated in the tables. They are shown as 'not applicable' as n/a. In such cases it is recommended that either the concrete is warmed to higher temperature, or the formwork is redesigned for a more suitable design value.
6. The tables are *not* applicable to underwater concreting.
7. Tabular information on Rates of Rise for WALLS up to 10 m high, and COLUMNS up to 15 m is given in Tables 17 and 18 of *Formwork—A Guide to Good Practice*.

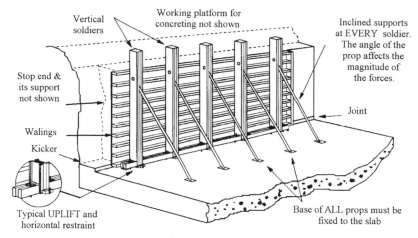

Figure 11.10   Typical arrangement of single-faced formwork.

11.5.2 Single faced vertical formwork

Single-faced formwork is frequently prone to problems (see typical example, Fig. 11.10).

The concrete pressure is calculated as for double-faced wall formwork, so the calculations are similar. The difference is that the entire force onto the single face has to be resisted by the external propping as there are *no* tie rods. When supporting props are inclined, the angle of inclination will generate uplift and a horizontal force, both of which require restraint. It is the restraint of this uplift force that generally causes problems. The forces can be large, for example a 2 m high form with propping fitted at say an angle of 45° will cause an uplift force of about 17 kN per metre length of wall.

A secure way to restrain uplift is to use vertical bolts secured into the base slab (see Fig. 11.10). An alternative solution is to use purpose made large 'A' frames with a single, angled bolt secured into the base slab. The exact fixing of this angled bolt requires care and attention as it is required to carry large forces.

It is worth reminding engineers that even on small stop ends, (as small as 300 mm high), the effect of inclined bracing members will also generate uplift forces, and once grout seeps under the forms, it may cause uplift and a tendency for the formwork to lift.

11.5.3 Striking wall, beam side and column formwork

The main criterion for the removal of vertical formwork from walls, beam sides and columns is a requirement to have achieved sufficient strength to avoid mechanical damage during the striking process. If the form face has a feature, then the draw and size of feature, as well as the direction of the strike, can affect the striking time. The following striking recommendations in Table 11.5 are taken from the CIRIA Report R136 on *Formwork Striking Times*. Striking time criteria related to the DTp (HA) classification of surface finish are now the industry norm.

Table 11.5 Striking criteria for vertical formwork.

Finish description	Striking criteria

Basic, plain or rough finish DTp Classes F1 and F2	Either: A minimum period equivalent to:
	   8 hours at 20 °C for unsealed plywood
	   6 hours at 20 °C for impervious forms.
	or Strike next morning if the mean air
	   temperature is above 10 °C overnight.

Fair, fine smooth quality finish DTp Classes F3, F4 and F5	Minimum *in situ* cube strength of 2 N/mm²

Reference should also be made to the contract specification for any limitations on the time of striking of vertical wall and column formwork. Tables in specifications may have different parameters; times presented in hours or in days; temperature related to surface temperature of the concrete, while others to the mean of the minimum and maximum daily air temperature.

Where formwork to free-standing reinforced concrete walls is to be struck, care is necessary to ensure that there is sufficient concrete strength at the time of striking to withstand the overturning moments from the wind loads.

In the case of sloping sections of walls, columns etc. in addition to the possibilities of mechanical damage during striking, the concrete has to support itself, and a minimum structural strength will be need to be achieved *before* striking can commence (see Section 11.7.4 on soffit form striking).

The heat loss through steel formwork in cold weather can increase the time for striking the forms. One solution is to increase the thermal insulation ('U' value) of the formwork, by attaching polystyrene sheets onto the steel face on the outside of the forms, thus effectively insulating the face and reducing heat loss from the concrete.

11.5.4 Soffit formwork

The design of soffit formwork is slightly different to that for walls—loads are usually much lighter and deflections more critical—as the spans can be larger. The principles are the same, the loads are followed through from the face to the supporting falsework (see Fig. 11.11).

There are many different soffit decking systems, from traditional plywood with timber bearers (some with aluminium bearers) to those with voided slabs, such as trough or waffle floors (for typical examples see Figs 11.4 and 11.11). Some proprietary systems have quick-strip arrangements allowing the expensive formwork face components to be struck early (normally at a minimum concrete strength of 5 N/mm²) to avoid damage, yet leaving the main slab supported until approval to strike is received. The time at which soffit forms can be removed, and the exact procedure for striking the forms, requires detailed consideration. If not, there is real risk of damaging the permanent works and possibly initiating an accident.

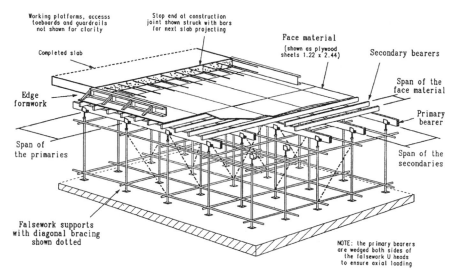

Figure 11.11   A typical arrangement of soffit formwork on a skeletal falsework.

The applied loads to the soffit formwork will include the weight of the structure being supported, plus an allowance of Service Class 2 loading (1.5 kN/m²) as a construction operations load to cater for operatives placing the concrete, plus, of course, the self-weight of the forms and any edge forms.

All slabs will have edge forms, stop ends or construction joints. The design of these is often left to the 'site' to complete. Particular care is necessary if stop ends are greater than 400 mm in depth. (For a typical Hy-Rib construction joint detail see Figure 11.3.)

11.5.5   Permanent formwork

An increasingly common method of construction is to use permanent formwork to create the soffit, with a structural concrete topping. The materials used include permanent steel sheets, thin precast concrete planks, thick precast concrete planks, glass-reinforced cement and other suitable, durable materials. If woodwool panels are used they may need to be channel reinforced for additional strength. They are generally designed as non-participating, and do not contribute to the strength of the final slab. If they are contributing then careful consideration needs to be given to suitability, buildability and durability.

The attention to detailing at the design stage by the PWD should reduce the common site problems of tolerances, placing and handling units of permanent formwork. The main industry reference is the the CIRIA Report *Permanent Formwork in Construction*, which gives detailed guidance on detailing, handling and the safety issues involved.

11.6   Scaffolding

11.6.1   General

The WAH Regulations require all working platforms to be fit for purpose. It also requires guard-rails fitted so that there is no unprotected gap exceeding 470 mm where a person may fall. All scaffolds shall be erected under competent supervision. The CDM Regulations impose duties on *all* designers, including scaffolding, formwork and falsework designers and all those site engineers carrying out 'on-site' designs.

The scaffold performance standard BS EN12811–1 is for all working scaffolds, in limit state terms and specifically excludes permissible stress design. It replaces BS 5973. Guidance for tube and fitting scaffolds is given by the NASC in TG20:08 *Guide to good practice for scaffolds with tube and fittings*. The WAH regulations require that *all* working platforms used for construction (including scaffold and tower places of work) and from where a person is liable to fall more than 2 m, shall be inspected:

- before being taken into use for the first time
- after any substantial addition, dismantling or other alteration
- after any event likely to have affected its strength or stability
- at regular intervals not exceeding seven days since the last inspection.

The inspection should be carried out by a competent person, who should complete a report before the end of the working period of the inspection. Although the inspection regulations do not apply to any mobile tower scaffold (unless it remains erected in the same place for a period of seven days) this should not preclude regular inspection of such structures.

There are many items that need to be checked on a scaffold, but very often it is the simple points that have been overlooked, particularly when it has been altered. Some of the more basic points are obvious. Several CITB books have helpful checklists for inspection and use of scaffolds.

Scaffolds require to be erected, altered and dismantled *safely*. To reduce the number of accidents caused by falls from height, the NASC have a detailed book (SG4: 2004) *Preventing Falls in Scaffolding and Falsework* which discusses the planning and safety hierarchy of prevention. The pocket-sized companion booklet *SG4: You,* is based on this document and is strongly recommended for site engineers.

Safety is everyone's responsibility: with careful planning, risk assessments, training and use of competent supervision, the number of scaffold accidents should be reduced.

11.6.2   Scaffold design

Basic principles

When erected against buildings, scaffolds are generally tied to the building for stability and have both an inside and outside row of vertical members, called standards. They are known as *independent tied scaffolds* (see Fig. 11.12 for typical arrangement). Where the scaffold is free standing, such as a tower or if ties to the building are not permitted, then its stability will rely on the width of the base and its self-weight to provide sufficient restoring moment to prevent overturning. Rakers may assist in this process.

Generally a very simple view is taken in most scaffold designs and imponderables, such as stiffness of couplers, are ignored and adequate safety allowed for in the factors chosen for the materials and components. Provided that the scaffold is erected within certain workmanship limits for verticality, joint connections and so on, the couplers are considered as giving no stiffness acting merely as connections. In practice, the couplers eccentrically load the standards, particularly where diagonal tubes connect to horizontals. Theoretically they torsionally load the standard: this is ignored, provided the diagonals are connected within a distance of 300 mm. **Keep it simple**: *If it is right then it will look right!* When thinking about the design of the scaffold, remember to:
Think vertical, think horizontal, then think horizontal *again*.

As with all structures, having established the loads, the three design requirements are to consider the strength and stability of the individual members, then the lateral stability of the structure, and finally the overall stability (see Fig. 11.12 for typical scaffold).

Element strength and element stability

The minimum factor of safety on the ultimate capacity of elements in scaffolding is 1.65.

Vertical standards are generally spaced about 2.1 m apart, (measured horizontally) and there will be horizontals spaced at about 2 m apart representing each lift. The transoms will be spaced to suit the permissible span of the chosen boards, but at a maximum span of 1.5 m. Designers should be aware of the effect of effective lengths on the load capacity in the members of the scaffold, covered in detail in Section 11.3.5.

Scaffold Designation

TG20 introduced in 2008 a scaffold designation that describes the Load Class (See 11.4.1) and the arrangement of boards. In order the three numbers are the Load Class, the number of boards between the standards and the number of inside boards, i.e. cantilevered in towards the structure as 0, 1 or 2. A suffix F is added to indicate that the inside boards are considered to have the same load as the main platform; otherwise it is assumed that inside boards are lightly loaded.

For example, a designation **3–5–2 F** indicates a Load Class 3 scaffold with five boards between the standards and two boards on the inside which a Fully loaded with Class 3 loading applied. Note that the TG20 configurations limit the maximum number of inside boards to two. Referring to Fig. 11.12, if it is assumed for light duty painting (Class 2) then its designation would be 2–5–0.

Lateral stability of scaffolds

Bracing is required in two directions to prevent overturning: at right-angles to the building this is called *ledger bracing*, and parallel to the length of the building, *façade bracing*. To resist the movement of the scaffold inwards and outwards from

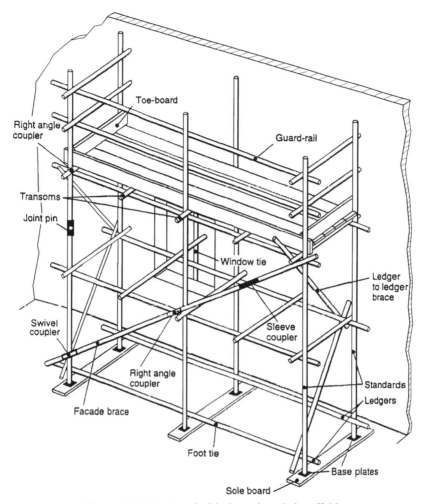

Figure 11.12   A typical independent tied scaffold.

the building adequate ties should be fitted. Braces fitted as diagonals are generally at an angle between 35 and 50. They make the structure into stiff triangles and transmit loads down through the scaffold to ties or to the ground—*bracing alone is not sufficient*—you must triangulate the bracing with horizontal lacing.

Ledger bracing is fitted up the scaffold at right angles to the building; on independent tied scaffolds it is fitted to *alternate* pairs of standards. If the bays are less than 1.5 m, then they are fitted onto every third pair of standards.

To stabilise an independent tied scaffold in the long direction, ties will connect to and stabilise the inside row of supports, but the outer row requires *façade* bracing to give longitudinal stiffness. Such *façade* bracing is fitted every six bays. On scaffolds higher than 8 m (four lifts) plan bracing fitted between adjacent tie positions are required at every four lifts, every twelve bays where the façade bracing is only fitted across single bays of the scaffold.

Tying scaffolding

The main way in which scaffolds are restrained is by tying to the building and using the permanent work for stability. The scaffold requires restraint in *four* directions: namely from moving *into* the building, *away* from the building and from longitudinal movement in the two directions parallel to the building (*left* and *right*).

There is also the consideration of the type work being executed—if fitting windows into a structure then there is need to remove some of ties in order to fit the windows, hence the concept of movable and non-movable ties. This does not mean that ties can actually be permanently removed, it simply classifies movable ties as those that can be *temporarily* removed—and then *replaced*. Scaffolds should be tied to the building façade at 4 m vertical centres (i.e. every other lift) and every other bay. This arrangement of tying is required for unclad, debrisnetted and sheeted scaffolds. Where larger lifts are required, such as a 2.7 m pavement lift, special provisions should be made. TG20 gives more details on the subject. The frequency of tying is also dependent on the exposure of the scaffold to the wind. The NASC Guide gives tables of the maximum safe height for various exposure conditions for basic scaffolds and specifies the capacity of the tie assembly required connected to the building.

The four main arrangements of ties used are the *through* tie, the *box* tie, the *reveal* tie and the *anchor* tie. Whichever arrangement is used, its connection to the scaffold needs to be satisfactory, and ideally should be within 300 mm of the standards and, if an independent tied scaffold, connected to *both* rows of standards. The NASC Guide specifies three load capacities of tie assemblies in tension as light duty (3.0 kN), standard (6.1 kN) and heavy duty (12.2 kN). It is important to use the correct capacity and type of tie assembly to suit your scaffold.

Remember that *adding* debris netting or sheeting to an existing scaffold will significantly alter the tie capacities required and the design should be checked.

Overall stability—free-standing scaffolds and towers of tube and fittings

Free-standing scaffolds and towers made of steel tube and fittings present special requirements for stability. Unless tied or weighted down, the only thing keeping them upright is their self-weight! The critical ratio is the height to the minimum base width to give stability. The height is measured from the foundations to the platform level (*not to the top guard-rail level*). The ratio depends on whether it is inside a building, protected from the wind, or whether it is outside, exposed to the elements. For example, a steel scaffold tube and fitting tower required inside a completed warehouse with a platform height of 6 m requires a minimum base width of 1.5 m (i.e. 4:1 ratio). In contrast, a similar tower, also 6 m high, when erected outside but adjacent to a building requires a minimum base width of 2.0 m (i.e. 3:1 ratio).

Overall stability—free-standing aluminium proprietary towers

Free-standing aluminium proprietary towers (Fig. 11.8) are easy to erect, but made with light components and having different rules for stability. Towers should not be left erected in exposed conditions, and preferably should be tied to adjacent buildings. BS 1139–3 specifies the requirements for prefabricated mobile and access working towers, by giving each Service Class loading a horizontal load allocation for design. It is the responsibility of the manufacturer to establish how these loads are to be restrained and his tower stabilised, e.g. with outriggers, stabilisers, weight boxes, etc. *Hence height to base ratios for PASMA towers are not appropriate.*

Placing sheeting onto any tower *significantly* increases the wind loads and structures used with sheeting must be appropriately designed.

11.6.3 Foundations to scaffolds and towers

Foundations to scaffolds and towers are important—a structure is only as stable as the foundations upon which it rests. They can vary from solid rock, pavements, to very soft ground. Often scaffolds are erected on poor, muddy locations, but with reasonable preparation an adequate base can often be provided. Always seek advice if you are unsure.

Each vertical member taking load should be fitted with a base plate to spread the load from the tube. On the majority of surfaces a sole plate of timber is also recommended to further spread the load. The minimum thickness of a sole plate is 35 mm and minimum width is 219 mm. Sole plates ideally should support *two* standards. Scaffold boards are often used as sole plates. Larger members, such as 225 75 mm C16 timber and railway sleepers 250 × 125 × 2600 long may also be considered.

11.7 Falsework

11.7.1 General

The responsibility for falsework safety and use is often left entirely to the contracting organisation, yet its effect on the final shape, location and any residual stress in the permanent structure can be significant. The *Government Advisory Report on Falsework 1976*, recommended tighter control on the procedures for falsework (subsequently reinforced by a BSCP in 1982) and made recommendations about minimum lateral stability of such structures.

BS EN12812 is exclusively in limit state terms. It gives much information about tube and fitting falsework, a type rarely used in the UK, so its relevance is limited. It classifies falsework into three classes A, B1 and B2, but gives no guidance on rules for class A falsework. It is based on the German standard DIN 4421.

The UK Code of Practice on falsework is BS 5975, updated in 2008 it, introduces procedures for *all* temporary works and covers the permissible stress design of falsework. It also gives rules for the design of simple class A falsework when used in the UK. It recognises that falsework has a different design philosophy to that used in the permanent works; it is loaded for a short time; rarely is it a billed item; often uses reusable components; has the unique structural requirement to be de-stressed under load (to remove it); it is not tied down and, therefore, relies on its own weight for stability; and is usually stressed to 90% of its safe working capacity.

BS 5975 is a permissible stress code with sufficient information, such as steel stresses from BS 449, contained in the appendices. It also has a comprehensive wind loading section based on the simplified method adopted for scaffolding in TG20, derived from BS 6399–2.

The code is one of only two BSI documents with procedures: this is because it was the management and control of the temporary works that was identified as being so important. One of the recommendations is to prepare the *temporary works design brief* (TWDB), listed and discussed in Section 11.1.3.

The other important code recommendation is that *all* sites should appoint a temporary work coordinator (*TWC*) to control the work. The site engineer should be aware of the name of the appointed TWC on the site. There are sixteen principal activities listed in the Code (Clause 7.2.5); they are not an onerous—they simply set down the actions that a responsible engineer should be carrying out. Some of the relevant site activities are:

- to ensure that those responsible for on-site supervision receive full details of the design including any limitations associated with it
- ensure any residual risks identified by the PWD are included in the TWDB
- ensure the temporary works design is independently checked
- to ensure that checks are made on site at appropriate stages covering the more critical factors
- to ensure that any proposed changes in materials or construction are checked against the original design and appropriate action taken
- to ensure that any agreed changes, or corrections of faults, are correctly carried out on site
- to ensure that during use, all appropriate maintenance is carried out
- after a final check, to issue formal permission to load if this check proves satisfactory
- when it has been confirmed that the permanent structure has attained adequate strength, to issue formal permission to dismantle the falsework.

Obviously inspecting and checking the falsework during erection and prior to use is vital, as well as being a legal requirement, but this should be a staged checking, it is not left until all the structure is erected. The CS's *Checklist for Erecting and Dismantling Falsework* is an essential tool for the site engineer.

11.7.2   Stability of falsework

The *Bragg Report* identified that failures occurred not because of *insufficient* bracing—but because there was *no* bracing. This prompted considerable interest in the industry—and the requirement was incorporated into BS 5975 Clause 19.4.1.1. Falsework fails either by the members buckling or simply falling over. To resist buckling the members have to have sufficient strength, this may require bracing to create stiff node points. To prevent freestanding falsework from falling over, it requires to be adequately stiffened, usually by bracing. There will be cases where the permanent works can be used to stabilise the falsework, such as by connecting to adjacent columns or existing walls. This is known as *top restrained* falsework. Stability now requires the involvement of both the PWD and the TWD, and more involved checking procedures are required. BS 5975 (Clause 19.3.2.2) gives guidance on top restrained falsework, and the required plate action of the soffit formwork to resist the lateral forces in order to stabilise the top of the falsework. BS 5975 recommends that for *every* falsework structure, three design checks are required, namely:

Check 1—Structural strength of members and connections

Obviously the designer will check for bending, shear, deflection, (occasionally torsion) and for stability of individual members and connections, such as welded or bolted joints.

The effective lengths of members in compression will need particular attention, especially long struts where the slenderness ratio is often the limiting criteria. Generally for all falsework with scaffold tube and fittings, the node points/couplers are considered as pin joints provided that the couplers are fitted within 160 mm of the node points.

Many proprietary falsework systems have patented joints which provide some moment restraint, and these can reduce the effective length factor below unity, thus increasing their load carrying capacity. Follow the suppliers recommendations. Particularly look out for unrestrained cantilever extensions on falsework at the head and bases. Often this will be the jack extension(s) and may have a crucial effect on the effective length to be considered for the adjacent lift of falsework.

The effective length (l) of unbraced cantilever extensions is considered to be that of the adjacent length of standard *plus* twice the length of the unrestrained cantilever (see Fig. 11.5).

Check 2—Lateral stability

The specified loads are the lateral stability loads specified in BS 5975 Clause 9.2.9.1. This minimum specified load (Section 11.4.2.2) is considered to act at the point of contact between the vertical load and the falsework.

Lateral stability also applies to the webs of steel beams at reaction points and at concentrated loading positions. BS 5975 Annex K.1 states that steelwork at *all* load transfer points require *web stiffeners* unless calculations show that such stiffeners are not required. Use web stiffeners unless calculations prove otherwise.

Check 3—Overall stability

Tall slender structures, and those exposed to high winds will have a tendency to overturn. This check requires that the *minimum factor of safety is 1.2* against overturning (BS 5975 Clause 6.4.5.1). As falsework is generally a gravity structure relying on its own weight for stability, the most onerous condition is often just after the soffit formwork is fixed, and just before the reinforcement is fixed, i.e. with little restoring force; once the concrete is placed there will be ample vertical load to resist overturning.

If the falsework is unstable, then use holding down bolts or kentledge.

Check 4—Positional stability

Lateral restraint to resist sliding of the structure should be considered. The assembly of formwork and falsework often rely on friction for transferring loads between members, in the absence of rigid permanent connections; as temporary works equipment is generally re-usable, holes and connections can reduce the strength and life of members. The recommended values of the coefficient of static friction of many of the combinations of components used in temporary works is given in BS 5975 Table 24 (Clause 19.4.5.1).

Remember it is important for users to be aware that the value of the frictional force is the value at which sliding is just about to occur. Thus, when lateral force is to be transmitted between the members a suitable factor of safety should be used.

It should also be noted that the orientation of materials does not affect the friction results, so that with the exception of the ground, surfaces can be top or bottom of the restraint system.

In temporary works design, the serviceability limit state is generally used for consideration of frictional restraint, such that the safe horizontal restraint force for a known vertical load is often required. Thus:

Safe lateral restraint force = (unfactored applied normal load) u divided by factor of safety.

It should be noted that BS 5975 Clause 19.4.4.1 recommends that the factor of safety used in the above equation should not be less than 2.0.

11.7.3 Continuity of members

Temporary works structures will have many redundancies, and the distribution of load into the verticals from the soffit formwork bearers will also be very indeterminate (see also arrangement of the bearers in Fig. 11.11). The face contact material, the secondary bearers, and often the primary members in the forkheads, will be continuous over several supports giving rise to increased reactions at internal supports from their elastic reactions.

The support reactions of beams change when they are continuous over the supports, such as a long scaffold tube or timber beam. The following examples assume that the distributed load on each span is 10 kN, and that the span between the supports are equal:

A worst case is a single beam continuous over two spans with three supports giving a central reaction of the static load 1.25 for continuity, i.e. a staggering 25% increase in load. BS 5975 Clause 19.3.3.2 accepts that in the case of falsework comprising random bearers, you design by working out the simple load on the area supported by the standard and *add 10%* for continuity—the 10% rule—but certain cases, for example over two spans, a more precise calculation may be justified. Certain proprietary systems incorporate simply supported beams where this *rule of thumb* would not be applicable.

11.7.4 Striking soffit formwork and falsework

The following striking recommendations are taken from the CIRIA Report R136 *Formwork Striking Times*—criteria, predictions and methods of assessment.

The time and procedure by which beam and soffit formwork can be struck should be carefully controlled and should be carried out in accordance with the requirements of the contract specification and drawings. The order in which supports can be removed is very important (see the procedures in Fig. 11.13). Always strike slabs by starting at the mid span position and working towards the supports, unlike cantilevers where commencement should be from the tip.

The criteria for striking soffits for flat thin concrete slabs are in the process of changing. This follows the recommendations from research carried out at ECBP in 1998. It found that a method of predicting striking times related to the control of the crack width in the concrete at critical areas, was a more realistic control of structural behaviour. The results of this work are outside the scope of this book, and for the striking of flat slabs, reference should be made to the *Guide to Flat Slab Formwork and Falsework.*

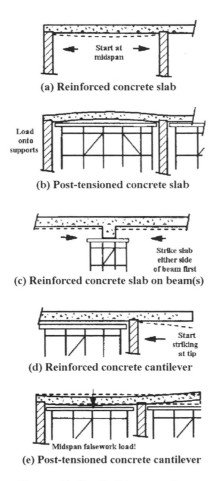

(a) Reinforced concrete slab

(b) Post-tensioned concrete slab

(c) Reinforced concrete slab on beam(s)

(d) Reinforced concrete cantilever

(e) Post-tensioned concrete cantilever

Figure 11.13   Striking procedures.

Table 11.6   Striking criteria for soffit forms, slabs and beams.

Finish description	Striking criteria
Basic, plain, rough finish or quality DTp all classes	Either: use specifications, codes of practice or tables (e.g. CIRIA R136) or: assess the concrete strength at the time of striking knowing the maturity, the concrete mix etc.
Flat slabs less than 300 mm thick	Use CONSTRUCT Flat slab guide based on control of crack width.

The basis of strength determination as a striking criterion is stated in more detail in *Formwork a Guide to Good Practice* (see Section 11.5.3; Table 11.6).

The main consideration in striking soffits is the early strength of the concrete to ensure that the member, when released, can support its own weight together with any imposed construction operations loads. All slabs should be struck completely and allowed to take up their deflected shape before applying further loading to the slab. If this cannot be achieved, then a calculation will need to be carried out in consultation with the TWC to assess the loading conditions. Additional considerations are the elimination of frost and mechanical damage, reduction in thermal shock, and limiting excessive deflections.

A main recommendation of the Concrete Society's *Formwork Guide* is that *the minimum concrete strength at time of striking be specified*, after taking into account the stage of construction. Calculations to justify a lower value of time and strength at striking should be approved by the permanent works designer. The TWD will need to assess the proportion of loading on the structure at time of striking to the PWD's design working load.

The load on the structure at the time of striking will include the formwork self-weight, the mass of the concrete, plus a minimum construction operations load for access of 0.75 kN/m^2. Where there are several other levels of construction, additional loads may also be required to be supported. In most cases, striking can be considered when the *in situ* concrete has obtained a characteristic strength at least proportional to the ratio of the loads multiplied by the grade of concrete. CIRIA Report *Formwork Striking Times* details two methods for determining the concrete strength required for striking, either using the proportion of load at time of striking to the design working load, or by analysing the actual section. The latter method is advantageous for lightly reinforced sections. The reasons for stating these two methods are due to the relatively *low* rate of gain of strength of pfa and ggbs mixes at early ages.

For the site engineer, the approval procedure and the permit to strike are very important considerations. It is already shown that it is the knowledge of the realistic strength of the concrete at the exact time of striking that is needed. Thus the assessment of concrete strength at an early age is vital to the striking procedures. BCA *Best Practice Guide No: 1* is an authoritative guide on the selection of test methods suitable for site use. The most reliable and favoured test from ECBP and other work is the *in situ* LOK test method, accepted in the UK as an established BS 1881 method of test.

11.7.5 Backpropping slabs in multi-storey construction

In multi-storey construction, the ratio of design service load on a slab to the applied construction loads needs careful assessment. When the supporting slab immediately below the level under construction has not gained full maturity, construction loads will require to be supported through several levels. This is known as *backpropping*. Ideally the supporting slab should have been struck and allowed to take up its instantaneous deflected shape. Note that repropping can be used at lower levels. The example shown at Fig. 11.14 indicates *two* levels of backpropping.

The order of installing, preloading, and removing of backpropping will affect the loads transferred into different floors. This should be discussed between the TWC and the PWD at an early stage of the contract. Once agreed, the procedure *must* be followed. The materials section highlights the importance of never mixing steel and aluminium back props in a structure. This is due to their different elastic properties under load.

The *Guide to Flat Slab Formwork and Falsework* identifies that due to the elasticity of the aluminium shores and the relative stiffness of the supporting floor, in some cases only 33% of the load is actually transmitted through the supporting floor to the backpropping, compared to over 50% conventionally assumed in BS 5975. In other words, over 70% of the applied construction loads will be taken by the supporting slab (see Fig. 11.14). Further, the earlier idea, on low-rise buildings, of *backpropping through to the ground* has been shown to be erroneous, as the supporting slab is still significantly loaded. Refer to the *Flat Slab Guide* for a fuller treatment.

The subject of backpropping is not understood by many at site level, and few PWDs understand the implications. It is a subject that needs careful consideration and control, because if the industry wants to build leaner more flexible commercial structures, then site engineers will have to understand the importance and significance of this subject for their particular contract.

11.8 Façade retention

11.8.1 General

Section 54 of the Town and Country Planning Act, 1971, empowered local authorities to give statutory protection to buildings of architectural or historic importance. This has led to the increase in the renovation and refurbishment of complete structures internally, yet requiring the facade to be undisturbed. CIRIA Report 111 *Structural Renovation of Traditional Buildings* is an ideal starting point for designers wishing to understand more about the nature and materials likely to be encountered in such work.

The retention of the building façade requires cooperation and coordination between the PWD and the TWD at all stages of the construction process to ensure that the temporary façade retention structure is able to transmit all the relevant forces safely. These forces arise from wind, out of plumb, accidental impact, etc. Prepared in a partnership with the HSE, the CIRIA book C579 *Retention of Masonry Façades— Best Practice Guide* gives detailed technical guidance on the subject. A user-friendly A5-size handbook C589 is also available.

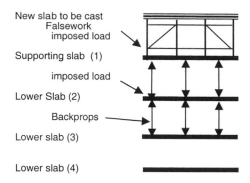

New slab to be cast
Falsework
    imposed load

Supporting slab (1)

        imposed load

Lower Slab (2)

        Backprops

Lower slab (3)

Lower slab (4)

Figure 11.14   Typical backpropping—two levels.

Where the façade structure has to be altered by the insertion of beams etc. the weight of the façade has to be carried by dead shores and temporary works or false-work—the design is covered by BS 5975. The procedure of inserting beams to support the façade is known as *needling*, the most critical aspect being the stability of the whole structure at each stage.

Where the façade remains on its own during renovation, it is often necessary to provide kentledge (or ties into the foundations) to resist uplift from wind. The retention scheme can either be fitted internally to the façade and support arranged to avoid clashing with subsequent construction or, more typically, arranged externally (see Fig. 11.15).

Whatever system is used, the connection of the façade to the retention system should be carefully considered. The usual method is to incorporate horizontal walings, pre-bolted through the façade. CIRIA C579 gives details of different fixings and connections. These connections must suit the sequence of operations and particular care is needed to ensure that these tension or compression forces can be adequately transferred and may also need to accommodate any envisaged movement. They will also need to be fitted *before* the internal structure is removed, so their location is often determined by consideration of the existing and proposed floor layout.

11.8.2   Loads on façade retention systems

Vertical loads

In a few cases, the façade retention system has to carry the vertical weight of the wall, in addition to overturning and stability requirements (see Fig. 11.15B). The mass of the wall will have to be estimated and CIRIA Report 111 gives some guidance on weights etc. In addition to the self-weight of the façade, there may also be loads arising on confined sites from site offices built into façade schemes. Some of these offices may be used for the storage of materials which will impose additional loads. During renovation of the façade, building work may have to take place involving the erection of working platforms (or scaffolding) within the façade retention scheme.

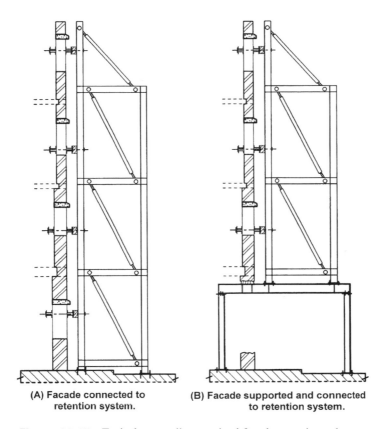

(A) Facade connected to
retention system.

(B) Facade supported and connected
to retention system.

Figure 11.15   Typical externally restrained façade retention schemes.

It should be pointed out that when a structure as illustrated Fig. 11.15B has the portal frame loaded asymmetrically, there will be consequential sway deformation of the steelwork which needs to be considered in the design.

Wind loads

The lateral load from the wind should be calculated from BS 6399–2. The retention system will often have to withstand the full wind force. With the wind blowing through the façade retention system, tension is generated in the outer legs, and kentledge may be required if anchorages are not possible. In the opposite direction, with the wind sucking out the façade, the self-weight of the actual wall façade may be sufficient to eliminate tension in any inside legs. The pressure of the wind should be calculated using a seasonal factor of $S_s = 1.0$ and, although the probability factor (S_p) may be modified for very short-term duration, it is recommended that $S_p = 1.0$ is used, as temporary façade structures are in use for periods longer than six months.

The effect of the façade (its number of openings etc.) will influence the value of the force factor ζ to be used. The total force should not be greater than that which would have been generated by an unperforated surface perpendicular to the wind, and a value of factor $\zeta = 1.0$ is recommended. The effects of sheltering or funnelling may also need to be considered.

During working operations there will be an operating limit for the cranes, considered at a wind speed of 18 m/s known as the *working wind*.

Accidental loads

It is likely that at some point in the construction process materials will be delivered to the outside of the structure, and then be crane handled over the façade into the building. Examples of this include: concrete skips, precast stair units, lift plant. Although this should be under controlled conditions, there will be a risk of accidental impact, especially to the arrangement of the retention system.

To allow for accidental loads, CIRIA C579 recommends that the temporary structure be designed to withstand an impact load of 10 kN in any direction. The lateral force should also include the working wind acting at the same time. This lateral force applied to the structure will cause sway and create a possibly critical deflection of the façade. To allow for a vehicle impact, the temporary façade structure should be designed to resist a 25 kN impact in any direction in the bottom 1 m of the structure.

Out of vertical loading allowance

An initial survey of the façade might identify the magnitude of the out-of-plumbness and allowances could then be made in the design for the additional overturning effect. Other effects to be considered include: settlement of the temporary works or foundations, and movements caused by the demolition of the existing structure (e.g. heave).

The specified horizontal load of 2½% of the vertical load used in falsework (see Section 11.4.2) is unrealistic in façade retention, and provided the plumb survey indicates a nominally vertical structure, CIRIA C579 recommends a lateral load of 1½% of the total vertical load (including the façade and the retention structure) when considering the overturning and the overall design of the temporary structure. When considering the design of the walings and the connections to the retention structure, the lateral load is increased to 2½% of the vertical load at the level considered. The effects of any out-of-plumb of the existing structure would also need to be considered. Table 8.1 in CIRIA C579 discusses in detail the principle lateral load combinations to be used in the analysis.

Note: 1½% represents 1:67 out-of-plumb, equivalent to nearly 300 mm on a 16 m frame.

11.8.3 Deformation criteria

When designing retention systems, one of the more critical aspects is the control of the deflections of the top of the façade under different loading conditions. The movement at the top will be caused in two ways: by differential settlement of the foundations of the retention scheme and also by the lateral deformation of the retention frame caused by horizontal forces such as wind, out of plumb and impact loads. All parties involved should be aware of the deflection limits and the regime of monitoring to be adopted. More onerous limits may need to be adopted for sensitive structures. Certain monitoring 'trigger' points may need establishing in consultation with the PWD.

Ground movements in response to partial demolition and reconstruction should be considered by the client's consultant as part of the new design. The additional deformation of the temporary structure caused by the wind, specified lateral loads, impact, elastic movement, etc. should be considered by the TWD.

CIRIA C579 recommends a lateral deflection limit of:

$$\delta = \frac{h}{750}$$

for the frame displacement, where: δ is the out of vertical horizontal dimension at the top and h is the overall height of the façade.

A typical aspect ratio for retention schemes is 1:5 for base width to height.

The precise loading combination for the limit needs stating in the TWDB (see CIRIA C579). Monitoring of deflections is important and the limiting deflection should be made clear, usually stated as a value in mm, not as a ratio. Thus on a 16 m frame the limit stated as h/750 represents 21.3 mm, which would be more practical stated as 'limit to 25 mm'.

Further reading

Note: The author particularly recommends those publications indicated in **bold**.

Standards and Codes of Practice

BS 5975: 2008 **Code of Practice for Temporary Works Procedures and the Permissible Stress Design of Falsework,** London, 195pp.

BS 5268–2: 2002 *Structural use of timber: Code of Practice for permissible stress design, materials and workmanship,* London, 176pp (includes Amendment No. 1 2008).

BS EN 12812: 2004 *Falsework—Performance requirements and general design,* London, 50pp.

BS EN 1065: 1999 *Adjustable Telescopic steel props—Product specifications, design and assessment by calculation and tests,* London, 35pp (replaces BS 4074).

BS EN 1004: 2004 *Mobile access and working towers made of prefabricated elements—Materials, dimensions, design loads, safety and performance requirements,* London, 2005, 12pp.

BS 1881–207: 1992 *Testing Concrete. Recommendations for the assessment of concrete strength by near-to-surface tests,* London, 20pp.

BS 1139–5: 1990 *Metal scaffolding. Specification for materials, dimensions, design loads and safety requirements for service and working scaffolds made of prefabricated elements,* London, 18pp (also known as HD 1000: 1988).

BS EN 12811–1: 2003, Part 1 *Scaffolds. Performance requirements and general design,* London, 2004, 58pp.

BS 2482–1: 2004 *Specification for Timber scaffold boards,* London. 2008.

BS 6399–2: 1997 *Loading for buildings. Code of practice for wind loads,* London, 118pp (reprinted 2002).

BS EN 39: 2001 *Loose steel tubes for tube and coupler scaffolds. Technical delivery conditions,* London 18pp.

BS EN 13377: 2002 *Prefabricated timber formwork beams. Requirements, classification and assessment,* London, 28pp.

BS EN 13374: 2004 *Temporary edge protection systems. Product specification, test methods,* London, 36pp.

BS 8110: 1997 *Structural use of concrete. Code of practice for design and construction,* London, 168 pp.

Other related texts

Formwork—A guide to good practice, 2nd ed.: Special Publication CS030, The Concrete Society, Slough, 1995, 292pp.

TG 20:2008 *Guide to good practice for scaffolding with tube and fittings* NASC, London, 250pp. *The control of blemishes in concrete,* Appearance Matters Series No. 3: British Cement Association, Slough, 1981, 21pp.

Plain Formed Concrete Finishes, The Concrete Society Technical Report No. 52, The Concrete Society, Slough, 1999, 48pp. (Catalogue ID TR52)

National Structural Concrete Specification for Building Construction, 3rd ed.: CON-STRUCT, Camberley, 2004, 76pp. (Ref: CS 152)

Concrete pressure on formwork, CIRIA, London, 1985, 32pp. (CIRIA Report 108)

Final report of the Advisory Committee on Falsework, Her Majesty's Stationery Office, London, 1975, 151pp.

Formwork striking times—criteria, prediction and methods of assessment: CIRIA, London, 1995, 60pp. (CIRIA Report R136)

Controlled Permeability Formwork: CIRIA, London, 2000, 100pp. (CIRIA Report C511)

Friction Resistance in Temporary Works: P.P. Pallett *et al.,* in Concrete, 36(6), 2002, Crowthorne, 12–15.

European Developments in concrete release agents: I. Hart, in Concrete, 34(3), 2000, Crowthorne, 18–21.

Guide to Flat Slab Formwork and Falsework: The Concrete Society, 2003, Crowthorne, 160pp.

Early age strength assessment of concrete on site: Best practice guides for in-situ concrete frame buildings, European Concrete Building Project, 2000, 4pp.

Checklist for Erecting and dismantling Falsework: The Concrete Society, Crowthorne, November 1999, 24pp, (Catalogue ID: 123).

Permanent formwork in construction: Joint Report CIRIA—Concrete Society, London, 2001, 175pp (CIRIA Report C558).

The designer's guide to the use of Expamet Hy-Rib, 2nd ed.,: Expamet, Hartlepool, 2002, 36pp.

Operator's code of practice for PASMA Aluminium Alloy Towers, 12th rev.: Leeds, 2005, 28pp.

Preventing Falls in Scaffolding and Falsework: SG4:05, NASC, London, 2005, 46pp.

User Guide to SG4:05 Preventing Falls in Scaffolding and falsework: in *SG4:You,* NASC, London, 2006, 52pp.

A guide to practical scaffolding, 5th ed.: CITB Publication No CE 509, Kings Lynn, 2005, 146pp.

Structural renovation of traditional buildings, CIRIA, London, 1986, 99pp (reprinted with corrections 1990). (CIRIA Report 111)

Retention of Masonry Facades—best practice guide: CIRIA, London, 2003, 332pp (CIRIA Report C579).

Construction joints and stop ends with Hy-Rib and ggbs concretes: P.F. Pallet, in *Concrete,* 34(10), The Concrete Society, Crowthorne 2000, 37–40.

Checklist for the Assembly, Use and Striking of Formwork: The Concrete Society, Crowthorne, 2003, 28pp (Catalogue ID 144).

Retention of masonry facades—best practice site handbook, CIRIA, London, 2003, 86pp (Publication C589).

Anchorage Systems for Scaffolding: 'NASC TG4:04', Construction Fixings Association, 2004, Sheffield, 8pp.

Health and safety in construction, 3rd ed.: HSE, HSE Books, Sudbury, 2006, 104pp (Series no. HSG 150).

Formwork for modern, efficient concrete construction: BRE, BRE Press, Bracknell, 2007, 24pp (Report 495).

Chapter 12   REINFORCEMENT

Chris Shaw and Neil Henderson

Basic requirements for good practice

- **Correct specification**

- **Cut bend and fix accurately**

- **Reinforcement to be free from loose scale or rust**

- **Store in dry condition if on site for prolonged period**

- **Welded steel fabric sheets to be sensibly flat**

12.1   General

Steel reinforcement is used in concrete to fulfil a number of functions all of which are essential to the efficiency of the member or structure. It is essential also that the design concept and criteria are maintained during construction. This being achieved by observance of the structural detail drawings and the establishment of a query and answer system with the structural designer.

All formwork and reinforcement must be checked by a competent engineer at critical stages during construction and steel-fixing, particularly before the formwork is completed and also immediately prior to the placing of concrete. It is essential that the checking engineer has practical and theoretical knowledge and experience of the type of work that he is examining so that he can detect discrepancies and their significance to the strength and function of the structure.

12.2   Details, drawings and schedules

12.2.1   Codes and Standards

Important changes have taken place in the scheduling of reinforcement. British Standard 8666: 2005 *Scheduling, Dimensioning, Bending and Cutting of Steel Reinforcement for concrete – specification*, replaced previous standards. There are significant differences between the new and previous standards, which can be summarised as follows:

- Notation: changes in the designation of the type and grade of steel reinforcement.
- Shape Codes: these have been increased in number to 35.
- Minimum diameters: of bending formers now vary according to the bar size.

Table 12.1   Notation used in BS 8666: 205 for type and grade of reinforcement.

Notation	Grade
H	Grade B500A, Grade B500B or Grade B500C conforming to BS 4449: 2005
A	Grade B500A conforming to BS 4449: 2005
B	Grade B500B or Grade B500C conforming to BS 4449: 2005
C	Grade B500C conforming to BS 4449: 2005
S	A specified grade and type of ribbed stainless steel conforming to BS 6744: 2001
X	Reinforcement of a type not included in the above list having material properties that are defined in the design or contract specification

Note: In the Grade description B 500 A etc., B indicates reinforcing steel.

The notation used in BS 8666 for the various grades of reinforcement is shown in Table 12.1.

A complete set of reinforcement drawings is to be available, together with the relevant bending schedules or computer print out sheets. Other reference documents which should be on site are:

- BS 8666: 2005 which gives bending shapes and codes, tolerances, and procedures for scheduling on standard bending schedule sheets.
- BS 8110: 1997 the *Structural use of Concrete*, which gives details of material and workmanship requirements. The relevant British Standards for highway and maritime are BS 5400–4: 1990 and BS 6349–1: 2000, respectively. These British Standards are being replaced by Eurocodes.
- BS 7973: 2001 *Spacers and Chairs for Steel Reinforcement and their Specification, Parts 1 and 2*.
- Relevant company standing instructions or procedures, for the operational control of the work.

12.2.2   Design and detailing

Detailing is the preparation of a drawing showing where reinforcement is to be placed in a structure. The reinforcement details must be checked to ensure that they allow for tolerances in the reinforcement, cover and for services, pockets, architectural features and all types of joints shown on the contract drawings.

12.2.3   Drawings and bending schedules

It is essential that site engineers, foremen and steel-fixers are familiar with the reinforcement detail drawings and schedules and that the significance of reinforcement in the various parts is understood e.g. main structural, distribution, anticrack, handling and stiffening.

The following are basic checks that should be carried out. More detailed instructions may be in your project quality plan or issued locally—if in doubt contact your quality manager.

- Check that the specified spacers and chairs have been supplied to BS 7973: 2001, Part 1. Non-compliant products are sometimes supplied.
- Check that the spacers and chairs are in the correct positions in accordance with the requirements of BS 7973: 2001, Part 2, and that they are providing the specified cover to the reinforcement. The cover is normally provided to the reinforcement nearest the surface of the concrete, i.e. to the links in beams and columns. The failure to achieve the specified cover is the most commonly occuring defect in reinforced concrete.
- Check that positions of cast-in items do not clash with main reinforcement. Drainage fittings are always larger than you think. Fixings with hooked tangs cause considerable problems. With beam sides and walls, fittings are often attached to shutters before 'offering up' to reinforcement cages already fixed. Remedial refixing of inserts or reinforcement may involve two trades on site and disruption of one or other is not always notified or corrected. In addition, refixing to new positions may be difficult and costly.
- Check that the designer has properly specified fixings and inserts and not just shown: 'Hilti type', 'Leibig type', 'Rawlplug type' etc. Make sure the designer has used the catalogue to get the correct type, length, fitting, material etc. and these are all adequately defined.
- Holding-down bolts for steelwork—fully appreciate the size of these complete with their associated bolt boxes/cones, plate washers etc. Ensure that the correct height is left projecting above the slab or foundation.

Ensure sufficient grouting space is left under column base plates. The positions of holding-down bolts relative to the structure and to each other are required to be set particularly accurately. Holding-down bolts must also be secured from movement during the concreting operation. It is necessary to locate the bolts by using a template which is normally made of plywood. Welding the reinforcement to ensure no movement during the placing of concrete must only be carried out when it is specifically permitted in the design.

- Check the reinforcement detailers' requirements where holding-down bolts, particularly in parapets, clash with reinforcement links.
- Check holding-down bolt 'clusters' are not deformed or bent and that the threads are not damaged prior to concreting.

Welding of reinforcement should be avoided wherever possible. Written permission must be obtained prior to welding, and welding must be carried out under carefully controlled conditions. Clause 1717 in the *Manual of Contract Documents for Highway Works* (Volume 1) and BA 40/93 of the *Design Manual for Roads and Bridges* gives advice. Sacrificial reinforcement may be used for welding purposes to support temporary works or bolt clusters (see also Chapter 14).

Note that the specified cover to the reinforcement is normally to the bar or welded steel fabric nearest to the surface of the concrete.

When a large area of slab is detailed over two drawings, check that no reinforcement has been left out on the 'join' line (especially when each drawing has been prepared by a different detailer). Check that the reinforcement fits, especially on slabs. Summate bar lengths, taking into account laps and cover and compare against the slab dimensions given on the concrete general arrangement drawings.

On reinforcement bending schedules:

- check that the job or contract title is stated
- check that the section, area or member is stated
- check that the correct revision letter is stated
- check that 'A' and 'B' dimensions are not transposed in error
- check that 'A' and 'B' dimensions for sheets of welded steel fabric reinforcement are not transposed in error (much wastage occurs when this happens in the case of 'rectangular mesh' sheets).

Steel fixers may be paid at a rate per diameter of reinforcement or for the tonnage stated on the bending schedule. Any other work, rebending, supplying and fixing additional reinforcement required but not scheduled, may be at daywork rates. Therefore, errors and omissions are costly and cause delays.

Check the critical dimensions of certain bars on the schedule, particularly links and 'U' bars with concrete faces either side. Ensure negative bending tolerances have been allowed for on the schedule. If necessary have bars re-scheduled to take account of negative tolerances. 'Flexible' detailing should always be used whenever possible.

The conversion of the detailers' two-dimensional information into a three-dimensional assembly of reinforcement can be critical in cantilevers and around columns in flat slab design. Steel fixers must use chairs to BS 7973: 2001 (usually continuous ones), and if there is no adequate supervision to ensure that top reinforcement is fixed at the correct effective depth required by the designer in critical areas, then potential problems are 'concreted in'.

Spacers and chairs should be specified by the designer in accordance with BS 7973: 2001, and specified on the reinforcement drawings. The chair height should take account of areas of slab with varying diameters of bottom reinforcement to support the chair and the varying diameters of top reinforcement the chair has to support. This can be rationalised in areas of less significance to reduce number of different height chairs.

In deep slabs it is good practice to detail a 'rider bar' on top of chairs to provide improved rigidity.

Check that the detailer has drawn out heavily reinforced column to beam junctions to a large scale before scheduling the reinforcement. 'Flexible' detailing should be used.

Look over the reinforcement detailing before the fixing starts and watch out for the following:

- Ribbed slabs—bars lapped in narrow ribs cause congestion and difficulties in achieving correct cover and adequate space for concreting.
- Heavily reinforced slabs—lapped reinforcement causes problems as in the note above.
- Two layers of reinforcement at right angles in one face lapped at same place causes problems as above, only doubled.
- Heavy slab reinforcement all curtailed in one line causes potential shrinkage crack points and provides points of sudden stress change in slabs when fully loaded. Check with the designer to stagger the ends of bars.

- Check columns and plan sections with gridlines in two directions. Square columns with asymmetric reinforcement must be cast with reinforcement fixed the correct way round.

Reinforcement checks prior to fixing

- Check bar dimensions.
- Prepare clear sketches for steel fixers.
- Check critical bar dimensions prior to fixing.
- Particular attention should be paid to the closed dimension of bars that fit a confined space e.g. Bearing/plinth links, capping 'U' bars.
- Plan storage—ensure a suitable area is chosen.
- Double check setting out for reinforcement using simple checks—e.g. measure up off the kicker, measure up off the base, and compare to previously poured area.
- Always double check the kicker, before, during and after pouring! This is particularly important where features are present.
- Spacer and chairs must be to BS 7973: 2001, Part 1, and of a single cover value to reduce the chances of misplacing.
- Aiming to set the blinding 5 mm low gives away structural concrete but avoids the cover problems associated with high blinding.
- Ends of bars are to be checked to ensure that they are not deformed due to bad cutting. Any bar which is deformed and encroaching into the cover area should be cropped back to give the required cover whilst still retaining any required lap.
- Double check that the specified spacers and chairs have been used—measure on site—don't take people's word for it. In particular, check that only single cover spacers are used.
- Do not cut or bend any pre-bent bars. Something is wrong if you are having to cut or rebend bent bars. If not sure, stop and resolve problem properly. Don't steal steel!
- Remember that when using couplers, this can lead to problems with cover as couplers are invariably thicker than rebar—make sure that the couplers fit!
- If using couplers, make sure that the coupler is fitted properly, the rebar is prepared properly and of the correct length to fit into the coupler.
- Check that link sizes fit and check that the links actually delivered are the correct size.

The CARES mark, 5 ribs, 1 rib and 2 ribs = barmark 5-12

Figure 12.1   CARES bar marks. The dot-dash-dot (shown on the left) is the CARES registered certification mark (number 2150684) and indicates that the product is CARES approved. The number of ribs between the next two dots after the CARES approval mark, indicate the country of origin. The number of ribs between the next two dots indicates the steel mill number. Details of steel mill numbers can be found in either the CARES certificate or the list of 'Firms Holding CARES Certificates of Approval'. Further details or confirmation of barmarks can be obtained from the CARES office. (Courtesy UK Certification Authority for Reinforcing Steels).

12.3 Delivery

12.3.1 Approved list

In the UK all the approved suppliers will normally be members of CARES. This is the UK Certification Authority for Reinforcing Steels, the third party certification body carrying out product conformity certification to British Standard Specifications.

Every reinforcing bar carries marks identifying 'CARES approved supplier', UK manufacture and mill number (see Fig. 12.1). Check the reinforcement on arrival and look for these marks. Full details of the marking system can be obtained from your quality manager or buying office.

12.3.2 Ordering

The process is logical:

- Define the order of construction.
- Define what reinforcement is needed to suit the order of construction.
- Call off the required bars from the bending schedules to suit the order of construction.
- Do not order reinforcement too early in a contract, if left lying around on site it may get too rusty or become contaminated with mud, oil etc.
- Ensure that the most up to date schedule has been used to order reinforcement.

For example, on a confined site where there is only room to store reinforcement for half a floor slab, it will be necessary to call up only the bars needed for each half of the slab. Don't forget to include the lap bars!

12.3.3 Supply

Cut and bent on site

Reinforcement can be delivered to site in stock lengths of 12 m and cut and bent on site. This system may be adopted:

- Where the site is too far from a cut and bend supplier.
- On large contracts where it may be more economical to buy bulk orders of reinforcement from the manufacturer.
- On some large civil engineering sites.

Delivered to site ready cut and bent

Specialist firms supply ready cut and bent reinforcement direct to site. This has become normal practice. The manufacturers cut and bend the reinforcement using computer-controlled machines. The reinforcement is bundled into groups of bars with the same number and labelled with the schedule number and Bar (or fabric) Mark.

12.3.4 Labelling

Whether bars are cut and bent on or off site they must all be labelled to identify them. The information required on the label is at least the *Bar (or fabric)Mark* and its schedule and drawing number. The diameter, shape code and location in the structure should also be given (see Fig. 12.2).

It is good practice to have spare labels on site and to label bundles of reinforcement for different areas to ease identification later.

12.3.5 Storage

The reinforcement must be stored on site in the following way:

* off the ground on sleepers, over hardstanding (concrete or hardcore)
* protected from dirt
* protected from the effects of the climate if it is to be stored for any length of time
* in such a manner that bars can be located when they are required without undue difficulty or delay
* in such a position that bundles may be easily lifted out by crane.

To ensure that location of the bars is possible, liaison may be necessary with the supplier to ensure the bars are delivered in the correct order. This is particularly important on very confined sites where reinforcement may be stacked on top of other reinforcement.
Note: Good storage achieves minimum handling. Thereby it reduces costs by reducing double and triple handling.

12.3.6 Delivery checks

Is the reinforcement visually as detailed and scheduled? For example:

* High bond deformed bar (high yield or high tensile) prefix H, A, B or C on drawings and delivery tags.
* Welded Steel fabric composed of ribbed high tensile wire (see also Table 12.4).
* Check that reinforcement is correct against the bending schedule in terms of both numbers and shapes. If discrepancies are not spotted at delivery stage, delays or extra costs will be incurred.
* Bars with two or more bends should be measured to ensure they have been bent correctly and within tolerance.

It is good practice to keep a copy of the reinforcement schedules and use these to record delivery dates and details of split bundles.

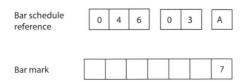

Figure 12.2   Form of bar or welded steel fabric label shown in BS 8666: 2005.

12.4   Checks before fixing reinforcement

12.4.1   Condition of formwork

Check that the formwork decking is ready to receive the loading from the reinforcement fixing operations and any statutory or contractual clearance for loading
has been received by the supervisor responsible. It is normal practice to have a
temporary works checker on site who will authorise loading.

The points mentioned in Chapter 11 (Formwork, Scaffolding, Falsework and
Façade Retention) should be checked when the formwork is clear and before reinforcement fixing operations commence. Delays and extra costs will occur if these
matters are left until the formwork is covered in a mass of reinforcement.

12.4.2   Condition of reinforcement

In accordance with good practice, light surface rust is acceptable, providing there is
no loss of section. Loose scale or flake must be removed and the steel checked for
shape, diameter, cross-section area and bond. In some cases additional reinforcement may be required. Bars that are pitted should be set aside until an assessment
of the degree of pitting related to the structural properties of the bar can be made.
Reinforcement which has scaling, laminations, splits or pitting on delivery should
be returned to the supplier.

Reinforcement is not to be coated with any material likely to affect the bond
adversely or to attack the concrete. Special details should be provided by the
designer for cases where the bond is to be eliminated, as in some types of joint.

12.4.3   Accuracy in bending

The bending dimensions and shapes (irrespective of where cut and bent) are to be
in accordance with the bending schedules and will normally be to the preferred
shapes as detailed in BS 8666. This is especially important in the case of links or
any other shape which directly controls the amount of concrete cover. Cutting and
bending tolerances are shown in Table 12.2. Bars greater than 16 mm size cannot
be 'adjusted' by hand during fixing. Attempts to do so are expensive and often only
succeed in distorting the formwork. Where bar shapes are intended to lie in more
than one plane, these planes should be mutually perpendicular.

Checks should be made to ensure that the critical dimension required to fix a bar
accurately has not been left as the 'run-off' in bending. Negative bending tolerances can seriously affect the ability to fix bars within given dimensions.

12.4.4   Services

Ensure that all service requirements and other openings and inserts are accounted
for. Never move or cut bars to allow passage for a pipe or similar detail. Refer back
to the designer.

Table 12.2 Cutting and bending tolerances.

Cutting and bending processes	Tolerance (mm)
Cutting of straight lengths (including reinforcement for subsequent bending)	+ 25, − 25
Bending:	
≤ 1000 mm	+ 5, − 5
> 1000 mm to ≤ 2000 mm	+ 5, − 10
> 2000 mm	+ 5, − 25
Length of wires in fabric	± 25 or 0.5% of the length (whichever is greater)

Note: For shapes with straight and curved lengths (e.g. shape codes 13 and 33) the largest practical radius for the production of a continuous curve is 200 mm and for larger radii the curve may be produced by a series of short straighter sections. The radius of bending for different bar sizes is given in BS 8666: 2005, Table 6. Tolerances for shape code 01, stock lengths, are subject to the relevant product standard e.g. BS 4449: 2005.

Where reinforcement has to be cut to allow for the inclusion of service pipes, anti-cracking bars should be fixed around the sides of the hole on both faces of concrete. Approval should always be sought prior to any cutting of reinforcement.

12.4.5 Joints and dowel bars

Movement and construction joints should be positioned and located on the drawings, and the reinforcement detailing should take these into account. A separate detail should be shown for a daywork joint, together with suggested 'best' positions. Any other positions required must be agreed with the designer before reinforcement fixing commences.

Dowel bars are often provided between two adjacent members with no other reinforcement. This invariably occurs at movement joints. Hence it is necessary for the dowel bars to be able to slide within the concrete. To allow this movement to take place, the dowel bars must be parallel to each other and parallel to the direction of movement.

De-bonding agents and ferrules on the bars must be used as specified. Dowel bars at movement joints are always plain round bars. Their ends should be sawn and not cropped as in most bar cutting machines. Dowels should be sufficiently fixed to prevent their movement and maintain their alignment during pouring of concrete. Tolerances for dowel bars are given in Clause 1011 of the *Manual of Contract Documents for Highway Works* (Volume 1) or as specified by the designer.

12.4.6 Reinforcement from previous work

Ensure that all starter and continuity reinforcement from previous lifts or sections of construction is correctly located. Any discrepancies should be referred back to the designer who will advise on remedial work where required. Also check lap lengths. Starter bars should be counted to check that the correct number have been fixed.

Bars protruding from previous pours should be cleaned, preferably while the concrete from the previous pour is still green, prior to fixing further bars.

12.4.7   Applied heat and welding

Reinforcement must not be heated or welded unless previously agreed with the designer.

12.5   Fixing reinforcement

12.5.1   Starters from previous work

Check that these are sound and in the correct position. Proprietary prefabricated cages are sometimes specified. A wide range of standard units are available covering high yield bars in the sizes 10, 12 and 16 mm which will produce allied rebates in the concrete up to 50 mm deep and 240 mm wide. Various special plastic strips cover such requirements as curved joints in slip-formed construction, starter bars for circular towers, connections to diaphragm walls and secant piling and bar anchorages for various cantilevered constructions. In all of these cases, the size and shape of both the reinforcement and removable plastic can be varied.

12.5.2   Patent mechanical connectors or splices

Many systems use sleeves that are of larger external diameter than the nominal bar diameter and it is necessary to consider the cover requirements. In most cases, the size space required for a single sleeve seldom exceeds the diameter of transverse reinforcement or links, so that special attention to the cover is not needed, provided the links are repositioned as necessary.

The types of connection fall into the following main groups:

Group A

For bars in tension or compression:

- threaded couplers
- swaged couplers
- swaged and threaded couplers
- metal grouted couplers
- welded joints
- bolted couplers—where nuts shear off—e.g. 'Ancon' type.

Group B

For bars in compression only:

- wedge sleeves
- bolted sleeves.

For examples of typical reinforcement connectors, see Fig. 12.3.

Agree all details with the designer before commencing work. This work should only be done by trained operatives following approved work instructions and using calibrated equipment.

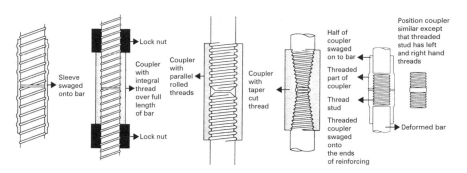

Figure 12.3 Typical reinforcement connectors.

12.5.3 Spacers, chairs and saddles

Before fixing of the reinforcement starts, ensure that sufficient quantities of spacers and steel wire chairs to BS 7973: 2001 are on site. Note that site-made spacers are prohibited (see Fig. 12.4). Special supports (Shape Code 98 in BS 8666: 2005) may be required for very thick slabs or foundations.

The main function of these is to ensure that reinforcement remains in its intended position, laterally and vertically throughout all subsequent stages of construction, particularly the pouring and vibrating of concrete. The most demanding application is for horizontal reinforcement as, in this case, the weight of the reinforcement has to be supported together with the weight of operatives, walkways and barrow runs.

Spacers and chairs must comply with the requirements of BS 7973: 2001, Part 1. Tables 1 and 2 give spacer and chair categories and design loadings. Spacers and chairs must be used in accordance with the requirements of BS 7973: 2001, Part 2. Figures 1 to 9 show layouts for slabs, beams, columns and walls, and can also be used for other members such as foundations.

Plastic spacers to BS 7973: 2001, Part 1 are generally suitable for all types of buildings. Cementitious spacers are used for supporting bars of 25 mm size and above, and where the surface of the concrete is subject to abrasion etc., for example the seaward face of a sea wall.

A further consideration is that the spacing of spacers may need to be significantly closer than the maximum centres given in BS 7973 in order to avoid applying high concentrated loads on the formwork which in turn could result in indentation or deformation of the surface of the formwork or overloading of the spacer.

Whichever type of spacer is used, it should provide the cover specified and be firmly held in position: clip-on types should be used only on bars of the correct size within the claimed range of the clip.

A particular problem is the lateral displacement of bars in narrow ribs in ribbed slab construction. To prevent this, special recessed saddle spacers are available.

Where cementitious spacers are being used, ensure no cracks are visible in them after placing reinforcement, particularly in bases or soffits of water-retaining structures.

Single cover 'A' spacer

This is a plastic 'A' spacer. It is 'state of the art' in plastic spacers, and the result of many years of development and use on actual projects. It is used for most purposes in building including foundations, columns, beams, slabs, and walls. It is designed for use with conventional formwork and 8 mm up to 20 mm size reinforcement. The spacer clips on to the reinforcement and overall is the most cost effective option because it does not need tying on with wire. The labour used in tying is expensive. It is manufactured for covers of 20, 25, 30, 40, 50, and 90 mm.

Soft substrate 'A' spacer

Where a plastic 'A' spacer has to rest on a soft substrate (i.e. not on conventional formwork) such as thermal insulation or a polythene damp proof membrane a spreader base is used. This clips to the base of the 'A' spacer and spreads the load carried by the spacer onto the insulation or polythene. The spreader base can also be used in vertical applications such as basement walls where, for example, a reinforced concrete wall is cast against a waterproof membrane, such as 'Sheetseal' or 'Bituthene'.

Soft formwork 'A' spacer

Soft formwork spacers are used where the formwork (which is usually permanent formwork for foundations) is made of polystyrene, cellular plastic or similar materials. The spacers are fixed to the links and reinforcement in foundation beams, pilecaps etc. They are manufactured for covers of 40 and 50 mm and bar sizes of 10, 12 and 16 mm.

End Spacers

End spacers are used at the ends of the wires of welded steel fabric and reinforcing bars to ensure the correct end cover. The left hand one shown in the picture is for the wires of the smaller sizes of welded steel fabric. The right hand one is for the wires of larger size welded steel fabric and for the smaller sizes of reinforcing bars. End spacers for the larger sizes of reinforcing bar are not currently available. In this situation the ends of the bars need to be bent through 90° and an 'A' spacer fixed to the short straight section at the end of the bar. End spacers also act as an 'anti-hazard' device until the concrete is poured.

Circular Spacers

Circular (or 'wheel') spacers have been used in the past, mainly on vertical concrete members such as walls and columns. However, they contain more plastic than is necessary for a single cover spacer and are therefore not a good use of resources. The 'A' spacer provides the same cover more efficiently, so circular spacers are normally no longer needed.

Pile Cage Former for CFA and Bored Piles

The pile cage former is used for continuous flight auger (CFA) and bored piles. It combines the functions of the spacer with two plastic rings to position the longitudinal reinforcement for the pile. The rings have pre-formed scallop shaped recesses on the inside edge for locating the main reinforcement, which can be up to 25 mm in size. The cover to the reinforcement provided by the spacers and rings is 75 mm. The spacer is used for piles from 300 mm to 450 mm in diameter.

CEMENTITIOUS SPACERS

Single cover cementitious spacer

Cementitious spacers, such as the one shown on the left, are used where the surface of the concrete may be subject to abrasion, e.g. in the seaward side of a sea wall. They require to be wired on to the reinforcement. The wire is traditionally 16 or 18 gauge soft iron wire but in marine environments stainless steel tying wire should be used. They are also used to provide cover to reinforcement of 25 mm size and above. This would include the end cover to the reinforcement at the edge of a slab or beam which should be bent at 90o to its longitudinal axis. Spacers made on site are not permitted in the Standard.

Cementitious line spacer

The spacers shown on the left are two types of cementitious line spacers. Either can be used to support reinforcement of 25 mm size and above. They are mainly used for bridge decks and heavily reinforced foundations.

They are manufactured in 1m lengths, and for covers of 25, 30, 40, 50, 60, 75, and 100 mm. They should be used in short lengths not exceeding about 350 mm, and the lengths should be staggered in plan on the formwork as shown in Figure 1(a) of BS 7973 -2: 2001. Spacers made on site are not permitted in the Standard.

WIRE CHAIRS

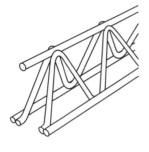

Lattice type continuous wire chair

There are two types of continuous wire chairs. One type is the lattice chair as shown on the left. The other type, the goalpost, is shown below. The Standard requires this type of continuous chair to be manufactured from three longitudinal steel wires of the same size in order to carry the design load. In practice, for a lattice support wire spacing of 200 mm the top wire needs to be of 5 mm size. Beware of non-compliant chairs with wires of a significantly lesser size. The chairs are manufactured in heights from 50 mm to 200 mm. Greater heights and larger wire sizes are also available upon request to the manufacturers.

Goalpost type continuous wire chair

On the left is shown a goalpost type continuous wire chair. The Standard requires this type of continuous chair to be manufactured from three longitudinal steel wires of the same size in order to carry the design load. In practice, for a goalpost support wire spacing of 100 mm the top wire needs to be of 3.5 mm size. Beware of non-compliant chairs with wires of a significantly lesser size. The chairs are manufactured in heights from 30 mm to 200 mm (standard duty) and up to 400 mm (heavy duty). Greater heights and larger wire sizes are also available upon request to the manufacturers.

Individual wire chair

Individual wire chairs can be used where there is no bottom reinforcement off which to support a continuous chair. For example, it can support the top reinforcement in a cantilever slab where there may be no bottom reinforcement. The Standard requires that the legs of the chair are cased in plastic protective tips for a distance of at least 40 mm where the chairs are required to support reinforcement off an exposed face, such as the underside of a cantilever slab. They are manufactured in heights from 75 mm up to 200 mm, and, with linked legs, from 175 mm up to 300 mm.

Figure 12.4   Examples of spacers and chairs.

12.5.4   Initial and periodic inspection

It is not sufficient to inspect the reinforcement at the completion of any section as some bars in deep or complicated formwork may be difficult to observe and any correction at that stage may involve dismantling the formwork or reinforcement cage. Collaboration with the steel-fixer at the start of a section sets the pattern and examination at subsequent intervals ensures that this is being achieved.

Reinforcement checks during fixing

- Ensure that the spacers and chairs are positioned in accordance with Figures 1 to 9 of BS 7973: 2001, Parts 1 and 2.
- Particularly ensure correct cover at kickers is maintained. This will avoid potentially insurmountable problems with future lifts.
- Always double check the kicker, before, during and after pouring! This is particularly important where features are present.
- The specified cover must be achieved to all bars nearest to the surface of the concrete prior to pouring the concrete.
- All ties on cementitious spacers should be tight: loose ties are not acceptable.
- Tying wire should go through the hole in the centre of the cementitious spacer.
- The ends of ties should be checked to ensure they don't encroach into the cover area.
- Use stainless steel wire on exposed faces if specified.
- Ensure all spacers and chairs are correctly positioned in accordance with Figures 1 to 9 of BS 7973: 2001, Part 2.

- Ensure lapped bars are tied at both ends to prevent one end becoming free and moving into the cover zone.
- Check lap lengths as work progresses.
- Where tape-ups have been given to the fixers on wall working, check correctness.
- Cracked, broken or deformed spacers and chairs are not acceptable. **Note**: if you are having to force in rebars using a crowbar, there is something wrong.
- Sufficient chairs must be used to ensure no movement of top reinforcement during concreting. See BS 7973: 2001, Parts 1 and 2.
- Ensure that reinforcement is tied in accordance with BS 7973: 2001, Part 2, section 5. (See BS 7973: 2001, Part 2, Figure 3.)
- Ensure that rebar is adequately supported where it cantilevers.
- Ensure that graded bars are placed correctly—very easy to get confused.
- Ensure that when placing bolt clusters, the correct cover is maintained to the rebar—it may be awkward but the cover must be maintained.

12.5.5 Additional reinforcement

Additional reinforcement may be required in some cases to stiffen or brace reinforcement cages during assembly, handling and the placing of the concrete. If the stiffening is to be temporary, timber or scaffold tubes can be used and removed later. If bracing is to be concreted in, it should be of the same specification as the designed reinforcement. 'Z' bars are often used to brace prefabricated beam reinforcement.

12.5.6 Prefabrication

The preliminary familiarisation with the drawings should indicate how much of the reinforcement can be prefabricated outside the formwork before lifting in and tying together. It may be found convenient to divide the work for this purpose between at or below ground, and above ground.

For work below ground, it is often found to be more convenient, especially where heavy bars are concerned, to build the reinforcement cage up *in situ*.

Where sections of reinforcement are to be prefabricated, it is important to check the orientation of bars.

12.5.7 Concrete cover

The reinforcement must be fixed with appropriate accuracy in the position specified by the designer to ensure proper structural performance, durability and fire resistance. Insufficient cover can result in corrosion of the reinforcement, leading to cracking and spalling of the concrete and eventually, if left unchecked, to a reduction in the cross-sectional area of the reinforcement and in the strength of the structure. The cover must also be sufficient to provide the required degree of fire resistance. Too little cover reduces the fire resistance; too much, unless precautionary measures are taken, can make the structure vulnerable to spalling, which again reduces the fire resistance. Incorrect positioning of the reinforcement can seriously reduce the load-carrying capacity of the structural member.

Unless otherwise specified, the concrete cover to the reinforcement should be within the following permissible deviations from the dimensions specified in BS 8110:

- bars up to 12 mm diameter: +5mm   −5mm
- bars over 12 mm up to 25 mm: +10 mm   −5mm
- bars over 25 mm: +15 mm   −5mm

For highway works, the actual cover achieved shall not be less than the required cover from the Explosive Class tables in BS 8500–1, and including any allowances for longer durability required under Clause A5 of BS 8500–1. The maximum cover achieved shall not be more than the nominal cover as defined in BS 8500–1, including the stated fixing tolerance Δ_c.

Normally, nominal cover is specified for which the above tolerances are acceptable. Where minimum cover is specified, ensure that this specification is complied with by providing spacer blocks to a size 5–10 mm over the minimum cover requirement. Further information can be found in the CIRIA Report (CIRIA C568, 2001) *Specifying, Detailing and Achieving Cover to Reinforcement* and the article *Cover to Reinforcement – Getting it Right*, in *The Structural Engineer*, volume 85, issue 4, pages 31–35.

12.5.8   Tying of reinforcement

The tying of reinforcement should comply with the requirements of BS 7973: 2001, Part 2, Section 5.

The most economic connection is the hand-made tie using 1.6 mm diameter (16 SWG) black annealed tying wire, although sometimes stainless steel wire is specified. It is essential that the ends of all ties are turned inwards away from the formwork.

It is essential to check which type of tie wire is required on an element of work. Generally, exposed faces require stainless steel wire and buried faces black annealed wire.

12.5.9   Congestion at splices

It is sometimes necessary, in order to avoid congestion at splice positions in columns or walls, for bars to extend above the formwork some 3m or so. These bars must not be allowed to wave about as this movement can crack the green concrete. Neither should they be temporarily bent down, as subsequent straightening of these bars will damage the newly cured concrete, and is sure to damage the bar by leaving a permanent kink. Long protruding bars should be supported by a type of scaffold. Congestion at splices may also be reduced by the use of patent connectors (see manufacturers' data sheets for further details).

12.5.10   Cleanliness

The general cleanliness of the works should be attended to, particularly the bottoms of beams, columns and walls. Start clean and clean out at intervals as required. Access doors should be built into the formwork where required.

12.6   Checks on completion of fixing and before concreting

Use a concrete pre-placement check and inspection sheet (see Fig. 12.5 for a standard form).

Reinforcement checks after fixing

- Check that the spacers and chairs comply with BS 7973: 2001, Part 2, Figures 1 to 9.
- Check that the gap between the reinforcement will allow concrete to fill the formwork.
- It is vitally important that the correct concrete cover is available at kickers otherwise there will be insurmountable problems when it comes to progressing the next lift. Always double check the kicker, before, during and after pouring! This is particularly important where features are present.
- The pre-pour check should pay special attention to the security of cast-in items and that the ends of the tying wires do not project into the cover zone.
- Any loose tying wire to be removed from the rebar.
- Use a timber of the correct size between the rebar and the formwork to check that there is the correct cover.
- Change concrete aggregate size if necessary, where steel is congested. Seek approval first.
- Carry out the checks listed in the Inspection and Test Plan.
- Check reinforcement is fixed properly around openings. Other trades will loosen/ cut ties to insert service pipes on temporary works ties etc.
- Do a final check on each of the following:

  - Ensure that there is sufficient lap protruding from kickers to tie in with the next lift.
  - Are the number, spacing and diameter of bars correct?
  - Is the cover to bars correct?
  - Are the splice/lap lengths correct?
  - Is the alignment of the bars correct?
  - Are the bars secured adequately with the correct type of tying wire? (Stainless steel wire?)
  - Are the spacers and chairs in accordance with BS 7973: 2001?
  - Is the reinforcement clean?
  - During concreting, dip through concrete with nail or similar object to check the cover to the top reinforcement. Ensure there are no localised dips in the hand-finished concrete surface.
  - Check that the kicker does not move during the concrete pour and ensure the correct cover to the rebar in the kicker before, during and after pouring.

Remember: All structural concrete must be checked with a cover meter after pouring, so ensure that the rebar is correct prior to pouring. If the rebar is displaced during concreting to afford access, then you must ensure that the rebar is put back to its correct position prior to finishing the concrete pour. If the requirements of BS 7973: 2001 are followed the cover will be correct.

Reinforcement

Type of steel, its location, cover, diameter and spacing to be in accordance with drawings and specification. Special attention should be given to the support of, and cover to, top reinforcement in a slab, beam, and especially a cantilever!

Laps and splices

Laps to bars, and splice bars between prefabricated sections to lie in correct position and plane. This can affect space available for the passage of concrete and vibrator.

Rigidity of reinforcement

To be adequate when considering placing and compaction of concrete. In slabs use additional numbers of spacers and chairs especially in those areas where walkways or barrow runs are planned.

Cleanliness

No mould oil or paint on bars. No mud or mortar from masonry operations. Cement grout, if clean and difficult to remove may remain. No water in bottom of form-work. All debris, particularly tying wire clippings to be removed from the formwork as they cause rust staining.

Column/wall starters

Use temporary links or wooden templates on column starters.

Column cranks

Cranks in vertical column reinforcement should have some links in their length to maintain cranks in their correct alignment. 'Flexible' detailing should be used in preference to cranted bars.

Surplus bars

Wherever possible check on any surplus reinforcement.

Concrete dimensions

Check overall dimensions, wall and slab thicknesses, and column cross-sections.

Props, struts and braces

These to be adequate for the size and span of members, and method of placing the concrete.

Distortion

Correct any displaced bars, distorted cages or formwork.

Day-work joints

Check that daywork joints are placed in the best position and firmly fixed.

Vibrator

Ensure vibrator is ready for use and its size suits the spacing of reinforcement, and that it works before placing the concrete!

12.7 Checks on completion of concreting

Starters

Positions and lengths of starter continuity bars: if these are to be exposed for any length of time, protect with a suitable covering such as cement grout, lean mix concrete or as instructed in the specification.

Finally check plumb of wall and column formwork.

Note: *Having taken all this care—ensure that the concrete is properly protected and cured.*

The cover can be checked with a cover meter after each pour to ensure that problems are discovered at the outset. Problems are very unlikely if the cover has been done in accordance with the requirements of BS 7973: 2001.

12.8 Alternative forms of reinforcement

12.8.1 Ferrous

Fusion-bonded epoxy coated rebar (FBECR)

FBECR is manufactured by bonding an epoxy, powder-based coating, under factory controlled conditions and decontaminated bars at high temperatures. FBECR is used to improve the corrosion resistance of reinforcement.

The following should be considered when using FBECR:

- The bending of already coated bars may result in damaged or potentially weak areas of the coating on the bend if the appropriate powder properties have not been selected. The fluidised bed dipping technique of coating may overcome this problem if appropriate reinforcement configurations can be produced prior to construction, however, transportation of the coated bars may be more difficult.
- Coating type and thickness will need to be selected with respect to the aggressiveness of the environment.
- Additional care and site supervision will be necessary, as special measures are required for handling and installation of the bars to prevent damage. Repair procedures for bars damaged in handling and installation will need to be established.
- Structural detailing will need to be modified to account for the different bond characteristics of FBECR, i.e. longer anchorage and lap lengths.

Galvanised reinforcement

The zinc coating used in galvanised steel is designed to act as a sacrificial coating by corroding preferentially (forming an anode) where steel is exposed. The corrosion products of the zinc migrate to the cathodic area, formed by the exposed steel to form a 'protective' barrier. Zinc coatings remain passive in carbonated concrete and the rate of corrosion is lower than for uncoated steel. This makes galvanised steel reinforcement suitable for use in concrete that is at risk from carbonation. Zinc is readily attacked by wet alkaline solutions, which occur naturally in concrete to produce zincates and hydrogen, and may result in poor bond characteristics and reduce the effectiveness of the coating. The galvanised reinforcement should be treated with a passivating coating before it is fixed in the formwork. The passivating coating prevents the zinc from reacting with the wet concrete, and may be applied in the factory during manufacture.

Stainless steel reinforcement

BS 6744: 2001, covers the specification of austenitic stainless steel as a reinforcement in concrete. Stainless steels comprise a family of alloys that derive their corrosion protection from the formation of a more stable, passive oxide film on their surface. The film is formed very rapidly by reaction of the alloying elements, principally chromium, in the steel with water and oxygen-bearing atmospheres. The durable nature of this protective surface film means that any corrosion attack that may occur under very aggressive conditions is usually in the form of pitting through the protective film, rather than general, overall attack found with conventional carbon steel. Reducing the risk of general corrosion means that expansive forces from the corrosion product's growth, which causes spalling of the concrete cover, are decreased.

Of the very wide range of possible alloys, only the following principal types have been investigated for their suitability as reinforcement for concrete:

• ferritic stainless steels
• austenitic stainless steels
• duplex stainless steels
• conventional reinforcing bars clad in stainless steel.

Austenitic and duplex steels offer a homogeneous solution to the problem of general corrosion and different grades have different degrees of corrosion resistance depending on their alloy contents. Even so, the type of stainless steel to be used needs to be chosen to suit the corrosive environment in question.

Stainless steel reinforcement is available in the form of plain and ribbed bars and as welded fabric and fibre reinforcements. These various forms of reinforcement can be used in the same way as conventional bars and fabric reinforcement.

12.8.2 Non-ferrous reinforcement

Non-ferrous reinforcements are generally fibre composites, and include glass fibre, carbon fibre or aramid fibres embedded in suitable resins. They are available in the form of rods or grids for use as reinforcement in concrete and as prestressed tendons.

	SITE:		CONTRACT No.:
CONCRETE PRE-PLACEMENT CHECK AND INSPECTION			POUR No.:
CONTRACTOR:		CONCRETE GRADE:	
LOCATION:		ADDITIVE:	
		TARGET SLUMP (mm):	VOL (m²)
		METHOD OF PLACEMENT:	
DRAWING NO.:		METHOD OF CURING:	

APPROVAL IS REQUESTED TO PLACE CONCRETE ON:

_____/_____/_____ of _____ hrs.

SIGNATURE: _____ (FOR CONTRACTOR)

DATE: _____/_____/_____

NB. 24 HOURS SHOULD BE GIVEN PRIOR TO CONCRETING

CHECKS	READ FOR INSPECTION		INSPECTION	DATE	REMARKS
	CONT.SIGN	DATE			
SETTING-OUT					
REINFORCEMENT CONTENT					
SPACERS AND CHAIRS					
REINFORCEMENT COVER					
JOINT DETAILS					
WATER BARS					
CLEANLINESS					
FORMWORK TIES					
BUILT IN ITEMS CIVIL					
M & E					

NOTES:

YOU ARE CLEARED TO PLACE THIS CONCRETE

SIGNATURE_____

(FOR CLIENT)

DATE _____ TIME_____

Figure 12.5 Concrete pre-placement check and inspection.

With the correct choice of fibre and resin, the composite material should be more durable than conventional reinforcement, particularly in a chloride environment. The composites have a higher ultimate strength (1500–2000 N/mm^2) than steel (300–700 N/mm^2), but the modulus of elasticity will generally be lower (approximately 50 kN/mm^2 for glass fibres and 150 kN/mm^2 for carbon fibres compared with approximately 200 kN/mm^2 for steel). However, the long-term durability of resin is still uncertain.

A comparison of the advantages and disadvantages of ferrous and non-ferrous reinforcement is shown in Table 12.3.

Table 12.3 Advantages and disadvantages of alternative forms of reinforcement.

Reinforcement type	Advantages	Disadvantages
Galvanised steel	Good performance in concrete at risk from carbonation	Poor performance in chloride-contaminated concrete
Fusion bonded epoxy coated steel	Good resistance to corrosion in chloride-contaminated concrete	Risk of damage to coating during handling, fixing and casting, leading to reduced durability
Stainless steel	Good resistance to corrosion in carbonated and chloride-contaminated concrete	Higher initial cost, although becoming more widely available.
Non-ferrous reinforcement	Good resistance to corrosion in chloride-contaminated concrete	Lack of long-term experience Lack of agreed Standards Limited supply. Cost.

12.8.3 Fibres

Fibres are generally based on polypropylene or steel and are added to the concrete at the batching plant or the ready-mixed concrete truck on site. In all cases, the fibres must be uniformly distributed throughout the mix. Sometimes, depending on the quantity and type of fibres added, adjustment to the basic concrete mix may be necessary to produce the workability and consistency required.

Fibres can be added to concrete at dosages as high as 60 kg/m^3 of concrete for steel fibres, and as low as 0.6 kg/m^3 for polypropylene fibres. Their main application is to control early-age cracking and improve impact resistance, toughness and abrasion.

Fibres are used to control early-age cracking, caused by high rates of shrinkage, which occurs in the first few hours after placing. They increase the tensile strain capacity of the concrete and reduce the frequency and size of cracks. Fibres for this application are invariably polypropylene and are added at 0.6–0.9 kg/m^3, depending on fibre diameter.

Impact resistance can be improved by the use of fibres. Both polypropylene and steel are used for this and addition rates 3–20 kg/m^3 for polypropylene and 20–40 kg/m^3 for steel are common.

The use of fibres in concrete can provide some residual strength by their ability to bridge across cracks as they develop. Fibres can delay the onset of fracture as concrete fails by crack propagation. Steel fibres are mainly used for this application but there are problems with achieving a good surface finish when using steel fibres in reinforced concrete. Although some polypropylene fibres have proved useful, relatively high addition rates are required.

The advantages of good quality durable concrete will be enhanced if the tendency to cracking can be reduced by the addition of fibres. In addition, the use of polypropylene fibres can improve the fire resistance of concrete.

Table 12.4 Standard fabric types and stock sheet size.

Fabric Reference	Longitudinal Wires			Cross Wires			Mass (kg/m²)
	Nominal wire size (mm)	Pitch (m²/m)	Area (mm)	Nominal wire size (mm)	Pitch (m²/m)	Area (mm)	
Square fabric							
A393	10	200	393	10	200	393	6.16
A252	8	200	252	8	200	252	3.95
A193	7	200	193	7	200	193	3.02
A142	6	200	142	6	200	142	2.22
Structural fabric							
B1131	12	100	1131	8	200	252	10.90
B 785	10	100	785	8	200	252	8.14
B 503	8	100	503	8	200	252	5.93
B 385	7	100	385	7	200	193	4.53
B 283	6	100	283	7	200	193	3.73
Long fabric							
C785	10	100	785	6	400	70.8	6.72
C636	9	100	636	6	400	70.8	5.55
C503	8	100	503	6	400	49	4.51
C385	7	100	385	6	400	49	3.58
C283	6	100	283	6	400	49	2.78
Wrapping fabric							
D98	5.0	200	98	5.0	200	98	1.54
D49	2.5	100	49	2.5	100	49	0.77

Note: Tolerances shall be in accordance with Table 4 of BS 8666: 2005. For standard welding fabric the type of wire shall be designated as a suffix to the fabric reference as illustrated in the example in Figure 3 of BS 8666: 2005. Standard lengths and widths shall be 4.8 m and 2.4 m respectively, giving a sheet area of 11.52 m².

Further reading

Standards and Codes of Practice

BS 4449: 2005 *Carbon Steel Bars for the Reinforcement of Concrete. Specification.*
BS 4483: 2005 *Steel fabric for the reinforcement of concrete. Specification.*

BS EN 1992-1-1: Eurocode 2 *Design of concrete structures*. Part 1: General rules for buildings.
BS EN 1992-2: Eurocode 2 *Design of concrete structures*. Part 2: Concrete bridges—design and detailing rules.
BS EN 1992-3: Eurocodes 2 *Design of concrete structures*. Part 3: Liquid retaining and containment structures.
BS 7973: 2001 *Spacers and chairs for steel reinforcement and their specification*.
BS 5896: 1980 *Specification for High Tensile Steel Wire and Strand for the Pre-stressing of Concrete*.
BS 6349–1: 2000 *Maritime Structures. Code of Practice for General Criteria*.
BS 6744: 2001 Stainless Steel Bar for the Reinforcement of and use in Concrete. Requirements and Test Methods.
BS 8110: 1985 to 1997 The Structural use of Concrete.
BS 8666: 2000 Specification for Scheduling, Dimensioning, Bending and Cutting of Steel Reinforcement for Concrete.

Other related texts

Cover to reinforcement—getting it right: IstructE, London, IstructE TSE Vol. 85 No. 4 2007, pp. 31–35.
Cover to reinforcement—getting it right: 6th International Congress on Concrete, Dundee, Application of Codes, Design and Regulations, 2005, pp. 147–154.
Standard Method of Detailing Structural Concrete: IStructE, 3rd Edition, London, IStructE, 2006.
Specification for Highway Works (Volume 2): Notes on Guidance on the Specification for Highway Works: HA, London, HA. (Date varies as Notes are updated.)
Economic Assembly of Reinforcement: a Review of Pre-fabricated Reinforcement and How it Results in Rapid On-site Installation of Reinforcement: D. Bennett *et al.*, BCA Crowthorne, UK, 1992.
Towards rationalising reinforcement for concrete structures: CS, Slough, UK, CS, 1999.
Coating Protection of Reinforcement: CEB, Lausanne, Thomas Telford, 1995.
Guidelines for the Use of Epoxy Coated Strand: PCI, Chicago, PCI, 1993.
Concrete on Sites (in 11 parts): BCA, Crowthorne UK, BCA, 1993.
Care and Treatment of Steel Reinforcement and Protection of Starter Bars: M. N. Bussell *et al.*, London, CIRIA Report R147, 1995.
Improving Development Characteristics of Reinforcing Bars: CERF, Washington, CERF, 1994.
Steel Reinforcement: J. Tubman, London, CIRIA Report SP 118, 1995.
Concrete Materials and Practice 6th ed.: L. J. Murdock *et al.*, London, Arnold, 1991.
Guidance for the Design of Steel-Fibre-Reinforced Concrete: CS, Camberley, CS, 2007
Guidance for the Use of Macro-Synthetic-Fibre-Reinforced Concrete: CS, Camberley, CS, 2007
Guide for the Design and Construction of Concrete Reinforced with FRP Bars: ACI, Farmington Hills, USA, ACI, 2001.
Specifying, Detailing and Achieving Cover to Reinforcement: E. S. King &, J. M. Dakin, London, CIRIA, 2001.
How Can We Get the Cover We Need? L. A. Clark *et al.*, London, IStructE TSE Vol. 75 No. 17, 1997.
Steel Reinforcing Bar Specifications in Old Structures: Concrete International, 1999.
The Product Certification Scheme for Steel for Reinforcement of Concrete: CARES Part 1, Sevenoaks, UK, 2004.
Manufacturing Process Routes for Reinforcing Steels: CARES Part 2, Sevenoaks, UK, 2004.
Properties of Reinforcing Steels: CARES Part 3, Sevenoaks, UK, 2004.
Fabrication of Reinforcement: CARES Part 4, Sevenoaks, UK, 2004.
Welded Fabric: CARES Part 5, Sevenoaks, UK, 2004.

Welded Pre-fabrication: CARES Part 6, Sevenoaks, UK, 2004.

Stainless Reinforcing Steels: CARES Part 7, Sevenoaks, UK, 2004.

Ancillary Products for Reinforced Concrete Construction: CARES Part 8, Sevenoaks, UK, 2004.

Information Technology in the Reinforcement Supply Chain: CARES Part 9, Sevenoaks, UK, 2004.

Note:

At the time of writing there has been a gradual change from British Standards to Eurocodes and this will need to be recognised in future work.

Chapter 13   CONCRETE

Chris Shaw and Neil Henderson

Requirements for best practice

- **Constituents: Adequate specification that clearly states performance or prescriptive**

- **Compaction: Method appropriate to the work in hand**

- **Cover: Accurate placing from rigid formwork to achieve specified value**

- **Curing: Appropriate method for the environment and the work in hand**

13.1 Materials

13.1.1 Basic requirements

First of all, *read the specified requirements* for the contract or project. The main controlling requirements for materials for concrete in the UK are given in:

- BS 8110: 1997 *The Structural use of Concrete*
- BS EN 206-1 *Concrete Part 1: Specification, performance, production and conformity*
- BS 8500 *Concrete*. Complementary British Standard to BS EN 206-1

Company standing instructions or procedures; general work instructions or the Quality Plan should give procedures to follow.

13.1.2 Cement

Cement should be obtained from a firm registered under the BSI Scheme for Firms of Assessed Capacity, unless other independent quality assurance procedures operate. However, note that the scheme does not cover blends of cement with fuel ash or ground granulated blast furnace slag (ggbs) made at the mixer, these being within the scope of the procedures applicable to the concrete mix as a whole. The term 'cement' includes the total cementitious content of a blend of materials.

The variety of common cement types is covered in BS EN 197-1. This Standard specifies several classes of Portland cement (PC) now available. The classes are designated 32,5; 42,5; 52,5 and 62,5 with the suffix N or R depending on the rate of early strength gain required for the concrete (N for normal strength gain and R for rapid strength gain). The strength classes commonly used by the UK ready-mixed concrete industry are 42,5 and 52,5.

To modify the characteristics of the concrete mix for technical or economic reasons, consideration should be given to the use of low heat or sulphate-resisting cements (SRPC) or cements blended with ggbs or fly ash.

It is wise to check the alkali content of all concrete, using the procedures given in BRE *Digest 330*. Cement certificates should contain data on alkali levels in cement.

In view of problems associated with the use of high-alumina cement (HAC) it is important that the guidelines set out in the Concrete Society Technical Report No. 46 *Calcium Silicate Cements in Construction—A Re-assessment*, are strictly adhered to. It should be noted that the building regulations ban the use of high-alumina cement for structural purposes although this material may be used for non-structural applications.

Note that cement is a hazardous material and is a well-known source of skin problems. Prolonged contact requires that suitable precautions are taken and protective clothing should be provided and used. There should be facilities for washing and changing clothes after work and for washing off dust, freshly mixed concrete and mortar.

If cement dust or mixture gets into the eyes, they should be washed out immediately with plenty of clean water. Suitable respiratory protective equipment should be worn during surface treatment of hardened concrete. For further detailed information see:

- HSE Construction Information Sheet No. 26, Revision 2
- HSE Guidance Notes: Environmental Hygiene (EH) EH 65/12, 1994

13.1.3 Additions

The two main forms of additions used in the UK are fly ash and ground granulated blast furnace slag (ggbs). These materials can be either inter-ground with Portland cement during manufacture to form a composite cement or they can be blended with Portland cement at the mixer, providing this is permitted by the specification. Other replacement materials such as silica fume or metakaolin may occasionally be specified.

When fly ash is considered as part of the total cementitious content of the concrete, it should conform to BS 3892-1. Likewise ggbs should conform to BS 6699. Again—check with the specification for your contract.

13.1.4 Aggregates

Coarse aggregate (greater than 4 mm particle size) is either gravel or crushed rock. Fine aggregate (less than 4 mm particle size) is sand or crushed rock. Lightweight aggregate is man made and is generally obtained from either clay or pulverized-fuel ash or pelletised slag. This is only to be used when specified. Recycled aggregates may also be specified

Inspect all aggregates visually before and during unloading for possible signs of poor grading, layer loading, and the presence of clay, shale or silt. Test and sample material for compliance with BS EN 12620 using PD 6682-1: *Aggregates–Part 1: Aggregates for concrete – guidance on the use of BS EN 12620*. Reject material with obvious defects at the time of delivery. If material has been discharged before deficiencies are observed, notify the supplier immediately. Quantity checks can be made at public or other weighbridges. All sampling and testing procedures are specified in the suite of BS 812 and BS EN 932 and BS EN 933 standards.

If marine dredged aggregates are used, give special attention to the chloride and shell content. Reputable suppliers will, on request, supply certificates that provide information on these values.

If in the position of having to choose aggregate, bear in mind the risk of subsequent deterioration of the concrete due to alkali–silica reaction (ASR). Damage from ASR can occur when alkalis (normally those present in Portland cement) react with certain forms of silica in the aggregate to form a gel which absorbs water, swells and exerts pressure that can cause the concrete to crack. In cases of doubt on any aspect of ASR seek further advice as necessary.

13.1.5 Water

Generally, water fit for drinking will be satisfactory for mixing concrete. If the quality of water from any source is suspect, sample and test in accordance with BS EN 1008: 2002 for impurities such as organic matter and dissolved salts before using. Samples should each be at least 2.5 litres in clean, uncontaminated containers. Totally immerse the closed container in the water and then remove the stopper. Before final filling, rinse the container several times, then completely fill and stopper firmly. Securely label to show source and date, and any other information considered relevant to variability in the supply. It will be necessary to take samples periodically if supplies are polluted suddenly, as for example by freak storms. Check sampling should be carried out immediately.

Do not use sea water for reinforced concrete as it will lead to corrosion of the steel reinforcement. If sea water is being used for non-reinforced concrete, it is necessary to add the chloride content to that of the aggregates if checking for ASR. For guidance on the quality of water for concrete see BS EN 1008: 2002.

13.1.6 Admixtures

Admixtures can enhance the properties of concrete both in the plastic state and in the final hardened state. There are five main types, namely water reducing agents, retarders, accelerators, air-entraining agents and superplasticisers. If admixtures are proposed, they should comply with BS EN 934-2. To aid compaction and finishing without loss of strength the use of water-reducing plasticiser may be beneficial. The use of excessive amounts of such admixtures should be avoided where the risk of unwanted or variable retardation could occur. Retarders, which increase the life of the fresh concrete, are of special benefit in hot weather conditions or when casting a large pour.

Permission to use admixtures usually requires the agreement of the designer and arrangements for obtaining this will normally be found in your contract specification. Confirmation must be obtained from the supplier that the admixture complies with the specification requirements.

13.1.7 Storage of materials

Pump bulk deliveries of cement or additions directly into silos. Take care that the specified type and correct quantity of material is delivered to the correct silo. Maintain silos in good weatherproof condition and clean filters regularly.

Store bagged cement in a weatherproof hut. If this is not available, store on a raised platform and cover with weatherproof sheeting. Use in strict rotation in the order of delivery, and discard if there is evidence of deterioration.

Store different sizes of aggregate separately on a slightly sloping hard standing with rigid partitions between the different sizes.

Store admixtures to avoid exposing them to extremes of temperature. Facilities might be needed to enable large drums of admixtures to be agitated and re-mixed prior to use. Ensure that drums of material are clearly identified.

13.2 Mix design

13.2.1 Types of specification

There are five methods of specifying concrete in BS 8500-1, namely:

- Designated concretes
- Designed concretes
- Prescribed concretes
- Standardized prescribed concretes
- Proprietary concretes

A *Designated Concrete* is produced in accordance with a specification given in BS 8500 which requires the concrete producer to hold current product conformity certification based on product testing and surveillance coupled with approval of his quality system to BS EN ISO 9001.

The *Designed Concrete* is the most commonly used. The purchaser is responsible for specifying the required strength and the producer is responsible for selecting the mix proportions to produce the required performance, subject to any restriction on materials, minimum free-water content ratio and any other properties required. Strength testing should form an essential part of the assessment of compliance with the specification.

For a *Prescribed Concrete* it is the purchaser who specifies the proportions of the constituents and is responsible for ensuring that these proportions will produce a concrete with the performance required. Even though the concrete performance has been specified and cube testing cannot be used to judge compliance, cubes are still required to be tested for compressive strength to ensure that quality is being maintained. A *Standardized prescribed concrete* is one selected from a list given in BS 8500 and made with a restricted range of materials.

Proprietary Concrete is concrete for which the producer assures the performance, subject to good practice in placing, compacting and curing, and for which the producer is not required to declare the composition.

13.2.2 Mixes required

The specification will normally list the specified strengths and maximum aggregate sizes, and may have a maximum water/cement ratio or minimum cement content specified. The other variable to specify is the workability (or consistence) of the concretes. Again, under the requirements of BS 8500, this is required to be specified by the purchaser. However, as consistence requirements will depend on the method and condition of placing, your company may choose to vary this.

Site personnel will not normally design mixes but may be involved in full-scale trial mixes. They should report any deficiencies in the mix, and make slight adjustments to mix proportions, as instructed by the mix designer. Refer to your company standing instructions or procedures or general work instructions, for procedures relating to material sampling testing and mix design.

13.3   Concrete production

13.3.1   Batching

The quality of the finished concrete is decided initially by the accuracy of batching materials. This accuracy requires the competent operation of well-calibrated plant. BS EN 206-1 requires that the accuracy of measuring equipment shall be within 3%. Your company standing instructions or procedures or general work instructions, will refer to the relevant operating and calibration procedures.

Weighing machines require regular attention and calibration to maintain their accuracy. The pointer of the weigh-dial must be adjusted to read zero when the hopper is empty. If it is found necessary to make frequent adjustment to the pointer, find out the cause and get it repaired. For volume batching, use a gauge box which is deep and narrow rather than wide and shallow: this gives less variation. Make allowance for bulking of damp sand. With a sharp sand, this can be as much as 25% with a moisture content of 5.0 to 6.0%. Hence, if no allowance is made, the mix may be undersanded by about 5% by weight of total aggregate.

13.3.2   Water

To achieve consistent quality of concrete, strict control of the quantity of mixing water is vital. This is particularly important when setting up a mixer on a new job or when the mixer is moved to a new location on site. Check calibrations of the water discharge must be carried out while the mixer is running.

Calculation of the water to add must allow for any free water in the aggregate, i.e.

Total water required = free water in the aggregate + water added

13.3.3   Mixing

- Ensure mixers with integral weigh hoppers are located on a firm level base and raised sufficiently on stands or supports to provide the required height to suit the site transport available.
- Ensure the weigh capsule is protected from damage by the spillage of aggregates building up around it.
- Check to ensure the full load of the hopper, with its ingredients, is carried directly on the capsule.
- Ensure that the weigh-hopper and dial are kept clean at all times.
- The sequence of loading the hopper is important. Start with coarse aggregate, particularly when cement is also loaded. This will result in a self-cleaning process avoiding a hardened layer material at the bottom of the hopper.

- *Do not* allow the hopper to drop heavily on the weigh capsule. *Always* check that the hoist ropes are slack to ensure that the loaded hopper bears on this capsule.
- *Do not* strike the hopper with a hammer or pick-handle. This leads to deterioration of the hopper and indicates that the shaker rollers are not properly adjusted.
- Check that the mixer is always being operated at the speed recommended by the manufacturer, since mixing times vary according to the mix and type of mixer.
- *Do not* discharge part batches in a number of small containers, e.g. wheelbarrows: this causes significant segregation. *Always* discharge the contents of a batch in one go.
- Protect the mixer drum, frame and axles etc. A wipe over with diesel oil followed by attention as necessary should keep the mixer free from adhering concrete and make it much easier to keep clean.
- Avoid an unsightly mess under the mixer. Two sheets of plywood or corrugated iron under the mixer drum catches and runs off all wet concrete droppings from the drum.
- When using batching plants, the procedure for calibration must be in accordance with the manufacturer's instructions and your company standing instructions or procedures.

13.3.4 Silos

- Ensur e substantial foundations, giving support over the entire base.
- *Do not* set up precariously balanced on oil drums or the like. The silo will be dangerously unstable and the silo will be grossly inaccurate.
- Ensure that silos are lifted only at lifting eyes provided or at the lifting trunnions at the centre of balance.
- Ensure that the height of the silo allows the dispensing hopper to clear and discharge into the mixer loading hopper at a suitable height.
- Ensure that the silo and mixer hopper are sited so that the pullway mechanism is kept clear of the hopper when it is raised or lowered.
- Ensure that cement dust is brushed off the shield and pullway of the silo at frequent intervals.
- Clean out the dispensing hopper at least once every hour to prevent the sticking and build-up of cement.
- Refer to the manufacturer's instructions on maintenance. It is best to have cement silos maintained by the manufacturer under his maintenance scheme. This should be done when the silo is first installed, and then at least every month afterwards. At the same time the manufacturer's fitter should check the calibration of the weighing mechanism, and provide the site agent with a properly completed service report. Any test weights used must be stamped by the local Trading Standards Officer.

13.4 Ready-mixed concrete

13.4.1 Enquiry and ordering procedure

Orders should only be placed with your company's or client's approved suppliers. These will normally be suppliers who are registered members of the Quality Scheme for Ready Mixed Concrete (QSRMC) or British Standards Institution (BSI), who carry out regular independent assessments of the plants and technical records of registered companies. Orders should be placed using your company's standard form.

Information on the current status of any QSRMC plant and its records system can be obtained from your Purchasing Manager, Quality Manager or QSRMC central records office (See www.qsrmc.co.uk for contact details).

Refer to your company standing instructions or procedures or general work instructions for procedures and standard forms. Check also what approvals are required by the contract's specification.

13.4.2 Control on site

Prior to the receipt of ready-mixed concrete check that the mixes quoted for agree in all respects with the specifications. Check that a purchase order has been placed covering all the specified requirements. Ensure that adequate testing facilities are on site, i.e. that suitable accommodation is ready for the curing tank and all the necessary testing equipment, and personnel are available.

If the concrete supplier is not a current valid member of QSRMC, additional costs might have to be allowed for to cover the additional testing and surveillance that might be necessary to assure compliance with the specified requirements.

All site-sampling of ready-mixed concrete must conform with the procedures referred to in your company standing instructions or procedures or general work instructions.

In the early part of a contract and subsequently at regular intervals, make physical checks on the quantities of ready-mixed concrete being delivered. The method of checking must conform to BS EN 206-1 and BS 8500.

Check that the details on the delivery ticket include all the key data related to the mix and that it agrees with the specification, in particular:

- cement types
- aggregate size
- admixtures
- minimum cement content
- maximum water/cement ratio
- consistence
- loading time.

Do not add water to the concrete to exceed the specified workability. If a higher workability is necessary for a particular purpose, this must be agreed with the designer responsible for the specification and the supplier notified accordingly. Water may be added on site under the conditions described by the contract providing that the water/cement ratio does not go beyond that specified. This procedure is only carried out under strict conditions.

Larger sites may have the luxury of their own testing laboratory which will be UKAS (United Kingdom Accreditation Service) accredited. However, smaller construction sites will have to carry out their own sampling and testing. It is usual the practice on smaller sites to send cubes to a UKAS laboratory for testing.

13.5 Transporting, placing and compaction of concrete

13.5.1 Transporting

This must conform with four main requirements:

• It must be rapid so that the concrete does not dry out, lose its workability or plasticity between mixing and placing.
• To avoid non-uniform concrete, segregation must be avoided, and any loss in fine material or cement and water prevented.
• Organise transport to avoid delays during the placing of any particular lift or section. These delays cause the formation of pour planes, cold joints, or poor construction joints.
• When using open vehicles in hot or wet weather, protect the concrete from the effects of sun, rain and wind, and wind-borne dirt.

Remember that some mortar from the first one or two batches of concrete transported nearly always sticks to the container. To overcome this, an extra 10% of cement and sand can be added to the first batch produced: this compensates for any loss of mortar in transit as well as in the mixer.

13.5.2 Placing concrete

Careless placing can cause displacement of reinforcement, pre-stressing cable ducts, ties for cladding and the movement and damage of formwork.

Segregation of the mix is associated with poor placing techniques. It occurs when concrete being discharged from a skip or chute is allowed to drop continuously and collect in one spot. Segregation can be very pronounced when concrete is discharged off the ends of conveyors without suitable funnel-shaped pipes or baffles.

Give the following points special attention during placing:

• Deposit the concrete as near as practicable to its final position. Do not deposit a large quantity at any one point and then allow to flow along the formwork. This causes segregation, honeycombing, sloping pour planes and poor compaction. Do not use a poker vibrator to move concrete, as this also causes similar problems.
• Deposit concrete in horizontal layers and compact each layer thoroughly before the next layer is placed.
• As far as practicable, place each layer in a continuous operation, the thickness depending on the size and shape of the sections, the concrete consistency, reinforcement spacing, method of compaction and the need for placing the next layer before the previous one has hardened. These layers should be 0.2 m to 0.4 m thick. In mass concrete, thicker layers 0.4 m to 0.6 m thick can be used. Several layers may be placed in succession to form one lift, provided they follow one another quickly enough to avoid *cold* joints.
• Make concreting continuous to avoid unsightly lift planes in the finished structure.
• Work the concrete thoroughly into position around reinforcement and embedded fixtures, and into corners of the formwork.

Pour rate

- Always begin pours at the downhill end if sloping.
- On wall pours check the pour rate with the temporary works supervisor. Higher than permissible pour rates can seriously affect the integrity of the temporary works.
- Use a concrete placement record form where appropriate, see your company standing instructions or procedures for details.

Pre-concrete checks

- Check that the specified concrete mix is approved and compatible with the spacing of the rebar, i.e. aggregate size and consistence. Will rebar congestion also influence the size of pokers?
- Establish the rate of pour, which must be sufficient to keep the face alive, but not to exceed the maximum rate of rise.
- Use a hopper and tremie pipes to avoid segregation when concrete has to be dropped from a height.
- Forewarn the concrete supplier of required quantity, rate and time. For large pours hold a meeting to confirm back-up supplies, numbers of trucks, possible traffic problems and stocks of aggregates, cement etc.
- Agree pour sequence, methods of placing, compacting, finishing, curing and protecting the concrete. Confirm on the Operation Plan and distribute to all involved.
- Ensure adequate hardstanding, access and egress for the chosen crane/concrete pump and lorries. Check that all areas of the pour are accessible using this method.
- Check that the specified finish is compatible with follow-on activities such as waterproofing.
- If vibrating screeds are to be used, ensure they are the right length and type.
- Check line, level and final cover of any temporary screeds.
- If a curing agent is to be used, check that it is compatible with final colour requirements and subsequent operations such as silane.
- Check that all the required equipment is available and will be present at the pour on time, e.g. pokers and vibrators, tamps, screeds, floats, rain covers etc.
- Check that the pokers work before starting to pour the concrete.
- Check that Permit to Load form has been signed as approved and received.
- Check that the correct mix and quantity has been ordered, with a hold volume given, to permit an accurate final figure to be determined.
- Check the weather forecast for adverse conditions and amend your plans accordingly (see Chapter 2).
- Ensure the delivery rate suits the placement rate of the concrete gang. This will prevent lorries standing with concrete starting to set.

Concrete pour checks

- Prior to discharge, check the delivery ticket for correct mix and location. Test for consistence/air content as appropriate, and do not permit use if results are outside acceptable limits.
- Place the concrete in accordance with the Operation Plan.
- Remember concrete must be placed within two hours of batching and any live face within the pour must be covered within 30 minutes.

- Do not heap concrete, especially on deck pours, and do not move concrete by vibration.
- The depth of concrete layers placed must not exceed the permitted maximum rate of rise or 500 mm maximum.
- Ensure that concrete is vibrated correctly:
  - The poker must be inserted quickly and extend at least 100 mm into the layer of concrete beneath to release the air correctly.
  - Take care not to touch the formwork or splash nearby works such as parapet rail or bolt threads.
  - The poker must be withdrawn slowly from the concrete as all the air voids are removed, taking extra care around pipes, ducts, bolts clusters, features and areas of congested reinforcement.
  - Over-vibration will result in aggregate segregation.
  - Do not leave the poker running when not in use.
- Ensure a 'Permit to Load' has been signed by the temporary works coordinator/ supervisor, prior to the concrete pour.
- Also carry out all checks detailed on the 'Inspection and Test Plan' prior to the pour.
- Check for possible movement of pipes, ducts, bolt clusters, formwork kickers, stop ends, etc. during concrete placement.
- Check for grout leakage. Wash off immediately if it occurs on permanent works such as bridge beams/steelwork.
- Look out for ties or spacers breaking and rebar springing free. Repair immediately.
- Ensure finished concrete is at the correct level. Beware of movement in suspended slabs that may affect level of screeds or stop ends. Do not overfill kickers.
- Check cover to top reinforcement and faces of kickers.
- Finish concrete to the specified requirements, with the right tools, i.e. wooden, or steel floats, or power finishing, brush finish etc.
- Remove all dummy screeds and fittings and ensure that any voids are properly filled, compacted and finished.
- Be aware of the volume of concrete placed and the void remaining. Give adequate notice if you need to revise the hold volume. Thereafter determine the final volume as accurately as possible.
- Ensure adequate protection and curing measures are taken, in accordance with the Operation Plan.
- Check that the requisite number of test cubes have been taken. Stripped cubes should be cured with the pour.
- Minimise wastage by having a home for any surplus concrete as nearby blinding/ kerb backing etc.
- Clean and tidy away all equipment immediately after use.

13.5.3 Compaction

The object of this is to remove air holes and to achieve maximum density. Compaction also ensures intimate contact between concrete and reinforcement and other embedded parts. To obtain maximum density, the mix design should produce a mix of adequate workability (consistence) to suit the dimensions of the section. On the other hand, it is important that the concrete is not too wet as this is liable to result in segregation, excessive laitance at the top of the pour, weakness and a lower density due to the volume occupied by the excess water.

When compacting concrete (by hand or vibration), do not displace the reinforcement or formwork. However, the concrete must be worked thoroughly around the formwork so that the finished surface will be even, dense and free from honeycombing or excessive blowholes. Good compaction is particularly necessary where the concrete may be subsequently exposed to aggressive chemicals in the ground or polluted atmosphere.

When using a poker vibrator the following points help in achieving a good result:

- Make sure the poker is of a diameter that will fit between the reinforcement.
- Move the poker frequently in the wet concrete and make sure the whole section is covered adequately.
- Do not let the poker touch the 'wet' side of timber formwork, especially if this is intended to leave a special finish.
- Do not allow a poker to touch reinforcement when part of that reinforcement is embedded in 'green' concrete, that is concrete which has stiffened but which is not yet one day old. The vibration might break the bond between reinforcement and concrete.
- When compacting a layer of concrete on top of another layer that is newly placed, push the poker at least 100 mm into the lower layer.
- Do not try to compact a layer of concrete more than 600 mm thick with a poker, as air has to be forced to the surface and the poker is inefficient over a greater depth than this.

There are four indications on how long to vibrate concrete:

- Air bubbles rise to the surface as the concrete is being vibrated; when these stop, it is generally a sign that not much more useful work can be done on the concrete.
- The surface texture also gives an indication as to whether compaction is complete or not. Do not allow an excess of watery material (laitance) to appear on the surface.
- The dangers from under-compaction are far greater than those from over-compaction, so there is no hurry to stop vibrating. Under-compaction will cause a serious loss in strength and durability, and will mar the appearance.
- Provided the concrete is still plastic enough to allow the poker to sink in under its own weight, it is perfectly in order to re-vibrate it. This is often necessary when a long run is being constructed; the previous layer has probably stiffened up somewhat and will need 'livening up' a little as the next layer is being vibrated. This will ensure a good bond between the two layers.

13.6   Curing concrete

13.6.1   Requirements

Curing may be considered as comprising two elements. While the concrete is hardening and gaining strength, there is firstly the process of preventing the loss of moisture. Secondly there is the control of the temperature within the concrete. The two processes may be dealt with together and the control of temperature generally is only required in *large* pours. Although large is not accurately defined, this may be taken to mean where the smallest dimension of a pour is greater than 500 mm.

The chemical action of the setting and hardening of concrete requires the presence of water, and is known as hydration. Hydration also generates heat which causes the temperature of the concrete to rise. A significant loss in water by evaporation may cause the hydration process to stop with consequent reduced strength development. In addition, evaporation can cause early and rapid drying shrinkage resulting in cracking. Methods of curing are therefore designed to maintain the concrete in a continuously moist condition over a period of several days or even weeks to prevent evaporation by the provision of some suitable covering or by repeatedly wetting the surface. Use your company concrete curing record form where appropriate.

Correct curing maintains the concrete at a favourable temperature for the time specified, and avoids excessive differences in temperature both within the concrete and in its surroundings. The extent and nature of curing will be affected by hot or cold weather. Thermal curing for large pours is discussed in more detail in Section 13.6.4.

Curing treatments commonly used can be divided into two main groups:

- large horizontal surface areas, such as roads and floors
- formed concrete, such as columns, beams and walls.

13.6.2 Large horizontal surfaces

These are usually very exposed and careful curing from the time of placement is necessary. The following methods are listed in order of effectiveness, but each have their drawbacks.

- Keep the concrete continuously wet by spraying with water. However, do not use this method in freezing conditions or when frost is likely to occur. The temperature of the curing water should be as near as possible to that of the concrete and, in any case, well above freezing.
- Spray with water as soon as possible after finishing the concrete and cover with polythene. Because of the poor insulating properties of this material, additional shading or other protection may be required to reduce the effects of solar radiation or hot dry winds.
- Lay damp hessian directly on the surface as soon as the concrete has hardened sufficiently not to cause damage or adhere to the concrete. Damp sand in a 50 mm layer is often used but both methods must be carried out properly with spraying or wetting at intervals, day and night.
- One of the most practical and economic forms of curing is the use of spray-applied curing compounds, provided the application is thorough and is carried out promptly. However, curing compounds must not be used on concrete where a bond between it and any form of topping such as granolithic concrete, cement-sand screed or rendering is required.

13.6.3 Formed concrete

Formwork generally provides adequate protection against drying out, but cover exposed surfaces immediately after casting, particularly in dry weather.

Stripping times for vertical formwork are related to the size of members and weather conditions, but striking times are increased in cold weather according to the reduced maturity of the concrete or when consideration is required to be given to the colour and appearance of the concrete. When uniformity of appearance is important, do not remove formwork until four days after casting the concrete. It should be realised, however, that formwork striking times are primarily based on the concrete having gained sufficient strength to be self-supporting. These times are likely to be less than the curing periods for vertical members.

Periods for curing before formwork may be struck may be included in the job specification. Guidance on these minimum periods is also available in BS 8110 (see also Chapter 11). The requirements may change with the forthcoming publication of BS EN 13670.

For large beams and columns where the mass related to surface area is large, retain formwork for as long as possible until most of the heat generated during hydration has dissipated, to obviate excessive temperature differentials within the concrete. As a general rule the temperature differential should not be permitted to exceed 20 °C.

Good curing to beam, column and wall vertical surfaces can be achieved by continuously sprayed water. This is seldom practicable particularly with vertical surfaces due to the nuisance of water dripping onto other work below. Small sections such as beams and columns may be cured most satisfactorily by spraying and wrapping in polythene sheeting as soon as the forms are removed. Alternatively, wet hessian in contact with the surfaces may be suitable, provided that arrangements are made to keep the hessian continuously wet.

For some members, curing membranes may be suitable, but the application of spray-applied compounds on large vertical surfaces can be very uneven and patchy if the spraying equipment is unsatisfactory and if the work is not carefully carried out. Windy conditions also create problems.

It is not usually practicable to wrap concrete walls in polythene sheeting. It is easier to drape wet hessian or polythene from the top of the wall, taking care that all edges are firmly held down to prevent drying winds getting in and evaporating the water out of the thin sections.

13.6.4 Thermal curing

The hydration process produces large amounts of heat in the first few days after pouring. Near the surface of the concrete the heat can be lost to the atmosphere; towards the centre of the pour, because of the poor thermal conductivity of concrete, heat can build up causing the temperature to rise. Problems are created when the heat generated is considerably greater than the heat lost at the surface, and a thermal gradient is created across the section. The thermal gradient causes different amounts of thermal expansion in different parts of the concrete. If different amounts of thermal expansion occur when the concrete is no longer plastic, but is rigidly set, the stresses set up can be sufficient to cause the concrete to crack.

Generally this situation only occurs when the thickness of the section exceeds 500 mm. Temperature rises can be very large. For example in a 1 m thick pour of concrete made with Portland cement, the core temperature can be more than 50 C above the ambient.

The potential for cracking is determined by a number of factors only over some of which we have control, and which include:

• type and amount of cement
• type and amount of cement replacement
• type of formwork (timber or steel)

- whether insulation is used
- ambient temperature conditions
- temperature of the constituents and concrete at mixing and placing
- temperature gradient (which is determined by the above factors)
- type of aggregate
- restraint to the concrete from earlier placed pours or other restraints such as piles or the ground.

To prevent cracking the usual procedure is to limit the temperature gradient across the section. This is often achieved by insulating the shutters or wrapping the pour in insulating blankets. Although this will cause the core temperature to rise more, it also reduces the temperature gradient by reducing surface heat losses. Temperature gradients may need to be measured. This is done by incorporating thermocouples (special bimetallic strips) into the core and around the edge of the pour.

Whenever you are involved in large pours, and particularly if the specification imposes limits on the temperature gradients or differentials, advice must be sought from your technical division.

Concrete curing checks

- Ensure curing of the concrete is carried out in accordance with the Operation Plan and does not affect concrete finish.
- If curing compound spray is to be used:
  - on a horizontal surface, apply as soon as possible after the initial bleed water has evaporated
  - on vertical surfaces, apply as soon as possible after the formwork has been struck
  - do not use on dry surfaces, wet with clean water and spray while moist
  - do not apply on top of water.
- If polythene, damp hessian or frost blankets are to be used, ensure that a chilling wind tunnel is not created. Any hessian must be kept damp at all times.
- Take care not to touch the formwork or splash nearby works such as parapet rail or bolt threads.
- If the formwork is to be retained during curing:
  - check formwork temperatures during cold or windy weather, it may be necessary to lag to retain heat
  - do not strip until the required strength has been achieved and a Permit to Unload has been issued
  - beware of cube strengths resulting from unrepresentative curing and remember that seven-day strengths are only indicative of final values.
- Before stripping formwork, agree a suitable laydown area or transport to the correct location. Do not place in a heap.
- Carry out all post concrete checks and cover surveys at the earliest opportunity, especially when similar pours are to follow.

13.7 Concreting in hot weather

In the UK, difficulties due to hot conditions are frequently overlooked until the sudden and unexpected arrival of a spell of hot sunny weather. In countries with hot climates, suitable precautions are carefully planned and fully implemented as a matter of course.

The precautions apply to both the fresh and hardened concrete. The main objective is to control the excessive evaporation of water, especially when the high temperature is accompanied by a low relative humidity and by warm drying winds.

To avoid adverse effects on the quality and durability of the finished structure, keep the temperature of the concrete when placed below 32 °C.

In hot weather, keep concreting materials cool:

- Shade stockpiles of aggregates from the sun. Periodic sprinkling with water is very good but avoid large variations in moisture content of the aggregates since this will affect workability of the concrete produced, unless allowed for.
- Keep mixing water cool by storing in tanks shaded or painted white. Bury supply pipelines or insulate them.
- In very hot conditions it may be necessary to cool the water by adding crushed ice, or by refrigeration. Requirements for cooling should be written into the method statement for the work.
- The use of hot cement should be avoided where possible, but this has less effect than the temperature of aggregates and water.

One of the problems of hot-weather concreting is rapid stiffening after mixing. The usual approach is to reduce the time of transporting the concrete. Other precautions are covering the concrete with damp canvas during transit, and spraying the transporting containers periodically with water to cool them and prevent water being drawn out of the concrete, or use a retarding admixture. However, the performance of these is sensitive to a number of factors, notably the type of cement. It is essential that trial mixes are made well in advance of the work in order to determine the effect on both the fresh and hardened concrete.

If transporting concrete over relatively long distances is unavoidable, the use of truck mixers is preferable. The concrete is protected in the drum, and mixing can be delayed until the discharge point is reached.

In hot climates, it is advisable to pour concrete in the early morning or evening. Working at nights is sometimes necessary.

Further information can be obtained from Clause 1710 of the *Manual of Contract Documents for Highway Works* (Volume 1).

13.8   Concreting in cold weather

As the temperature of concrete decreases, the rates of hardening and development of strength slow down; below freezing point the chemical process of hardening stops. When the temperature rises again, hardening is resumed. Therefore ensure that:

- water in newly placed concrete does not freeze
- concrete is protected at an early age, i.e. until it will withstand cycles of freezing and thawing without deterioration
- strength development is maintained even though at a lower rate than at higher temperatures
- hot water in the mix from the batching plant is requested to raise the temperature of the concrete.

Unless some form of heat is applied and the concrete can be protected, stop concreting when the ambient temperature reaches about 2 °C on a falling thermometer. Concreting can be continued at freezing temperatures if certain precautions are taken. These include heating the materials so that at the time of placing the concrete temperature is greater than 10 °C. Together with ensuring that no ice or frost is present on the formwork or reinforcement and keeping the concrete warm with insulating blankets for several days after placing will normally be sufficient to prevent damage to the concrete and to maintain a satisfactory strength development. The temperature of the concrete when deposited in the formwork must be in no case less than 5 °C *until it has hardened*. Refer to your company standing instructions or procedures or general work instructions, for specific procedures.

13.9 Sampling and testing

13.9.1 General

There are three series of concrete test methods:

- BS EN 12350: Testing fresh concrete (currently seven parts)
- BS EN 12390: Testing hardened concrete (eight parts)
- BS EN 12504: Testing concrete in structures (four parts)

In most cases the tests are very similar to the BS 1881 tests they replace.

BS EN 12350–1: 2000. Testing fresh concrete

Part 1:	Sampling
Part 2:	Slump test
Part 3:	Vebe test
Part 4:	Degree of compatibility
Part 5:	Flow table test
Part 6:	Density
Part 7:	Air content—pressure methods
Part 8:	Self-compacting concrete—slump-flow test
Part 9:	Self-compacting concrete—V-funnel test
Part 10:	Self-compacting concrete—sieve segregation test
Part 11:	Self-compacting concrete—J-ring test

BS EN 12504–1: 2000. Testing concrete in structures

Part 1:	Cored specimens—taking, examining and testing in compression

BS EN 12390–1: 2000. Testing hardened concrete

Part 1:	Shape, dimensions and other requirements for test specimens and moulds.
Part 2:	Making and curing test specimens for strength tests
Part 3:	Compressive strength of test specimens
Part 4:	Compressive strength—specification of compression testing machines
Part 5:	Flexural strength of test specimens
Part 6:	Tensile splitting strength of test specimens
Part 7:	Density of hardened concrete
Part 8:	Depth of penetration of water under pressure

13.9.2 Sampling

The procedures for sampling are covered by BS EN 12350–1. Clause 4 provides for samples to be taken from either a ready-mixed concrete truck or from a heap. Samples should be taken from the concrete using a standard scoop. Samples are either composite (consisting of a number of single scoops distributed through a batch of concrete and throughly mixed together), or spot samples (consisting of a number of single scoops taken from part of a batch of concrete). A shovel must not be used for sampling because larger stones have a tendency to roll off. Both a slump test and a pair of cubes would need four scoopfuls.

Each scoopful should be taken from a different part of the batch. If sampling from a ready-mixed concrete truck, the very first and last part of the discharge should be ignored. The scoop should be passed through the whole width and thickness of the falling stream in a single operation. Spot samples from a heap should be distributed through the depth of the concrete as well as over the exposed surface.

13.9.3 Consistence tests

Several methods of determining workability (or consistence) are given in BS EN 12350–1, the one most commonly used is the slump test for concretes of medium to high consistence. In practice this means concrete having a measured slump of between 5 and 175 mm. These are the slump classes as given in BS EN 206–1:

Specified slump class	Requirement			
	For composite samples taken in accordance with BS EN 12350-1		For spot samples taken from initial discharge	
	Not less than	Not more than	Not less than	Not more than
S1	0	60	0	70
S2	40	110	30	120
S3	90	170	80	180
S4	150	230	140	240
S5	210	—	200	—

Dimensions in mm

The rest of this section is confined to slump tests. In simple terms, the concrete is thoroughly mixed with a shovel on the sampling tray by building a cone of concrete and then turning the cone over and rebuilding another cone. This is done three times. The final cone is then flattened by repeatedly inserting the shovel vertically into the top of the cone.

The equipment, which is standardised, comprises a truncated cone, optional funnel, tamping rod and a tray. The cone is placed on the level tray with its narrowest diameter at the top. The cone is filled in three layers with each layer being tamped to its full depth. The rod should not forcibly strike the tray through the bottom layer, but should just penetrate the layer below for the top two layers. After tamping the final layer there should be a surplus of concrete above the top of the cone. The surplus is removed with a sawing and rolling motion of the tamping rod level with the top of the cone. After cleaning away the surplus concrete from around the base of the cone, the cone is carefully lifted from the tray by raising it vertically. The entire operation of filling the cone and lifting it should be completed within 2.5 minutes.

Immediately after the cone is removed, measure the height of the fallen pile of concrete below the height of the top of the cone. This is the value of the slump. The measurement is taken by placing the cone alongside the concrete and placing the tamping rod horizontally across the cone, over the pile of concrete. The height should be measured and recorded to the nearest 10 mm.

According to BS EN 206–1, compliance with the slump requirements will be considered to have been met if the slump values are within the following limits:

Specified target slump	Tolerance	
	For composite samples taken in accordance with BS EN 12350-1	For spot samples taken from initial discharge
≤ 40	−20, +30	−30, +40
50 to 90	−30, +40	−40, +50
≥ 100	−40, +50	−50, +60

Dimensions in mm

The frequency of slump testing depends on the specification, the quantity of concrete being produced and the variability of the material used. Tests should however be carried out at least at the start of each day's concreting operations, each time concrete is sampled for making test cubes or whenever adjustments are made to mix proportions. Observations should be made of the effects on the slump of temperature and length of haul from the mixer to the point of placing. Do not accept concrete with a measured slump in excess of that specified or designed, having taken into account the specified or allowed tolerance.

13.9.4 Test cubes

Test cubes are used to determine the compressive strength of the concrete. See company standing instructions or procedures or general work instructions, when issued, for procedures on making, storing and despatching test cubes to testing laboratories.

BS EN 12390–2 covers the making and curing of specimens for strength tests. Cubes from dry-lean concrete require to be compacted by vibration and pressure. A local independent testing laboratory may be used provided it holds current UKAS accreditation and, in the case of contract work, meets with the approval of the employing authority.

All cubes for testing at an age of 7 days must be accompanied by their companion cubes for testing at 28 days. Local testing laboratories must not test 7-day cubes unless accompanied by their companion 28-day cubes.

The number of cubes made from each sample will depend on the compliance requirements of any particular specification. Normally at least three cubes should be taken allowing for at least one test at 7 days and two at 28 days. It may be prudent to make a fourth cube for possible testing at 56 days in the event of non-compliance when tested at 28 days. There may be an advantage in determining strength at ages between 7 and 28 days, in which case additional cubes will be required.

Standard cube moulds are used for making the cubes, and these can be either 150 mm or 100 mm. As a general rule, 150 mm cubes should be used. 150 mm mould should be filled in three approximately equal layers. Each layer is compacted with a special tamping bar which has one square end to be used for the tamping.

Tamping should be done uniformly over the whole layer until the concrete appears to be fully compacted. The number of strokes required will depend on the type of mix, but at least 25 strokes per layer are required. Generally it will be found that more than the 35 strokes will be required to compact the concrete fully.

After the top layer has been compacted, smooth it level with the top of the mould using a steel float. If the mould is too full after floating, do not scrape the surface. Rather, some of the top concrete including some coarse aggregate, should be removed, and the top should be re-compacted in the normal way.

All specimens should be cured before they are tested. They should be left in the moulds for 24 hours, covered with damp sacking and wrapped in plastic sheeting. They should not be disturbed during this period, and should be stored at a temperature of about 20 °C. The moulds are stripped at 24 hours. The base plate is removed first and then the sides are removed. Remember that the concrete will be very weak and susceptible to damage. Any pieces of concrete knocked off or cracked will affect the strength readings. Straight after demoulding, the specimens should be indelibly marked with a unique reference and then submerged in a water bath. The water should be thermostatically controlled at a temperature of 20 °C ± 2 °C. The moulds themselves should be cleaned of all traces of concrete, re-oiled, reassembled and carefully stored for re-use.

Daily maximum and minimum temperatures should be recorded throughout the whole curing period.

Normally, compliance with strength requirements is assessed on the basis of the average strength of the two cubes made from the same sample and tested at 28 days.

This constitutes a 'test result'. In the absence of other explicit requirements of the specification, refer to the requirements of BS 8500 and BS EN 206–1.

This means that, on all sites, it is necessary to maintain continuous profiling of each individual test result and the average of four consecutive test results as each test is received. This may be done in either tabular or graphical form. Simple computer programmes are being produced to assist in this monitoring.

It can be seen that if a test result falls below the specified characteristic compressive strength this does not necessarily mean that concrete production is out of control, but it can be an indication that batching techniques require improvement or the mix design requires adjustment.

In the event of non-compliance with the specification, immediate steps must be taken to improve the situation. Action regarding any suspect concrete already cast is determined by the designer. His instructions can range between adjustments to mix design or batching techniques, to removal and replacement of the non-compliant concrete.

13.10   No-fines concrete

All general sampling, testing and construction requirements given for dense concrete also apply to no-fines concrete. Precautions against frost action before hardening need to be more fully applied owing to the more open texture of the material.

13.11   Pre-cast concrete

Unless otherwise overridden by a particular contract specification and expressly agreed in writing, all orders placed for the supply of pre-cast concrete products must conform with the procedures contained in your company standing instructions or procedures or general work instructions, and all of the relevant standards.

Further reading

Standards

BS EN 1008: 2002 *Mixing Water for Concrete.* Specification for sampling, testing and assessing the suitability of water, including water recovered from processes in the concrete industry, as mixing water for concrete.
BS EN 450: *Fly ash for concrete. Definitions, requirements and quality control.*
BS EN 196: *Methods of testing cement.* Parts 1–7.
BS EN 15167–1: *Grand granulated blast furnace slag for use in concrete, mortar and grout. Definitions, Specifications and conformity criteria.*
BS EN ISO 9001: *Quality Management and Quality Assurance Standards.*
Note: This Standard is in several sections dealing with such topics as principal concepts and applications; design/development; production, installation and servicing, and final inspection and test.
BS 8110: 1997 *Structural use of Concrete.*

Note: This Standard is in several parts and deals with such topics as design, construction and special circumstances.

BS EN 197–1: 2000 *Cement*. Composition, specification and conformity criteria for common cements.

BS EN 206–1: 2000 *Concrete*. Part 1: Specification, performance, production and conformity.

BS EN 934–1: 2008 *Admixtures for Concrete, Mortar and Grout*.

BS EN 12620: 2002 *Aggregates for Concrete*.

BS 8500: 2006 *Concrete. Complimentary British Standard to BS EN 206–1: 2000*.

Part 1: Method of specification and guidance for the specifies.

Part 2: Specification for constituent materials and concrete.

Other related texts

Alkali-silica Reaction in Concrete: BRE, Garston UK, BRE Digest 330 2004.

Minimising the Risk of Alkali-silica Reaction: Alternative Methods: BRE, Garston UK, BRE IP1/02, 2002.

Volume 1: *Specification for Highway Works*: HA, HMSO, Norwich, UK, 1998, as amended.

Volume 2: *Notes on Guidance on the Specification for Highway Works*: HA, HMSO, Norwich, UK, 1998, as amended.

Concreting in Hot Weather: D. E. Shirley, C&CA, Wexham UK, C&CA Report, 1980.

Calcium Aluminate Cements in Construction: A Re-assessment: CS, Crowthorne UK, CS Report TR 46, 1997.

Construction Materials Reference Book: D. K. Doran ed., Oxford UK, Butterworth, 1992.

Construction Materials Pocket Book: D. K. Doran ed., Oxford UK, Butterworth, 1994.

Structural Effects of Alkali-silica Reaction: IStructE, London, IStructE, 1992.

Concrete and its Chemical Behaviour: M. S. Eglington, London, Telford, 1987.

Microsilica in Concrete: CS, Slough UK, CS Report TR 41, 1993.

Non-structural Cracks in Concrete: CS, London, CS Technical Report No. 22, 1986.

Alkali-silica Reaction: Minimising the Risk of Damage to Concrete: CS, Slough UK, CS Report No. 30 3rd ed., 1999.

Formwork: A Guide to Good Practice: CS/IStructE, London, CS/IStructE, 1986.

The Design Life of Structures: G. Somerville ed., Glasgow, Blackie, 1992.

Design of Liquid Retaining Concrete Structures 2nd ed.: R. D. Anchor, London, Arnold, 1992.

Properties of Concrete 4th ed., A. Neville, London, Longman & Wiley, 1995.

Handbook of Coatings for Concrete: R. Basssi & S. K. Roy (eds), Latheronwheel UK, Whittles Publishing, 2002.

Concrete Mixes—Planning and Design for Transporting, Placing and Finishing: G. G. T. Masterton *et al.*, London, CIRIA Report No. 165, 1997.

The Chemistry of Cement and Concrete 4th ed.: P. C. Hewlett (ed), London, Edward Arnold, 1998.

Self Compacting Concrete: CS, Camberley, UK, CS Technical Report No. 62, 2005.

Chapter 14 STRUCTURAL STEELWORK

Tony Oakhill

Basic requirements for best practice

- **Plan the small things as well as the big ones**
- **Careful attention to choice of erection method and plant**
- **Preparation of method statement before erection commencement**
- **Regular flow of materials**
- **Accuracy in setting out**
- **Site splices preferably to be bolted rather than welded**
- **Well-maintained lifting equipment**

14.1 Introduction

One advantage of structural steelwork over its competitors is that of quality assurance. Every stage of its manufacture from the raw material through to the finished frame is controlled to achieve a product that can be guaranteed. Until the erection stage is reached, all the processes involved have been carried out in factory conditions and within closely prescribed Standards and Codes of Practice, even where necessary to the stage of trial shop erection.

In order that the benefits of this careful process can be passed on to the client, it is necessary that everything possible must be done by the erector to overcome the often largely uncontrollable effects of site conditions and weather on his efforts.

If a structure is to be erected stage by stage, safely, within the shortest possible time and at the lowest possible cost, then positive action is required from all the parties concerned. The first step is for the designer to have clearly in mind the means by which it is to be erected. The second step is to convey these ideas to the fabricators and the erectors through an initial method statement.

The capability of structural steelwork to be readily able to conform to closely defined tolerances of dimension and integrity is one of its main assets. When it is remembered that the tolerances required of the finished structure are considerably finer than would be the aggregate of the permissible tolerances of its components then the task facing the erectors can be seen. Attention paid to accuracy at the very start of the erection process will always help those subsequent operations to achieve their tolerances with a minimum effort in terms both of time and money.

Careful attention to the setting out will pay generous dividends when the structure comes to be checked later on for line, level and plumb. Likewise, careful attention to the choice of erection method and the plant and equipment to be used will ensure a smooth erection process which enables the job to be finished in the optimum time and at optimum cost.

14.2 Information and procedures

14.2.1 Legislation

Statutory instruments are legal documents including Acts of Parliament and subordinate regulations approved by the Secretary of State, which detail, among other matters, the Health and Safety Standards necessary to comply with the law.

Health and Safety at Work Act 1974

Refer also to Chapter 19. This important piece of legislation lays down the general duties of employers and employees at work. This means it shall be the duty of every person while at work:

- to take reasonable care for the health and safety of himself and others who may be affected by his actions or omissions at work
- to cooperate with his employer or anyone else where it is necessary to enable them to comply with their statutory duties
- not intentionally or recklessly to interfere with or misuse anything provided under a statutory requirement in the interest of health and safety at work.

14.2.2 Management systems and method statements

A management system is the overall control of the project through the design, drawing, fabrication and erection stages and is often embodied in a quality assurance scheme. It covers the ability of all those all those involved in the project to carry out their individual tasks properly, and sets down their various areas of responsibility.

For the steelwork erector the most important aspect is to have a method statement before commencing erection work. Under Section 2 of the 1974 Act employers are required to supply safe plant and systems of work, provide adequate information, training and supervision, provide safe access and egress and a working environment without risks to health.

As a means to achieve these aims, an important part of the planning for a safe system of work is the provision of a method statement. This will set out the proposed erection scheme and the necessary procedures. Employees must follow these procedures which may consist of standard sheets for simple jobs up to extensive requirements for more complex work.

The method statement should take account of the following:

- the erection scheme and procedures to be adopted
- the drawings to be used
- the accuracy of the information given
- the design checks required prior to erection.

Other items covered should include the required degree of inspection and erection tolerances, the recording of any modifications made on site, and the services and facilities either received from, or supplied to, other site operatives.

14.2.3 Types of drawings

Drawings convey information between the designer, the fabricator and the erection team. They consist of:

- General arrangements—plans, elevations and sections showing the relationship between each member. These are sometimes over printed with erection marks and thus referred to as marking plans. Each steel member should be numbered.
- Assembly details—showing how the members are connected together.
- Lists and schedules—giving details of all the bolts, brackets and individual members being delivered to site.

14.2.4 Accuracy of information

Check with the method statement that all the appropriate and latest drawings are available on site before commencing erection work.

Check each drawing to ensure it contains all the necessary information to erect the components safely.

Check for any special features and connection details to ensure they fit properly and the correct size and type of fasteners are available to secure individual pieces and to achieve stability at each stage of the erection process.

If it is found that there are discrepancies or misunderstanding about the drawing detail or erection notes then the site supervisor should be consulted so that, if necessary, the matter can be brought to the attention of the designer or detailer for clarification.

14.2.5 Design aspects

Components should be delivered to site in the correct order so that the erection process can be followed. Weights should be clearly marked on all members, bundles of components and sub-assemblies of one tonne and over.

The following items concerning the site conditions need to be clarified prior to commencement of erection:

- the provision and maintenance of suitable hardstanding for cranes and general access to the site
- the investigation of soil conditions and identification of hazards affecting the safe operation of cranes
- the definition of any anticipated special environmental and climatic conditions that may affect the work on or use of the site.

Also especially important is the resistance to overturning which a single column or column assemblies can safely tolerate. This should be critically assessed and recorded, highlighting the design assumptions made. Where holding-down bolts are to be used, a sufficient factor of safety should be provided against pull-out. Holding-down bolts in addition to those necessary for the permanent structure may be required during erection.

14.2.6 Selection of steelwork contractor

Pre-selection

The following comments apply to major steelwork contracts. For smaller contracts, site management can select those items for checking appropriate to the nature and magnitude of the contract. The steelwork contractor's commercial standing and professional competence should have been vetted as a matter of course at tendering stage but, sometimes, the steelwork contractor is appointed by others. In many projects the steelwork contractor acts as a subcontractor to the main contractor. Often, the steelwork fabrication and the subsequent site erection is carried out by the same firm, although sometimes these two activities can be split between two different organisations.

A considerable number of steelwork contractors are members of the *Steel Construction Certification Scheme* (SCCS). The operation and management of the SCCS is carried out by the British Constructional Steelwork Association (BCSA) under an independent Certification Board with representatives appointed from related sectors of the industry. All certification assessments and surveillance monitoring visits are conducted by qualified auditors experienced in the various design, procurement, materials control, fabrication and erection management activities relevant to steelwork contracting. SCCS certification should give confidence to customers in their selection of steelwork contractors as specified in the National Building Specification and the National Structural Steelwork Specification for Building Construction (NSSS). Several of the companies certified by SCCS have integrated their quality, environmental and occupational health and safety management systems.

An increasing number of steelwork contractors have also joined the Register of Qualified Steelwork Contractors (RQSC). This too is administered by BCSA to enable steelwork contractors to demonstrate that they can do what they say they can do. The RQSC is twofold: it gives the *category* of steelwork that registered companies may undertake and it gives the *class* of project that companies are registered to undertake. It applies equally to fabrication and to the erection of steelwork.

- *Category* refers to the type of steelwork for which a company is registered.
- *Class* refers to the size of the steelwork portion of a project for which the company is registered to undertake and is dependent on financial criteria.

Assessment

It remains an important requirement that free access to the steelwork contractor's works is obtained so that the client or contractor can inspect the fabricated steelwork during the progress of the contract. Visit the works, get to know the layout and the people involved and check their methods of quality control and traceability. All unpainted steel supplied to the fabricator's works will have been cut slightly over length, and will have the strength grade stamped and colour coded for easy recognition on one end. This also applies to plate when the grade and colour code is generally in one corner.

Check how the steelwork contractor keeps track of the member. During the fabrication process the stamped mark and the colour coding often disappear in cutting, welding, grit blasting etc. Similarly, with plate, if the stamped corner is cut off to make a stiffener, is the remaining piece of plate re-stamped so that it can be identified for future use?

Both mild and high-strength steels can be used in the same design. Check how the fabricator ensures that these are identified correctly in the works to prevent transposition. The method of *traceability* must make it possible for a piece of steelwork to be traced back to a Material Test Certificate for a specific casting of steel.

Visit the steelwork contractor's design or drawing office. Check the fabrication drawings:

• Are there sufficient dimensions?
• Are the sizes of all welds and welding techniques given?
• Are the holing sizes and dimensions shown?
• Is the steel grade shown?
• Are the grades of bolts given?
• Is the paint treatment and preparation shown?
• Have the drawings been checked and signed off ?

Check the method of cross-referencing the fabrication and the general arrangement drawings. The number of drawings on a big job may run into many hundreds and easy cross-referencing is essential.

Ask the fabricator for a copy of his welding procedures and check that the welders comply with them. Ask to see the coded welders' certificates. Ask for copies of Mill Certificates, Procedure Qualification and Welders' Certificates, and any welding test results by an independent testing laboratory. Ask to visit the cutting, plating, drilling, grit blasting and painting shops.

Generally ask about deliveries of steelwork to site. Will they be by road, rail or a combination of each? If by road, are transport vehicles in house or subcontracted? What is the size of vehicles and maximum load?

During a works visit one can gain an overall impression of the steelwork contractor. He must be able to satisfy you against the specified requirements. Until he can do this he should not be allowed to proceed.

14.3 Steelwork delivery, handling and storage

A job can only proceed smoothly on site with a regular flow of the materials required for its construction. The preparation of a delivery schedule based on a compromise between expected outputs, possibly from more than one fabricator, and the demands of the site programme is an early priority for the management team who must also consider the need to ensure early stability of the frame.

The almost inevitable lack of correlation between these demands often makes the provision of a stockyard on the site a necessity. A stockyard is additionally a buffer area which can absorb, for example, variations in transport or shipping times due to outside influences which, without an adequate stock of material, would leave the erection teams unemployed. Typically, as much as four weeks' erection work may have to be provided by the stockyard capacity.

It may well be found that the demands of the site and its sequenced delivery requirements will be beyond the capability of any one fabricator. Planning the schedule for all the material requirements is hence another early role of management. Typically this will involve a purchasing department using specifications provided by the designers and detailers, working to the demands of the programme and the original cost estimate for the supply of each item required within the financial limit, to define delivery requirements. The day-by-day monitoring of the actual performance of the suppliers in terms of the quality of work, its adherence to the specifications of workmanship and tolerance and adherence to programme is a management responsibility.

Materials delivered to site that are not required for immediate use should be off-loaded and stored in a suitable area. This area must be firm and level with wooden sleepers or other suitable material placed on the ground at regular intervals to act as bearers. It is essential to avoid placing heavy loads on top of underground services such as electrical cables, culverts and drains, which may be damaged due to the weight of steelwork resting on the ground.

Steelwork should be stacked safely and in the correct sequence so that components that are required to be erected first are easily accessible. Where possible, members should be stacked so that water does not collect in webs and hollows and behind stiffeners. Ground bearers and timber battens should be used to separate stacked material so that slings can pass around components when they are to be lifted clear of the stack. Green timber should not be used as timber battens. Instead use properly dried timber. Timber must be large enough to allow the chain hook or the eye of the sling to pass under the material. At all times leave sufficient access ways between stacked material for slinging and crane movements.

There are several methods of stacking steel which will give stability and optimise space. The actual method used will depend largely on the uniformity of the steel and the overall dimensions of each component.

Ensure that individual members are not subjected to excessive forces or risks of corrosion during handling and storage. Suitable protection should be provided at sling pick-up points to prevent deformation, buckling and damage to paint. With storage of galvanised members such as cold rolled 'Z' sections, allow maximum air circulation around individual members. Keep the duration times for storage to a minimum to prevent 'white rust' formation in badly ventilated conditions. Rolled purlins are usually delivered to site in tightly bundled and banded packs. Opening these out for storage with air circulation may take too much space so consult suppliers on the bundled storage life before the onset of 'white rust' and take the necessary precautions.

14.4 Setting out and tolerances

14.4.1 Types of tolerances

'Tolerance' as a general term means a permitted range of values. Other terms that need definition are given below:

• *deviation*—the difference between a specified value and the actual measured value, expressed vectorially (i.e. as a positive or negative value)

- *permitted deviation*—the vectorial limit specified for a particular deviation
- *tolerance range*—the sum of the absolute values of the permitted deviations each side of a specified value
- *tolerance limits*—the permitted deviations each side of a specified value,

  e.g. ± 3.5 mm or + 5 mm − 0 mm.
  For structural steel there are three types of dimensional tolerance:

- *manufacturing tolerances*—such as plate thickness and the dimensions of rolled sections
- *fabrication tolerances*—applicable in the workshops
- *erection tolerances*—relevant to work on site.

14.4.2 Values for erection tolerances

The values for erection tolerances are given in Tables 14.1 and 14.2. Each of the specified criteria should be considered and satisfied separately. The permitted deviations should not be considered as cumulative, except to the extent that they are specified relative to points or lines that also have permitted deviations.

These values represent current practice at the time of writing and are taken from the fourth edition of the NSSS. However, in any case, the latest relevant standards should always be used. The clause numbers referred to in Table 14.2 are clause numbers in the NSSS which should be referred to for further information.

14.4.3 Setting out and tolerances

Before steelwork setting out or erection commences, carefully study the specification, the general arrangement drawings and as many fabrication drawings as are available. This in-depth study and discussion with the engineering staff on site is to decide a method of dimensional control starting from the ground. Study the critical points, for instance a transition at first floor cantilevering beyond the ground floor columns and then rising vertically from the end of the cantilever. The critical point here is the alignment and level of the end of the cantilever rather than the ground floor columns.

Elastic compression in the stanchions is very significant in tall steel structures. Consult the designer on this matter, as floor datum levels can vary considerably as the structure is loaded out, and allowances are needed if this is critical to fit up. Similarly on bolted structures, *take-up* of clearance holes in adjacent bolted connections, combined with deflection of the surrounding steelwork, can affect the level of floors and especially cantilever beams when loaded out.

The steelwork subcontractor must provide the main contractor with a method statement of setting out and controlling the structure so that all concerned know the job is being controlled. It is normal on most contracts for the main contractor to provide the main grid lines and setting out points. The specialist steelwork subcontractor normally carries out his own detailed setting out based on these.

Ensure that the steelwork contractor's setting out is properly checked. The essentials in setting out are:

Table 14.1   Permitted deviations for foundations, walls and foundation bolts.

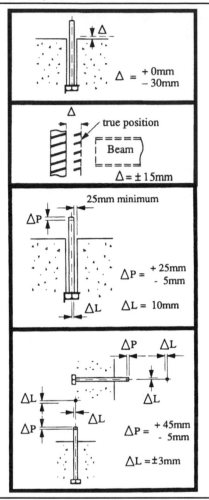

Foundation level

Deviation from exact level.

$$\Delta = \begin{array}{l} + \, 0mm \\ - \, 30mm \end{array}$$

Vertical wall

Deviation from exact position at steel-work support point.

$$\Delta = \pm 15mm$$

Pre-set foundation bolt or bolt groups when prepared for adjustment

Deviation from the exact location and level and minimum movement in pocket.

$$\Delta P = \begin{array}{l} + \, 25mm \\ - \, 5mm \end{array}$$

$$\Delta L = 10mm$$

Pre-set foundation bolt or bolt groups when not prepared for adjustment

Deviation from the exact location, level and protrusion.

$$\Delta P = \begin{array}{l} + \, 45mm \\ - \, 5mm \end{array}$$

$$\Delta L = \pm 3mm$$

- accuracy
- line
- level
- plumb
- tolerances.

The effort involved in achieving greater accuracy involves greater cost. Before job commencement, it is important to agree the required degree of accuracy with all parties. A statement of this should then be incorporated in the contract documents. Keep records of any company/employer agreements together with site records of actual levels, setting out dimensions and plumb elements of the building. This will help to resolve any claims at hand-over concerning inaccuracies due to elastic compression, take-up of bolted connections and other points of contention.

Table 14.2   Permitted deviations of erected components.

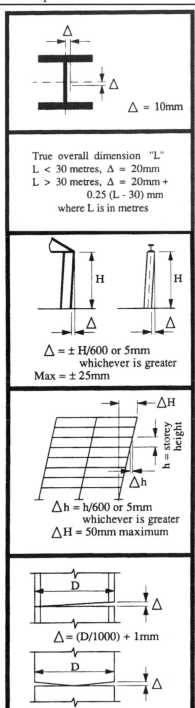

Position at base of first column erected Deviation of section centreline from the specified position.	$\Delta = 10\text{mm}$
Overall plan dimensions Deviation in length or width.	True overall dimension "L" L < 30 metres, $\Delta = 20\text{mm}$ L > 30 metres, $\Delta = 20\text{mm} +$ $0.25\,(\text{L} - 30)$ mm where L is in metres
Single storey columns plumb Deviation of top relative to base, excluding portal frame columns, on main axes. *See clauses 1.2A (xvii) and 3.4.4 (iii) regarding pre-setting continuous frames.*	H H $\Delta = \pm\,\text{H}/600$ or 5mm whichever is greater Max $= \pm\,25\text{mm}$
Multi-storey columns plumb Deviation in each storey and maximum deviation relative to base.	ΔH h = storey height Δh $\Delta\text{h} = \text{h}/600$ or 5mm whichever is greater $\Delta\text{H} = 50\text{mm}$ maximum
Gap between bearing surfaces (See clauses 4.3.3 (iii), 6.2.1 and 7.2.3).	D Δ $\Delta = (\text{D}/1000) + 1\text{mm}$ D Δ

Alignment of adjacent perimeter columns
Deviation relative to next column on a line parallel to the grid line when measured at base or splice level.

Beam level
Deviation from specified level at supporting column.

Level at each end of same beam
Deviation in level.

Level of adjacent beams within a distance of 5 metres
Deviation from relative horizontal levels (measured on centreline of top flange).

Beam alignment
Horizontal deviation relative to an adjacent beam above or below.

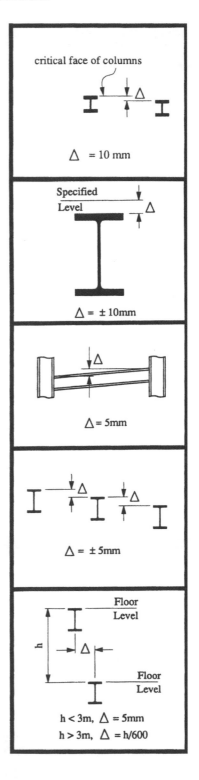

Table 14.2   (continued)

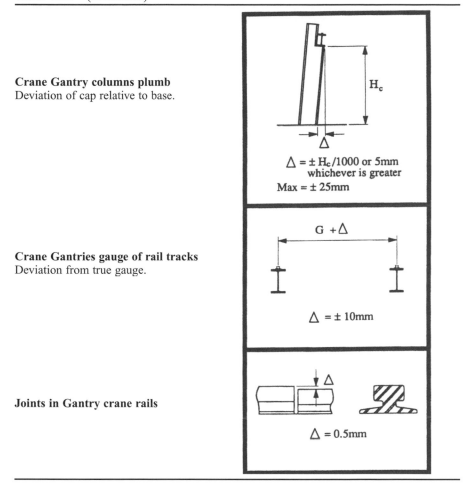

Crane Gantry columns plumb Deviation of cap relative to base.	H_c $\Delta = \pm H_c/1000$ or 5mm whichever is greater Max $= \pm 25$mm
Crane Gantries gauge of rail tracks Deviation from true gauge.	$G + \Delta$ $\Delta = \pm 10$mm
Joints in Gantry crane rails	Δ $\Delta = 0.5$mm

Steelwork is generally erected speedily. An unnoticed early stage-error can require drastic and costly corrective action: possibly dismantling and re-erection. It is therefore important to check and re-check key items as work proceeds.

- All substructure baseplates should be levelled from a common fixed Temporary Benchmark (TBM).
- Avoid levelling one baseplate from another.
- Be aware that rolling margins affect the nominal thickness of baseplates.
- Check that baseplates are set on shims of varying thickness to give the exact level at the underside of the baseplate.
- Use epoxy mortar to fix shims in position. Movement of them during erection may be disastrous.
- Check that baseplates are exactly aligned to main grid lines. A stanchion that is level and plumb will still give trouble if it is twisted in plan as connections will not fit.

• If the subcontractor dot-punches the centre line of the stanchion on the baseplate edge then it will be easy to align these onto pre-laid lines on the substructure.

Steelwork setting out is mostly achieved on the ground. Once the work is being erected (even if fabrication is correct) there is little that can be done to correct for setting out errors. Tolerances set out in NSSS and the specification already allow for:

• rolling margins
• fabrication deviations
• shrinkage in erection welding
• setting out tolerances.

Before and during the erection programme it is important to bear in mind other major components that will interact with the main structure. These include floors (particularly pre-cast concrete units), cladding (pre-cast concrete units, curtain walling), masonry and mechanical plant. It is important to check that these items have sufficient built-in adjustment to fit properly without requiring site modification to steelwork constructed to agreed tolerances. If in doubt discuss with the engineer.

Be aware that designers may confuse *serial* sizes with *actual* sizes of beams and columns. They may also overlook the fact that splice plates have thickness and depth and that bolt heads may protrude into the work of following trades. As a rule of thumb an axial stress of 50 N/mm^2 causes 2.4 mm of elastic shortening in a 10 m column. Temperature effects are very significant on large structures such as bridges, e.g. a 100 m long structure changes in length by 12 mm for a 10 °C temperature change.

14.5   Anchorage and foundation bolts

As noted in Section 14.4 the finished work must be within a designated range of dimensional tolerances, and the best way of complying with these is to make sure that accuracy is achieved from the commencement of the job. Setting out the foundations and the holding-down bolts is often done at a stage when close tolerances and the thick mud on site seem incompatible. However, time spent on marking the centre lines and the levels of the concrete bases accurately is time well spent.

• Check setting out positions and levels for conformity with drawings and tolerance requirements.
• Delay drilling until all checks are to the satisfaction of the designer.
• Cast holding-down (HD) bolts in sleeves or boxes to allow bolt movement for final stanchion base location.
• Rotate HD bolts periodically within sleeves or boxes while concrete is green to prevent rigid set.
• Use accurate templates for setting.
• Provide adequate reinforcement *outside* of the HD bolt boxes where these occur at the edge of a base or foundation slab (see Figs 14.1 and 14.2).

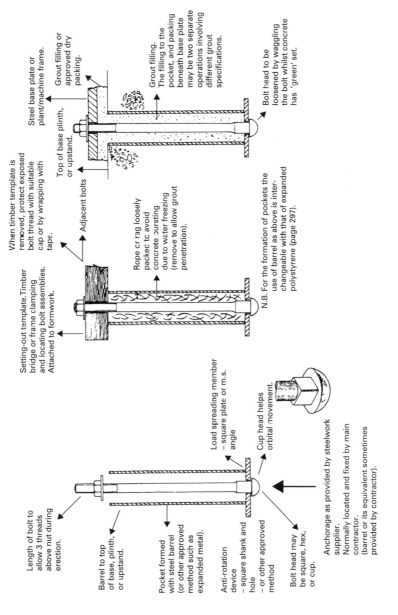

Figure 14.1 Holding-down bolts and anchorages.

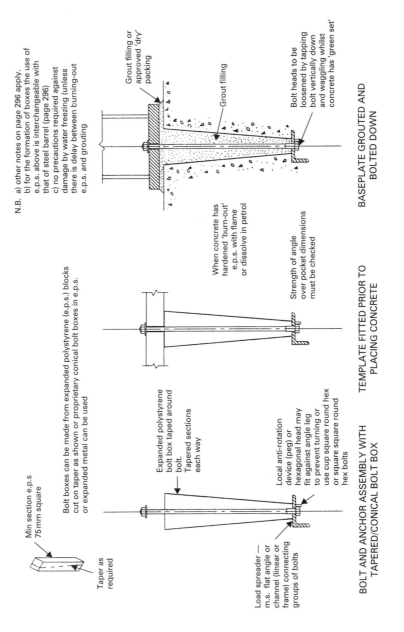

Figure 14.2 Holding-down bolts and anchorages.

Min section e.p.s
75mm square

Taper as
required

Bolt boxes can be made from expanded polystyrene (e.p.s.) blocks
cut on taper as shown or proprietary conical bolt boxes in e.p.s.
or expanded metal can be used

Expanded polystyrene
bolt box taped around
bolt.
Tapered sections
each way

Local anti-rotation
device (peg) or
hexagonal head may
fit against angle leg
to prevent turning or
use cup square round hex
or square square round
hex bolts

Load spreader —
m.s. flat angle or
channel (linear or
frame) connecting
groups of bolts

BOLT AND ANCHOR ASSEMBLY WITH
TAPERED/CONICAL BOLT BOX

When concrete has
hardened 'burn-out'
e.p.s. with flame
or dissolve in petrol

Strength of angle
over pocket dimensions
must be checked

TEMPLATE FITTED PRIOR TO
PLACING CONCRETE

N.B. a) other notes on page 296 apply.
b) for the formation of boxes the use of
e.p.s. above is interchangeable with
that of steel barrel (page 296)
c) no precautions required against
damage by water freezing (unless
there is delay between burning-out
e.p.s. and grouting

Grout filling or
approved 'dry'
packing

Grout filling

Bolt heads to be
loosened by tapping
bolt vertically down
and waggling whilst
concrete has 'green set'

BASEPLATE GROUTED AND
BOLTED DOWN

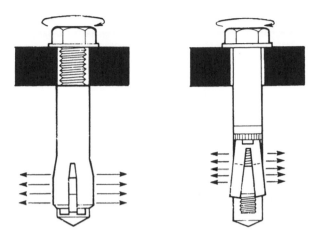

Figure 14.3 Heavy-duty anchors for concrete and masonry.

For lighter loads a heavy-duty anchor of the type shown in Fig. 14.3 may be used. The available size range is M8 to M24 with working pull-out loads in the range 6.9 kN to 45.5 kN when used in conjunction with 30 N/mm^2 concrete. Manufacturers' instructions should be carefully followed with regard to shear loading and combined shear and tension loading. All such arrangements should be agreed with the engineer. The following should be carefully noted:

- Anchorage is obtained by the wedging action of an expanding sleeve which is forced radially outwards against the surrounding concrete as the bolt is tightened with a torque wrench.
- Accurate positioning and sizing of holes is vital as no repositioning is possible after drilling.
- Templates must be used.
- Oversize holes will reduce the compression between the bolt and the surrounding concrete thus reducing the efficacy of the fixing.
- The depth of hole is also important since the fixing must penetrate into the base material to at least the minimum depth required.
- Manufacturers' instructions with regard to base plate thickness must be strictly followed.
- The surface of the base material must be sensibly flat and sound as no grouting will be possible.
- Because of compression stresses in the base material there are limitations on the minimum spacing of anchors, the acceptable distance of an anchor from the edge of a base and to the thickness of the base.

So-called chemical anchors may be used in certain circumstances. For pull-out resistance they rely mainly on the shear strength of the grout injected into a hole drilled to receive the anchor.

For the preparation of stanchion bases, refer also to Fig. 14.4 and Fig. 14.5

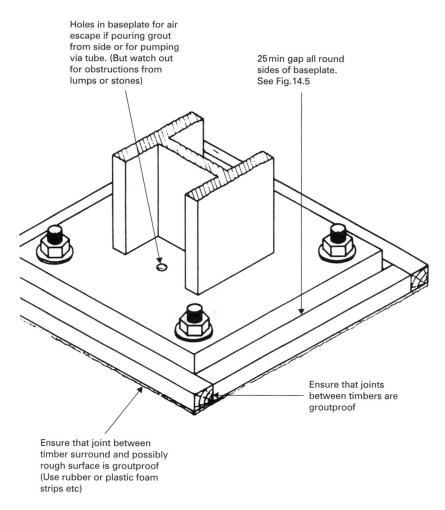

Holes in baseplate for air
escape if pouring grout
from side or for pumping
via tube. (But watch out
for obstructions from
lumps or stones)

25 min gap all round
sides of baseplate.
See Fig.14.5

Ensure that joints
between timbers are
groutproof

Ensure that joint between
timber surround and possibly
rough surface is groutproof
(Use rubber or plastic foam
strips etc)

Figure 14.4   Preparation of stanchion bases for grouting.

14.6   Erection

14.6.1   General

Steelwork erection is potentially a very hazardous task, and it is most important
to make proper provision for the safety of the steel erector. The Health and Safety
Executive and Statutory Regulations lay down the requirements of the law regarding the provision of equipment and arrangements, which will minimise the risk of
an accident.

The regulations stipulate requirements for people working in places where they can
fall more than 2 m. Provision must be made for safe access to and from the working
place, as well as the security of the working place itself. The width of the walkway,
the height of the handrail and means of preventing small tools being kicked off

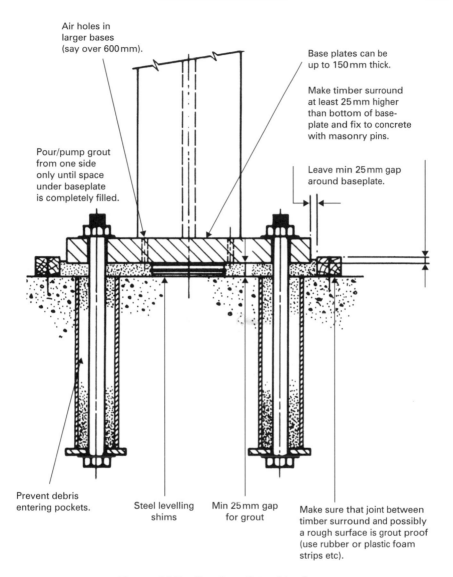

Air holes in larger bases (say over 600 mm).

Base plates can be up to 150 mm thick.

Make timber surround at least 25 mm higher than bottom of base-plate and fix to concrete with masonry pins.

Pour/pump grout from one side only until space under baseplate is completely filled.

Leave min 25 mm gap around baseplate.

Prevent debris entering pockets.

Steel levelling shims

Min 25 mm gap for grout

Make sure that joint between timber surround and possibly a rough surface is grout proof (use rubber or plastic foam strips etc).

Figure 14.5 Grouting of stanchion bases.

the platforms are all defined. Cranes, slings and lifting devices in general are all subject to inspection and to testing at regular intervals.

The main aim during erection is the preservation of the stability of the structure at all times. Most structures which suffer some form of failure, or even collapse, do so during erection, and these failures are very often due to a lack of understanding on the part of one person of what another has assumed about the erection procedure.

It is important that the provisions of the erection method statement are followed at all times. Failure to follow this advice could result in an incorrect erection sequence leading to instability during construction. Frequent reference to the relevant working drawings and instructions will ensure stability is achieved at each stage.

14.6.2 Planning

Planning for safe erection should start at the design stage. The design engineer should consider how the structure will be erected, and the need during the erection to ensure safe and adequate working conditions for those engaged in each stage.

Detailed planning is necessary for the safe erection of steel structures. It is important that all contractors and subcontractors are informed fully about this planning information. The steelwork contractor must ensure that at least all of the following aspects are addressed:

- Prior to the commencement of steel erection, the erector will, after consultation with the design engineer, the steelwork fabricator (if a different company), and main contractor, determine and prepare a schedule that will be followed throughout the project. All the erector's employees shall be briefed and made aware of the erection method statement.
- On a site plan, the various types, position and coverage of the intended erection cranes should be marked, as well as such locations as unloading points, storage areas (if any), and lifting positions, irrespective of whether cranes are provided by the main contractor or the steelwork erector.
- The sequence of proposed erection, to take into account such items as:
  - access to work areas
  - location of workers in respect to other trades
  - restricted areas.
- The required crane usage in the overall plan.
- The method of handling various components, e.g. heavy members or pre-assembly of members.
- The stability requirements of all items of the structure.
- List of ancillary equipment proposed and their adequacy for purpose.
- Site organisation chart with names and experience of key personnel including steel erectors.
- Outline of how it is proposed to set out the structure in its correct position and level, and to maintain accuracy, all within specified requirements, throughout the erection phase.

14.6.3 Craneage

General

All crane drivers shall be appropriately certificated in accordance with the relevant statutory regulations. An adequate communication system between the ground, work areas, the working level, and the crane driver shall be established by the party providing the cranes. The erector is to check boom clearances to ensure there is unimpeded operation throughout the erection period.

Tower cranes

Where a tower crane is to be used for lifting and positioning the steel, then the crane supplier shall in addition to the general requirements above, plan for the following:

- The positioning of the crane in order that it can lift all the steel required on each level into final position.

- The crane's compliance with statutory regulations and the testing regime.
- The stabilising of the tower of the crane, which is to be carried out in accordance with the manufacturer's instructions.
- Ensure that the frame of the building erected at any stage of the construction can adequately support the crane tower structure.
- The final removal of the crane upon satisfactory completion of all steelwork erection.

Note: crane base should be designed by a chartered civil or structural engineer.

Mobile cranes

Where a mobile crane is to be used to lift and position the steel, the crane supplier shall, in addition to the general requirements above, plan for the following:

- The positioning of the crane on the site after taking into consideration its reach, the weight of the steel to be lifted, and the safe working load at the required radii.
- Check the ground conditions and spread loads on outriggers by using mats if necessary to ensure stability during erection.

14.6.4   Means of access to working places

Types of accesses to working places with conditions of use, comments and examples are given in Table 14.3. All accesses should either comply with statutory requirements or have the approval of local regulatory authorities.

The sequence of erection should be planned so that the permanent structure can be used as much as possible to provide safe access ways and working places, with little or no adaptation.

Long-term construction access such as permanent staircases, ladders, floors and walkways should be provided as the erection progresses together with suitable edge protection as required, thus simplifying access to working places. Such accesses should be restricted to persons actually engaged in work in that area.

In choosing a particular method of *temporary access* to the working place consideration should be given to:

- the type of operative required to undertake the work, i.e. competent steel erectors and supervisors used to working at heights, or alternatively, other employees
- ground conditions—a great variety of access aids, for example, scissor lifts, can be used if suitable hardstanding is available
- any permanent access routes, such as stairways, that could be erected as construction progresses
- permanent floors, if these are constructed as the structure progresses
- the height and accessibility of connections
- the frequency and number of times access is required including any necessary supervision and inspection
- the type of activity, the tools to be used, and how long it will take
- the availability of, and the clearances required for, any plant or equipment to be used
- the need to move access ways as the job progresses to form the shortest possible safe route
- any abnormal site conditions (e.g. unloading restrictions, difficult access, limited hours of working, unusual environmental and climatic conditions).

Table 14.3 Access and working places.

Method	Examples	Some Conditions of Use
Walkways and Stairways	General Access	Edge protection must be adequate. Grating or flooring units must have flush finish and be securely fixed. They must be maintained in a fit condition and checked regularly.
Ladders	General access to height, sometimes in confined spaces.	Permanent fittings should be securely fixed before use.
• Vertical ladders	Access up columns.	Fix securely along length. At edge of structure, and for excessive heights, a proprietary fall protection system on the ladder should be used.
• Inclined ladders	General access to heights for bolting up of joints.	Securely fixed and founded. Slope to be 1 in 4. A person at bottom to stabilise ladder.
Tower scaffolds	Making connections. Permanent and welded connections.	Firm foundation, sound, even surface. Erected by qualified persons only. Height to base width ratio can be critical and should be checked for stability. Safe means of access to working places.
Suspended cradles	Remedial work on completed structure.	Outriggers to be of adequate length and correctly counterweighted.
Common or proprietary scaffolds	Construction of ladder access towers. Providing a working platform.	Properly constructed under the supervision of a competent person. Inspected regularly by an authorised person. Should be capable of being dismantled safely.
Lightweight staging	General access, working place. More than one unit may be required to provide sufficient width for intended use.	Must be adequately supported. Edge protection should be provided. Safety harnesses or belts may be required if working below guard rail.

14.6.5 Erection stability

Bracing

Where practicable, erection should start in a braced bay in order that the structure can be plumbed and made self supporting by having the necessary internal or in-built restraints. Such a stable and self supporting bay can then be used to support subsequently erected steelwork. If it is not practical to commence erection at a braced bay, some temporary support to the steelwork may be required, generally in the form of guys or temporary bracings.

Building stability

The erection of any element or sub-assembly should start only if all the necessary equipment and tackle is on site to enable the stability of the structure to be maintained at all times. This includes the provision and appropriate use of sufficient temporary guys or bracing to ensure the stability of all parts of the structure as well as the structure as a whole. Added care shall be taken to ensure that all such temporary guys or bracing are always safely anchored. To avoid accidents, guys shall be identified by coloured markers or similar, especially in areas of plant and vehicle movement.

The stability of the building structure and the effectiveness of all temporary guys, bracing, and supports shall be checked at the beginning and end of each working shift, or before further erection begins.

Column stability

The erector should be advised by the main contractor that the concrete in the bases has reached the specified strength before erecting the columns. The erector should give consideration to the use of tightly fitted steel packers or steel wedges driven under the edges of the column base plate to provide added stability especially when less than four anchor bolts are being used to anchor the column.

14.6.6 Lifting steelwork

The gross weights of all members to be lifted, including the weight of any protective coatings, shall be provided to the steelwork erector who will ensure that correctly designed lifting gear of the appropriate capacity is being used. Before lifting any steelwork, the erector must ensure that the members are safely and suitably slung and, where necessary, guidance ropes and pulling lines are fixed to their ends. When transferring lifts from a horizontal to a vertical position care must be taken to avoid the unrestrained movement of the lower end. The use of lifting beams may be necessary during lifting and positioning of some members to ensure member stability. Multiple lifts (e.g. a beam for each level, for several levels) and bundles of steel members can be lifted, provided adequate safety measures are taken. These include:

- specifically designed lifting slings to avoid steel members becoming entangled
- cradles for bundles of steel.

14.6.7 Steel components

The suggested methods that are covered in the following clauses for making steel-work connections are provided as minimum standards. Other methods may be acceptable provided that they meet these minimum standards.

Columns

Consideration must be given to plumbing and supporting the column at height, and the required craneage capacities. All single columns or column assemblies should be designed to remain stable by bolting the column to the column below or the holding-down bolts in the concrete foundation. Once the column has been securely anchored and stabilised against overturning, the column lifting sling or device may be released (see Fig. 14.6).

Where possible, the lifting sling or device should be released from the working accessed level by the use of long slings, remote release shackles or other suitable devices, that have the necessary approval. If access is required to the top of the column, competent erectors can carry out these duties using suitable access ways or working platforms and by attaching themselves to the structure by wearing a safety harness.

Connection of primary beams to columns or concrete core

The initial landing and bolting of the beam may be made from:

- ladders, provided they are supported and lashed or otherwise fixed to the column, and the erector is attached to a fixture on the column or core
- a purpose-built platform supported from a column or adjacent member

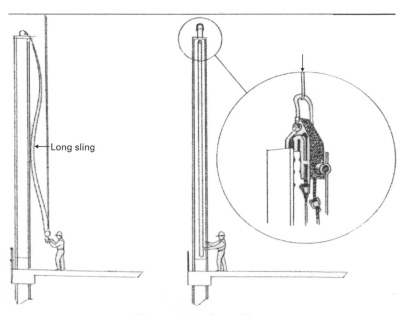

Figure 14.6 Long slings.

- a power-operated mobile platform
- an access cradle suspended from a crane.

Connection of secondary and in-fill beams

The initial landing and bolting may be made from a position straddling the primary or previously fixed beam. Competent erectors can carry out these duties using suitable access ways or working platforms and by attaching themselves to the structure by wearing a safety harness. Where possible, the lifting sling or device should be released from the working accessed level by use of long slings, remote release shackles or other suitable devices, that have the necessary approval.

Metal decking

Prior to landing the metal decking on the required working level, packs are to be placed in the correct orientation and location to suit laying. The erector must ensure that there is a safe anchorage to the beam being straddled while landing the bundle of metal decking. Any pack or single sheets must be secured when work ceases, to prevent them from blowing away.

By using the bundle as a datum or base the top sheet can be placed on to the beams adjacent. After positioning the first sheet at the point where the deck laying is to commence, square up and fix the first deck to the beams. This can then be repeated with all subsequent sheets, and always laying the deck back towards the bundle. The already laid sheets then provide a platform for deck fixing, side lap fastening and stud welding.

Shear studs

The decked area to be shear studded must be enclosed by perimeter-type handrails along all external and atrium edges. The area immediately under the deck area being studded must be roped off to prevent access because of the danger of hot sparks or displaced shear studs falling off. For safety reasons it is generally preferable for the stud welding to be carried out by the deck-laying gang simultaneously with the deck laying.

14.7 Bolting

Ensure on delivery that bolts are in accordance with drawings and specification. All bolts must be fitted with washers under the bolt head or nut, always under the turned part. Fit taper washers at all sloping face connections to ensure an all-round bearing of bolt head and nut.

The bolt length must be at least two full thread projections beyond the nut after tightening. In bearing-type connections the threaded parts of bolts must not protrude into bearing members being joined.

In connections consisting of larger bolt groups always start the sequence of tightening from the stiffer part or from the centre of connection and proceed outwards. On completion of tightening, check bolt tensions and repeat the tightening procedure as necessary.

In structures carrying vibrating machinery, bolts must either have spring washers, lock nuts or similar to prevent loosening, or be of the pretensioning type such as High Strength Friction Grip (HSFG) bolts, or Tension Control (TC) bolts.

Previous familiar bolting standards BS 3692 and BS 4190 have been replaced by a range of European Standards. Although the new European Standards have been published their adoption has not yet occurred and bolt manufacturers still supply bolts to British Standards.

It should be noted that bolt suppliers have been supplying fully threaded bolts for some time and steelwork contractors are using them in increasing numbers. They are ordinary bolts in every respect except that the shank is threaded for virtually its full length allowing a more rationalised range of bolts to be used. Thus the usual variable of bolt length (grip nut depth washer minimum thread projection beyond nut) can be replaced by a variable projection beyond the tightened nut. This reduces the number of different bolt lengths required.

14.7.1 Installation of bolts

The diagrams in Fig.14.7 illustrate the method of bolt installation.

14.7.2 Structural bolts

These fall into two groups:

- black bolts—usually used in shear and tension
- HSFG (high strength friction grip) bolts.

14.7.3 Black bolts

Used in holes with 2 mm clearance over shank diameter. The two grades most used in structures are grade 4.6 Black Mild Steel (BS 4190 and BS 4933) and grade 8.8 HT bolts and nuts (BS 3692), see Fig. 14.8.

Strength grade designation

This is indicated by two figures separated by a full stop (e.g. 4.6). The first figure is one-tenth of the minimum ultimate strength in kgf/sq. mm. The second figure is one-tenth of the percentage of the ratio of minimum yield stress to minimum ultimate. Thus *4.6 grade* means minimum ultimate stress of 40 kgf/sq. mm and yield stress of 60% of this. For high-tensile products where the yield point is not clearly defined, the stress at a permanent set limit is quoted in lieu of the yield stress.

The single grade number given for nuts indicates one-tenth of the proof load stress in kgf/sq. mm and corresponds with the bolt ultimate strength to which it is matched. Thus an *8 grade nut* is used with an *8.8 bolt*. It is permissible to use a higher strength nut than the matching bolt. For example *grade 10.9 bolts* are supplied with *grade 12 nuts* as *grade 10 nuts* do not appear in the BS series.

Podger spanners should be used to tighten bearing bolts supplied to BS 2583. Ensure that assembled parts are correctly drawn together on completion of tightening and that, where used, spring washers are fully closed. Check that bolts, nuts and washers all match.

(a) Members are brought together and holes aligned with drift pins.

(b) Enough bolts with their appropriate washers and load indicators are partially tightened.

(c) Fit bolts into remaining holes. Tighten until average gap on each load indicator is not more than the figure quoted in Table 14.5. Work from the centre of the joint outwards towards the free edges.

(d) Knock out drift pins and replace with further bolts and tighten. To avoid trapping bolts by joint slip, it is important that they are tightened before the drift pins are knocked out.

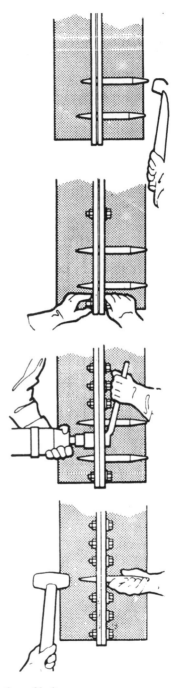

Figure 14.7 Installation of bolts.

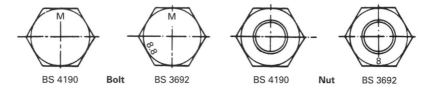

Figure 14.8   M or ISOM on bolt heads indicates ISO threads.

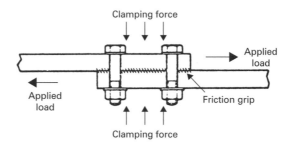

Figure 14.9   Load held by friction between plates.

14.7.4   High strength friction grip bolts

High strength friction grip bolting is an economic and efficient way of joining steel elements. The procedure is simple, but it is important that the principles of their use is well understood. It is important to give erectors adequate instructions to see they appreciate the differences between black bolts and HSFG bolts. Unless the correct tools and techniques are use it will be impossible to obtain the correct bolt tension.

Friction grip depends on tightening each HSFG bolt to a minimum shank tension if the correct clamping force is to be obtained. This enables the applied load to be carried by friction between matching faces rather than relying on the shear strength of the bolt. For adequate friction to be developed, the matching surfaces must be clean bare metal or, in the case of a waiver, in accordance with the designer's instructions. It is essential that the faces are not painted (see Fig. 14.9).

BS 4604 allows the slip factor to be taken as 0.45 for grade bolts through members whose contact surfaces are free from paint or any other applied finish, oil, dirt, loose rust, loose scale, burrs or any other defect which would prevent solid seating of the parts or would interfere with the development of friction between them. If any surface treatment is employed, or if higher grade bolts are used, the slip test fully described in BS 4604 should be carried out. It is generally recognised that shot or grit blasting the surfaces can improve the slip factor to a higher figure than 0.45, whilst paint, galvanizing and some primers can lead to a reduction in the slip factor.

Even when the applied load is in the direction of the bolt axis, tightening to a minimum shank tension is still required. The compression induced in the joining plies by the clamping force prevents the plies separating and thus there is very little or no increase in bolt tension so long as the applied load is less than the bolt preload (see Fig. 14.10).

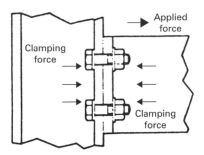

Figure 14.10   Load held by bolts in tension.

The security of HSFG bolts depends on correct tightening. There are three methods of tightening:

Torque method

This uses a *calibrated torque wrench* or a *torque cut-out pneumatic power wrench* to give a torque measurement. However it must be appreciated that a torque wrench only measures a resistance to turning. Because of varying conditions of nuts and bolt threads the relationship between torque and induced tension is dependent on those conditions.

When threads are dry, dirty, damaged or lacking in lubrication, friction will be high and, at a given torqure setting, bolt tension will be low. If threads are well lubricated and free running, induced tension at a similar torque setting will be high. A scatter of ± 40% can readily be obtained in different conditions using a pre-set torque wrench and a load meter.

Part turn method

The *part turn method* was devised to overcome concern caused by these factors. The principle behind this system is to tighten bolts into or approaching yield by first snug tightening and then applying an additional part turn, the amount depending on the grip length of the bolt. This method needs a great deal of operator reliability and supervision. It can only be inspected during tightening or by reverting to the already suspect accuracy of the calibrated torque wrench method. It should be noted that the BS for the use of higher grades states that the part turn method is not permitted.

Direct tension method

The *direct tension method* gives a direct indication of the load induced along the axis of the bolt. A load indicating device is fitted into the grip of the bolt assembly thus eliminating the unpredictability of the torque measurement. It should be noted that the standards specify tension not torque.

The *Coronet* load indicator is a typical direct tension measuring device comprising a hardened washer with protrusions on one face (see Fig. 14.11). As the bolt is tightened, the protrusions are flattened and the gap between the load indicator face

General grade (Metric series)

Higher grade

Figure 14.11 'Coronet' load indicators.

and the bolt head is reduced. The gap is measured by a feeler gauge (see Fig. 14.12) and, at the specified gap given in Table 14.4, the shank tension will not be less than the minimum specified in the appropriate standard.

Sometimes it is more convenient to fit the *Coronet* load indicator under the nut. If so a special nut face washer should be placed between the protrusions of the load indicator and the nut (Fig. 14.13).

If the bolt head has to be turned in order to tension this assembly then an additional flat round washer should be placed under the bolt head. In turning the bolt head with the load indicator under the head, use a nut face washer assembled in the same way but underneath the head. No further washers are required in this application. With taper flanges, use taper washers (Fig. 14.14 and Fig. 14.15).

Components

Bolts, nuts, load indicators and washers are clearly marked according to grade. Check that bolts, nuts and washers are correctly matched (Fig. 14.16).

Corrosion

HSFG bolts, washers and load indicators will corrode if not properly protected; they should be kept in a clean, dry well-ventilated store and should only be issued for immediate use. Moisture absorbent hessian storage bags should be avoided as they encourage rapid deterioration of zinc and cadmium coatings. Final assemblies should be painted as soon as possible after erection: this is particularly important in marine or other corrosive environments. Research indicates that susceptibility to stress corrosion and hydrogen embrittlement increases with tensile strength. In certain conditions a metallic coating may reduce resistance to stress corrosion and a specifier should be aware of the site environment before deciding on a coating.

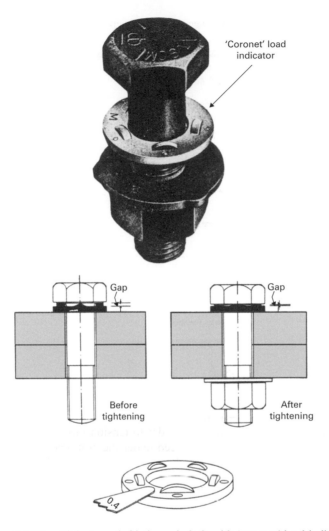

Figure 14.12   High strength friction grip bolt with 'coronet' load indicator.

Tools

Hand wrenches may be suitable for tightening small-diameter bolts but the use of power tools or torque multipliers for diameters in excess of M22 is recommended. Impact wrenches must have adequate capacity to tighten a bolt in 15 seconds. Prolonged impacting may damage a bolt assembly or fracture a bolt.

Tools with a torque output in excess of that required to tighten the largest bolt should be selected. This will leave a safety margin for wear, air loss and help to overcome the energy absorbed by higher than usual thread friction or *springy* joints. Most air wrenches require 1 m³/min at a pressure of 690 kN/m² at the tool for optimum performance. To maintain this delivery it is recommended that 20 mm diameter air line is used and that the delivery line is kept as short as possible.

Table 14.4   Clearances required for tightening bolts.

Size of bolt	Grade of bolt	BS 4395 and BS 4604	A mm	B mm	C mm	D Power tools mm	D Hand tools mm
M12	General grade	Part 1	23	27	30*	–	150
M16	All grades	Parts 1, 2 & 3	30	46	60	650	350
M20	,,	,,	30	46	,,	,,	,,
M22	,,	,,	33	52	,,	,,	,,
M24	,,	,,	36	65	,,	,,	,,
M27	,,	,,	41	71	70	750	,,
M30	,,	,,	45	78	,,	,,	,,
M33	Higher grade	Parts 2 & 3	50	90	100	,,	420
M36	General grade	Part 1	54	97	,,	,,	,,

*Hand tool only—allow 60 mm where power tools are used.

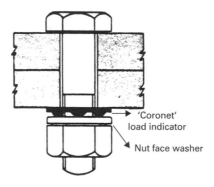

'Coronet' load indicator

Nut face washer

Figure 14.13   Use of nut face washer.

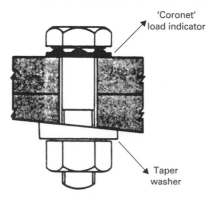

'Coronet' load indicator

Taper washer

Figure 14.14   Use of taper washer (1).

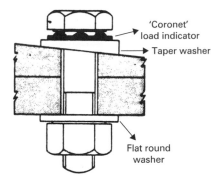

Figure 14.15   Use of taper washer (2).

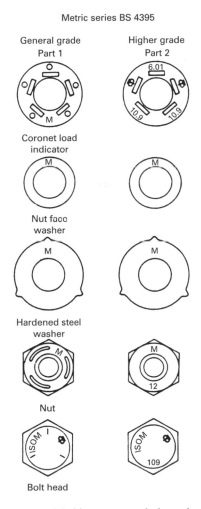

Figure 14.16   Markings on nuts, bolts and washers.

If tightening in 15 seconds is not achieved then the following checks should be made for:

- damaged or dry bolt and nut threads
- poor fit of joint plies
- bad alignment of holes causing bolt trap
- insufficient air supply due to leaks, lack of capacity or restrictions to air line
- excessive length of air line
- blockage of air inlet filter*
- incorrectly calibrated torque spanner**
- incorrect application of torque spanners or other equipment.

* Can be cured by cleaning with paraffin then lubricating with light oil, say, 5–10 SAE.

**These should be re-calibrated at least once per shift.

Installation tips

- Check all bolts, nuts, load indicators are of required grades.
- Ensure bolts fit freely.
- Load indicator protrusions must always bear against bolt head or nut face washer: **no other way will do!**
- Do not allow bolt head to spin on load indicator protrusions.
- Take care with inserting bolts with the load indicator fitted under the head through the hole without using force.
- If necessary, clean and lightly lubricate dry, rusty or dirty bolt threads.
- Do not get lubricant on ply contact surfaces.
- Tighten until the *average* load indicator gap complies with figure quoted in Table 14.5. Check gap with feeler gauge (see Fig. 14.17). If gap is smaller than specified, decide whether the error is sufficiently serious to need a change of bolt.
- When gap is not uniform, the *average* as measured by the gauge should be checked against Table 14.4.
- After erection and final checking carry out painting procedures as soon as possible to avoid corrosion of steel.
- Regard Coronet and other load indicating devices as precision items and resist the wish to tamper with them in any way.
- Do not assume that the presence of a load indicator adds to the difficulty of tightening bolts. If correct procedures have been followed no difficulties should occur.
- Remember BS 4604, in effect, states *IF AFTER FINAL TIGHTENING A BOLT IS SLACKENED OFF FOR ANY REASON THEN THE WHOLE BOLT ASSEMBLY MUST BE SCRAPPED AND NOT RE-USED.*

Inspection

- Check that the average load indicator gaps are not larger than those shown in Table 14.6 (for general grade bolts) or Table 14.5 (for higher grade Part 2 bolts). For further confirmation of bolt tightness, some assemblies may be tightened in the same manner as in a load cell.

Table 14.5 Load indicator average gaps after tightening.

Load indicator fitting	BS 4395 Metric series		
	General grade Part 1	Higher grade Part 2	
Under bolt head Black finish bolts	0.40 mm	*Max.* 0.50 mm	*Min.* 0.40 mm
All platings except spun galvanised bolts	0.40 mm	0.50 mm	0.40 mm
Spun galvanised bolts	0.25 mm	0.35 mm	0.25 mm
Under nut with nut face washer Black and all nut face washer finishes	0.25 mm	0.35 mm	0.25 mm

At the average gaps shown the shank tension will be:
For general grade bolts, not less than minimum given in Table 14.6.
For higher grade Part 2 bolts within the tension limits given in Table 14.6.

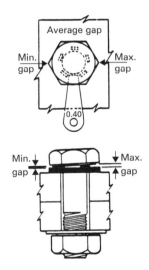

Figure 14.17 Use of feeler gauge.

- Check that the minimum requirements for shank tension, at the average gap, given in Table 14.6 complies with BS 4604 Part 1 and BS 3294 (for general grade bolts) and BS 4604 Part 2 (for higher grade Part 2 bolts).

Inspection notes

- Load indicator gaps are rarely uniform. The average gap will normally indicate that correct bolt tension has been achieved.

Table 14.6 Shank tensions.

Bolt diameter	BS 4395 Metric series		
	General grade Part 1 Min. proof load	Higher grade Part 2	
		Minimum 0.85 × min. proof load	Maximum 1.15 × min. proof load
mm	*kN*	*kN*	*kN*
M16	92	104	140
M20	144	162	219
M22	177	200	271
M24	207	233	316
M27	234	303	409
M30	286	370	500
M33	–	459	621
M36	418	–	–

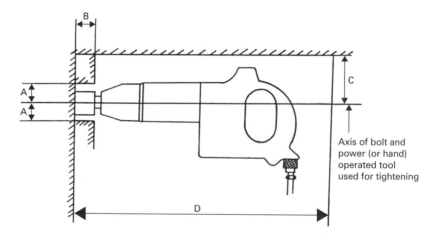

Axis of bolt and power (or hand) operated tool used for tightening

Figure 14.18 Clearances required for tightening bolts (see Table 14.4).

- Calibrated torque wrenches are not accurate for checking these bolts. They only measure resistance to turning and *must not be used* in any circumstances.

Clearances for tightening

The clearances required for the operation of some tools in general use are given in Table 14.4. The dimensions referred to are illustrated in Fig. 14.18.

14.7.5 Tension control (TC) bolts

The tension control bolt is a complete system employing a different approach to that of the conventional type of friction grip bolt. The bolt is replaced by a button head pin with a grooved shank and an additional spline or tail. These bolts are installed and tightened in a given sequence and after completion of tightening with a special wrench, the sheared off tail can be discarded. This leaves the pin stressed to not less than the required minimum.

14.8 Site welding

Site welding is only allowed with the express approval of the designer. Where site welding has been specified special consideration must be given to the planning of the erection of the structural frame. Some means must be provided for temporarily aligning adjacent components which are to be welded together, and of holding them in position until they are welded. The particular methods adopted to cope with the need for alignment may have to be able to carry the weight of the components, and in some cases they may have to be able to carry a substantial load from the growing structure.

Safe means of access and of working must be provided for the welder and his equipment. Since wind, rain and cold temperatures can all adversely affect the quality of the weld being produced, working platforms may have to incorporate weather protection.

Provision must be made in the initial setting and positioning of the components in order that the weld shrinkage, which takes place as the joint cools, will not result in the loss of the required dimensional tolerances across the joint.

Unless the contract documents state otherwise, ensure that site welding preparation, fit-up, pre-heating requirements, electrode storage etc. comply with BS EN 499. Welding procedures and welders should be tested in accordance with the series of welding standards BS EN 287 and BS EN 288.

The extent of quality control of site welding depends on the contract specification. Generally some degree of non-destructive testing (NDT) will be necessary. Where the contract documents do not specify otherwise, visual checks of welding could be deemed to be adequate. It is important and necessary to identify and correct any visual faults including heat affected zone (HAZ) cracking, bead cracking, undercutting etc. Any corrective work required must be specified by the designer. Particular attention should be given to welding steel of exceptional thickness (generally high strength steels over 25 mm thick), or large panel sizes.

It is recommended that welding specialists' advice should always be sought on the best welding procedure and pre-heating requirements.

14.9 Protection of steelwork

The maximum amount of protective treatment should be applied at the workshop in enclosed conditions. Generally it is convenient to apply at least the final paint coat at site after making good any erection damage.

Steelwork in a warm, dry interior will not corrode such that the structural stability of a building would be threatened during its design life (generally taken as about 30 years). When the steelwork is *hidden and dry* (e.g. fire protection boards, suspended ceilings) it is unnecessary to apply corrosion protective coatings for a normal building schedule. When intumescent or spray-on lightweight fire protection is to be used, advice should be sought from the manufacturers. Coatings are normally applied to *visible* steelwork only for decorative reasons.

When steelwork is exposed to moisture, corrosion will occur at a rate depending on the aggressiveness of the environment. *Hidden* steelwork which is cased, either for decorative or fire protection reasons, is not subject to mechanical damage. It may be impossible to carry out maintenance but there are no aesthetic requirements. Again, coatings are normally applied to *visible* steelwork only for decorative reasons.

When structural steelwork is embedded in concrete, it is recommended that the provisions of BS 8110–1 *The Structural Use of Concrete* should be followed. The concrete should have the correct composition and compaction with a depth of cover of at least 20 mm.

Correct surface preparation is essential for the adequate performance of coatings. Careful inspection and thorough cleaning of grease and dirt, removal of rust and loose paint must be carried out prior to the application of site coats. Where galvanised steelwork is affected by *white rust* (wet storage stain) this should be removed with a stiff brush and washed with water before subsequent pre-treatment and coating.

Incorrect storage on site prior to erection can accelerate the deterioration of coatings. Steelwork should be supported off the ground with items separated by timber battens allowing free circulation of air. It is essential to avoid standing water by laying down sections to ensure adequate drainage—refer to Section 14.3. Care in handling to minimise mechanical damage is critical to the performance of the protective system. The responsibility for the repair of damaged coatings should be clearly defined.

Other aspects of applying site protective coatings are:

• Ensure that paint storage and application are in accordance with the manufacturers' recommendations and specified requirements.
• Remove surplus bolt lubricants before painting.
• Make good to the satisfaction of the designer and purchaser painting and galvanising damaged in transit or during erection. Check the compatibility of repair materials with the original primer and paint coating.
• For general guidelines on painting refer to BS EN ISO 14713 *Protection against Corrosion of Iron and Steel in Structures*.
• Certain combinations of metal with steel will result in increased bimetallic corrosion.
• Where connections between dissimilar metals is unavoidable they should be insulated from each other and if possible kept dry.
• Give careful consideration to eliminate gaps and pockets in fabricated steelwork. Moisture can enter these causing intensive corrosive attack known as 'crevice corrosion'. This often results from lack of fit and the worst corrosion occurs in gaps only 0.75 mm wide.
• If possible, tightly close or pack out gaps and seal properly in a manner compatible with the protective paint system.

When making site-bolted joints using black bolts, where shop primer has been applied no further treatment is required before assembly. For site-made joints which have not been primed in the shop, the appropriate primer should be applied prior to assembly. Nuts and bolts should be cleaned and the full paint system applied. Galvanised or sheradised nuts and bolts should be used on galvanised steelwork.

If high strength friction grip (HSFG) bolted joints are being assembled at site it is essential to check that the faying surfaces at the time of assembly are free of any contaminant which would prevent the development of the slip factor required of the joint. After tightening and inspection of the bolt groups, the crevices around the load indicating washers, bolt heads, edges of the faying surface etc. can be sealed if necessary with a compatible paint or mastic. The nuts and bolts should then be cleaned and the full coating system applied.

For application of paint coats to site welded areas, particular attention should be paid to the removal of welding flux residues and weld splatter. Site welds should be treated in such a manner as to achieve the original standard of preparation and coating.

14.10 Site fabrication

Although steelwork fabrication is normally carried out in a covered workshop or factory under regulated and controlled conditions there will be occasions when certain fabrication activities need to be carried out on the site. Such situations may include:

- the attachment of new members to existing steel members
- strengthening beams and columns
- strengthening connections
- adding new floors to a building
- increasing the loading capacity of a floor by inserting additional beams
- creating new service openings
- the repair of damaged members

It is well known that structural steel provides maximum adaptability for changes in building use because structural alterations can be accommodated with relative ease. Also, where additional members are required, connections can be made to the existing frame with minimum disturbance and cost.

Steel construction generally consists of using standard rolled sections that are sized on the basis of the highest moments or forces existing in the member. Inevitably, there is a 'reserve' in strength at many points in the structure which can be utilised if, for example, holes are to be formed. Holes can be *flame cut* without affecting the performance of the material. The cutting of steel members on site is also a relatively straightforward procedure.

Because of the nature of the material, steel members can be man-hauled into position and can be connected to each other by welding or bolting or clamping. All of these techniques can be carried out *in situ*, although bolting or clamping are generally preferred to welding, as site conditions are not always conducive to achieving the best quality control and accuracy.

The use of modern portable magnetic site bolt clamps has improved the accuracy of on-site drilling, making the formation of connections much easier. The clamps are secured to the steel by an electromagnet. They weigh from 11.3 kg to 25 kg with a magnetic adhesion force between 750 kg to 1000 kg, and cutting depths up to 110 mm and 52 mm diameter. Because of the geometry of the drill, holes cannot be formed at less than 44 to 47 mm from an adjacent face or element.

The cutters, unlike a twist drill that only has two cutting edges and removes all of the hole, have four to sixteen edges. These chip away at the material leaving a central slug which is ejected by a spring loaded pin on completion of the drilling. The cutting area is thus reduced, making the process faster and more cost effective. Site drilling machines can be adapted to use twist drills and can produce counter-sunk holes with special cutters. Portable drills can also be clamped using special brackets onto pipes and tubes.

In recent years site welding equipment has become a lot lighter and more readily portable to site and easier to use. Normal manual metal arc welding (MMA) techniques are usually employed for site welding, although pre-heat may be required for some types of steel. Overhand or downhand welding is preferred because of the difficulty, and hence the slow speed, of welding underhand. There may be situations where overhead welding is impossible, but these cases should be minimised at the outset through careful design.

Cutting of steel can be undertaken by using oxy-acetylene flame or by mechanical disc cutter. Care must be taken at all times to ensure that sparks or molten metal do not impinge on flammable material or expensive cladding or other material finishes.

Purchased materials, fabrication tolerances and painting must be in accordance with the specified requirements. To ensure that steel materials conform, Mill Certificates should be obtained from the supplier or stockist with each order.

Fabrication of building steelwork should comply with the requirements set down in BS 5950–2. Detailing of steelwork should be in accordance with *Steel Detailers' Manual* or the BCSA Handbook *Metric Practice for Structural Steelwork*.

14.11   Crane rails

Normally these are specified by the crane manufacturer but supplied and installed by the steelwork contractor. These must be installed, bedded and fixed in the approved manner and to the specified tolerances. Ensure that this is done, as misalignment can cause excessive and speed wear on the crane wheels which could lead to break-downs. The placing of resilient pads under the rails are usually specified to prevent vibration and excessive loading on the steelwork structure.

14.12   Cladding and services

Ensure that cladding fixings are as indicated on the detail drawings in type, number and location. The type and spacing can vary over the surface covering of a building depending on localised wind effects.

Check that laps are in accordance with the manufacturer's instructions and have sufficient coverage. Also check that the laps are facing away from the prevailing wind direction to prevent moisture entrapment and to minimise any chance of storm damage.

A watch should be kept for problems with the use of dissimilar metals placed in contact with each other, e.g. the steel sheeting rails, metal cladding and metal fixings. If these are found to be dissimilar metals, then each may be insulated by placing an inert sheet or washers between.

Responsible supervision and regular inspection is essential to ensure structural integrity, satisfactory performance, acceptable appearance and quality in general. Deviations in steel frame construction, including such members as rafters side purlins and roof purlins, can often be greater than those acceptable for the cladding material, especially at critical bearing positions such as end laps or joints in composite panels. The steel framework should be surveyed prior to handover to the cladding contractor. Any deviations in line, level and plumb must be acknowledged by all parties and the necessary adjustments made to suit the cladding requirements before starting the installation.

Loading and off-loading cladding packs by crane or fork-lift should be carried out with care to avoid damage to the outermost sheets or panels in the pack. Never off-load with chains but use only wide soft slings for lifting. Use lifting beams if recommended by the manufacturer or supplier. Stacks should be carefully positioned and stored on site to prevent damage or deterioration. Some sheets are supplied with a protective plastic film on the weatherface to help prevent minor damage to the coating. This must be removed as soon as possible after the cladding has been installed because if it is left in place for long periods the film will become very difficult to remove. Individual manufacturers' instructions should always be followed.

Site cutting is generally avoided by the use of sheets cut to length in the factory. However, where site cutting is necessary nibblers and reciprocating saws (jig saws) should be used. Abrasive wheel cutters must not be used because they generate heat which will damage the coating. For optimum durability, site cut edges and end laps of steel sheets should be painted.

As in all building work, good safety standards are essential to prevent accidents on site. The erection of cladding is one such critical operation. In particular, construction of the roof is one of the most hazardous operations because of the potential for falls or material dropping onto people below. The contractors must plan and document a safe system of work before starting construction. This would especially take the fragility of the cladding materials into account. Whilst fully fixed metal sheeting is regarded as non fragile, rooflights and profiled metal liners must be treated with more care.

In addition to the basic safe system of working the following specific precautions should be taken when using metal cladding:

- Take care when handling sheets or panels to avoid cuts from the edges of the sheets. Wear gloves to protect hands.
- Take normal precautions when lifting heavy awkward objects to avoid lifting injuries.
- When cutting, wear goggles and dust masks.

For all other aspects of general site safety reference should be made to Chapter 19.

14.13   Model checklist for erection of steelwork

Note: This can be tailored to suit a particular contract. Points illustrated can be used as an aide-memoire by project managers, main contractors and subcontractors.

14.13.1   Preliminaries

- Access and working area
- Storage and security for materials
- Is pre-erection builder's work completed?
- Are safety, health and welfare facilities organised?

- Is power laid on as required?
- Have overhead obstructions and electrical and other services been removed or diverted?
- Has an inspection been carried out of the fabricator's facilities?
- Have any test results been received from the fabricator?

14.13.2 Setting out

- Are elastic compression calculations available?
- Is subcontractor's method statement available?
- Are main grid lines established?
- Has subcontractor's setting out been checked and is it being constantly re-checked?
- Is setting out to the accuracy required by the specification?
- Are adequate records being maintained to indicate accuracies and variations from specification?
- Are baseplates accurately aligned and levelled?
- Are following trades adversely affected by deviated steelwork?

14.13.3 Handling and storage

- Is storage of delivered components dry, well ventilated and secure?
- Are components stored in a manner that will not cause long term distortion?

14.13.4 Anchorages and foundation bolts

- Are foundation bolts accurately positioned with respect to line and level and have they been approved by the designer?
- If anchorages require to be drilled, are the concrete foundations accurately positioned to receive drill holes?

14.13.5 Erection

- Are legal Health and Safety requirements in place?
- Is security adequate?
- Are there safeguards to ensure that no final bolting or welding takes place until structure is properly aligned and approved by the designer?
- Do all steel components comply with specification?
- Is all wind and other bracing correctly positioned?
- Are sag rods properly positioned and correctly tightened?
- Is grouting-up of stanchion bases proceeding to specification?
- Has required testing been carried out (e.g. ultrasonic tests on welds)?

14.13.6 Site connections

- Do bolts comply with specification?
- Are bolt lengths sufficient to allow a two full thread projection beyond the nut after tightening?
- Is correct tightening procedure being followed? This is particularly important when HSFG bolts or similar components are being used.
- Have mating surfaces between components been properly prepared?

- Has paint or other surface treatment been applied as soon as possible after erection?
- Where support steelwork for vibrating machinery is being erected are correct bolts being used?
- Has required testing been carried out?
- Is site welding to specification and are welders suitably qualified?

14.13.7 Protection of steelwork

- Has surplus lubricant been removed before painting?
- Is the preparation to specification?
- Is painting to specification?

Further reading

Standards and Codes of Practice

The following list of current British Standards represents some of the more commonly used documents in structural steelwork construction. It should always be ensured that the most current version of each standard is used for reference.

General

BS 8888: 2002 *Technical Product Documentation* (TPD).
BS EN ISO 1660: 1996 *Technical Drawings. Dimensioning and Tolerancing of Profiles.*
BS EN ISO 7083: 1995 *Technical Drawings. Symbols for Geometrical Tolerancing.*
BS EN ISO 4157: 1999 *Construction Drawings. Designation Systems.*
BS EN ISO 6284: 1999 *Construction Drawings. Indication of Limit Deviations.*
BS EN ISO 8560: 1999 *Construction Drawings. Representation of Modular Sizes, Lines and Grids.*
BS EN ISO 9431: 1999 *Construction Drawings. Spaces for Drawing and for Text, and Title Blocks on Drawing Sheets.*
BS EN 1261: 1995 *Fibre Ropes for General Service. Hemp.*
BS EN 696: 1995 *Fibre Ropes for General Service. Polyamide.*
BS EN 697: 1995 *Fibre Ropes for General Service. Polyester.*
BS EN 698: 1995 *Fibre Ropes for General Service. Manila and Sisal.*
BS EN 699: 1995 *Fibre Ropes for General Service. Polypropylene.*
BS EN 700: 1995 *Fibre Ropes for General Service. Polyethylene.*
BS EN 701: 1995 *Fibre Ropes for General Service. General Specification.*
BS 4278: 1984 *Eyebolts for Lifting Purposes.*
BS 6166–3: 1988 *Lifting Slings.*
BS EN 12385–3: 2004 *Selection, Care and Maintenance of Steel Wire Ropes.*
BS 6994: 1988 *Steel Shackles for Lifting and General Engineering Purposes.*
BS 7121–1: 2006 and BS 7121–5: 1997 *Safe use of Cranes.*
BS 4604–1: 1970–2: 1970 *The use of HSFG Bolts in Structural Steelwork.*

Materials and products

BS 4–1: 1993 *Specification for Hot-rolled Sections.*
BS EN 10111: 1998 *Steel Plate, Sheet and Strip. Carbon and Carbon-manganese Plate, Sheet and Strip. General Specification.*

Note: This Standard is in several parts dealing with such topics as, hot and cold rolled steel plate; sheet and strip; formability; strength; heat treatment and general engineering purposes.

BS EN 10162: 2003 *Specification for Cold-rolled Steel Sections.*

BS EN 10225: 2001 *Specification for Weldable Structural Steels for Fixed Off-shore Structures.*

BS EN 10020: 2000 *Defintion and Classification of Grades of Steel.*

BS EN 10025: 1993 *Hot-rolled Products of Non-alloy Structural steels. Technical Delivery Conditions.*

BS EN 10027: 1992 *Designation systems for steels.*

BS EN 10029: 1991 *Specification for Tolerances on Dimensions, Shape and Mass for Hot-rolled Steel plates 3 mm Thick or Above.*

BS EN 10048: 1997 *Hot-rolled Narrow Steel Strip. Tolerances on Dimensions and Shape.*

BS EN 10051: 1992 *Specification for Continuously Hot-rolled Uncoated Plate, Sheet and Strip of Non-alloy and Alloy steels. Tolerances on Dimensions and Shapes.*

BS EN 10052: 1994 *Vocabulary of Heat Treatment Terms for Ferrous Products.*

BS EN 10079: 1993 *Definition of Steel Products.*

BS EN 10095: 1999 *Heat Resisting Steels and Nickel Alloys.*

BS EN 10113: 1993 *Hot-rolled Products in Weldable Fine Grain Structural Steels.*

BS EN 10130: 1999 *Specification for Cold-rolled Low Carbon Steel Flat Products for Cold Forming. Technical Delivery Conditions.*

BS EN 10137: 1996 *Plates and Wide Flats made of High Yield Strength Structural Steels in the Quenched and Tempered or Precipitation Conditions.*

BS EN 10142: 2000 *Specification for Continuously Hot-dip Zinc Coated Low Carbon Steels, Strip and Sheet for Cold Forming. Technical Delivery Conditions.*

BS EN 10155: 1993 *Structural Steels with Improved Atmospheric Corrosion Resistance. Technical Delivery Conditions.*

BS EN 10163: 1991 *Specification for Delivery Requirements for Surface Condition of Hot-rolled Plates, Wide Flats and Sections.*

BS EN 10164: 1993 *Steel Products with Improved Deformation Properties Perpendicular to the surface of the Product.*

BS EN 10210: 1994 *Hot Finished Structural Sections of Non-alloy and Fine Grain Structural Steels.*

BS EN 10219: 1997 *Cold Formed Welded Structural Sections of Non-alloy and Fine Grain Steels.*

BS EN 10258: 1997 *Cold-rolled Stainless Steel Narrow Strip and Cut lengths. Tolerances on Dimensions and Shape.*

BS EN 10259: 1997 *Cold-rolled Stainless and Heat Resisting Steel Wide Strip and Plate/ Sheet. Tolerances on Dimensions and Shape.*

Design

BS 449–2: 1969 *Specification for the Use of Structural Steel in Building. Metric Units.* [Now partially replaced by BS EN 1993–1: 2006]

BS 2853: 1957 *Specification for the Design and Testing of Steel Overhead Runway Beams.* [See also BS EN 1993–6: 2007]

BS EN 5400–1: 1988 *et al. Steel, Concrete and Composite Bridges.*

BS EN 5400–1: 1988 *Structural use of Steelwork in Buildings.*

Bolting

BS 2583: 1955 *Specification for Podger Spanners.*

BS EN 7371–12: 2008 *Specification for Electroplated Coatings on Threaded Components.*

BS 3692: 2001 *ISO Metric Precision Hexagon Bolts, Screws and Nuts.*

BS 3693: 2001 *ISO Metric Black Hexagon Bolts, Screws and Nuts.*
BS 4320: 1968 *Specification for Metal Washers for General Engineering Purposes. Metric Sizes.*
BS 4395: 1969 *Specification for High Strength Friction Grip Bolts and Associated Nuts and Washers for Structural Engineering.*
BS 4604–1: 1970–2: 1970–3; 1973 *Specification for the Use of High Strength Friction Grip Bolts in Structural Steelwork. Metric Series.*
BS 4933: 1973 *Specification for ISO Metric Black Cup and Countersunk Head Bolts and Screws with Hexagon nuts (obsolescent).*

Welding processes

BS 499–1: 1991 *Welding Terms and Symbols. Glossary for Welding, Brazing and Thermal Cutting.*
BS EN 22553: 1995 *Welding Terms and Symbols. European Arc Welding Symbols in Chart Form.*
BS 638–4: 1996; –5: 1988; –7: 1984; –9: 1990/BS EN 60974: 1996–2003 *Arc Welding Power Sources, Equipment and Accessories.*
BS 4872–1: 1982; –2: 1976 *Specification for Approval Testing of Welders when Welding Procedure Approval is not Required.*
BS EN 287: 1992; –1: 2004 *Approval Testing of Welders for Fusion Welding.*
BS EN 288: 1992 *Specification and Approval for Welding of Metallic Materials. General Guidance for Arc Welding.*
BS EN 22553: 1995 *Welded, Brazed and Soldered Joints. Symbolic Representation on Drawings.*
BS EN 24063: 1992 *Welding, Brazing, Soldering and Braze Welding of Metals. Nomenclature of Processes and Reference Numbers for Symbolic Representation on Drawings.*

Welding consumables

BS EN 440: 1995 *Welding Consumables. Wire Electrodes and Deposits for Gas Shielded Metal Arc Welding of Non-alloy and Fine Grain Steels. Classification.*
BS EN 499: 1995 *Welding Consumables. Covered Electrodes for Manual Arc Welding of Non-alloy and Fine Grain Steels. Classification.*
BS EN 756: 1996 *Welding Consumables. Wire Electrodes and Wire-flux Combinations for Submerged Arc Welding of Non-alloy and Fine Grain Steels. Classification.*
BS EN 757: 1997 *Welding Consumables. Covered Electrodes for Manual Arc Welding of High Strength Steels.*
BS EN 758: 1997 *Welding Consumables. Tubular Cored Electrodes for Metal Arc Welding with and without a Gas Shield of Non-alloy and Fine Grain Steels. Classification.*
BS EN 760: 1996 *Welding Consumables. Fluxes for Submerged Arc Welding. Classification.*
BS EN 1599: 1997 *Welding Consumables. Covered Electrodes for Manual Arc Welding of Creep-resisting Steels. Classification.*
BS EN 1668: 1997 *Welding Consumables. Rods, Wires and Deposits for Tungsten Inert Gas Welding of Non-alloy and Fine Grain Steels. Classification.*

Surface preparation and coating

BS 5493: 1977 *Code of Practice for Protective Coatings of Iron and Steel Structures Against Corrosion (obsolescent).*
BS 7079–A3: 1989 *Preparation of Steel Substrates before Application of Paints and Related Products. Visual Assessment of Surface Cleanliness.*
BS EN 10143: 1993 *Continuously Hot-dip Metal Coated Steel Sheet and Strip. Tolerances on Dimensions and Shape.*

BS EN 12329: 2000 *Corrosion Protection of Metals. Electrodeposited Coatings of Zinc with Supplementary Treatment on Iron and Steel.*

BS EN 12330: 2000 *Corrosion Protection of Metals. Electrodeposited Coatings of Cadmium or Steel.*

BS EN ISO 1461: 1999 *Hot dip Galvanized Coatings on Fabricated Iron and Steel Articles. Specifications and Test Methods.*

BS EN ISO 8503–1: 1995 *Preparation of Steel Substrates Before Application of Paints and Related Products. Surface Roughness of Blast-cleaned Steel Substrates.*

BS EN ISO 14713: 1999 *Protection against Corrosion of Iron and Steel in Structures. Zinc and Aluminium Coatings. Guidelines.*

Inspection and testing

BS 709: 1983 *Methods of Destructive Testing Fusion Welded Joints and Weld Metal in Steel.*

BS 3683–2: 1985 *Glossary of Terms Used in Non-destructive Testing. Magnetic Particle Flaw Detection.*

BS 6072: 1981 *Method for Magnetic Particle Flaw Detection.*

BS EN 571: 1997 *Non-destructive Testing. Penetrant Testing. General Principles.*

BS EN 970: 1997 *Non-destructive Examination of Fusion Welds. Visual Examination.*

BS EN 1330–3: 1997 *Non-destructive Testing. Terminology. Terms Used in Industrial Radio-graphic Testing.*

BS EN 1330–4: 2000 *Non-destructive Testing. Terminology. Terms Used in Ultrasonic Testing.*

BS EN 1330–5: 1998 *Non-destructive Testing. Terminology. Terms Used in Eddy Current Testing.*

BS EN 1435: 1997 *Non-destructive Examination. Radiographic Examination of Welded Joints.*

BS EN 1714: 1998 *Non-destructive Examination of Welded Joints. Ultrasonic Examination of Welded Joints.*

BS EN 10045–1: 1990 *Charpy Impact Tests on Metallic Materials. Test Method (V-and U-notches).*

BS EN 10160: 1999 *Ultrasonic Testing of Steel Flat Product of Thickness Equal or Greater than 6 mm (Reflection Method).*

BS EN ISO 12706: 2001 *Non-destructive Testing. Terminology. Terms Used in Penetrant Testing.*

Swedish standards

SIS 05 59 00 *Pictorial Surface Preparation for Painting Steel Surfaces.*

Construction

BS 5531: 1988 *Code of Practice for Safety in Erecting Structural Frames.*

BS 9999: 2008 *Fire Precautions in the Design, Construction and Use of Buildings.*

BS 5606: 1990 *Guide to Accuracy in Buildings.*

Health and Safety documents

See also Chapter 19 and the HSE website *www.hsebooks.co.uk.*

Guidance Note GS28–1 *Initial Planning and Design.*

Guidance Note GS28–2 *Site Management and Procedures.*

Guidance Note GS28–3 *Working Places and Access.*

Guidance Note GS28–4 *Legislation and Training.*

Note: GS28 is under revision.

Guidance Note GS15 *General Access Scaffolds.*

Guidance Note GS42 *Tower Scaffolds*.
Guidance Note PM30 *Suspended Access Equipment*.
Guidance Note GS31 *Safe use of Ladders, Step Ladders and Trestles*.
Guidance Note HS(G)19 *Safety in Working with Power Operated Mobile Work Platforms*.

Building Research Establishment documents

See also the BRE website *www.bre.co.uk*
Digest 234: *Accuracy in Setting Out*.

CIRIA

See also the CIRIA website *www.ciria.org.uk/ciria/*
CIRIA Report 174: 1997 *New Paint Systems for the Protection of Constructional Steelwork*.

BCSA publications

See also the BCSA website *www.bcsa.org/publications*
Guidance notes: *Safer Erection of Steel-framed Buildings*.
Erectors Manual: Ref. 16/93.
Structural Steelwork Erection: Ref. 20/89.
Health and Safety Guide for Managers and Supervisors: Ref. 29/00.
National Structural Steelwork Specification for Building Construction (NSSS): Ref. 203/94.
Commentary on NSSS: Ref. 209/96.
Acceptance Inspection of Grade 10.9 Structural Fasteners for Controlled Tightening: Ref. E64/91.
Slip Factors of Connections with HSFG Bolts: Ref. E37/84.
HD Bolt Systems for Steel Stanchions: Ref. 8/80.
Metric Practice for Structural Steelwork (3rd ed.): Ref. 5/79.
Historical Structural Steelwork Handbook: Ref. 11/84, W. Bates.
Structural Steel Fasteners and their Application: Ref. July 1978.
Manual on Connections: Ref. 19/88.

Institution of Structural Engineers publications

The Operation and Maintenance of bridge access gantries and runways (2nd ed.): October 2007.

Books

Civil Engineer's Reference Book (4th ed.) (L. S. Blake ed.): W. H. Arch, Oxford, Butterworth, 1989.
Construction Materials Reference Book: D. K. Doran (ed.), Oxford, Butterworth-Heinemann, 1991. [2nd ed. anticipated in 2010.]

Chapter 15   TIMBER

Paul Newman
revised by Andy Pitman

Basic requirements for best practice

- **Accurate specification**
- **Adequate control of moisture content**
- **Dimensionally accurate scantlings**
- **Defect-free timber**

15.1 Scope

This chapter deals with timber used as formwork, falsework, carcassing, structural components and joinery. For comprehensive guidance on the design, specification and erection of timber-framed buildings the reader is referred to guides published by TRADA (see references).

15.2 Material properties

Timber is a natural material; therefore, unlike other manufactured materials, its efficient utilisation is dependent on some form of selection and grading. An understanding of the characteristics and properties of timber as a raw material will enable the designer or user to ensure that it is used to best effect.

15.2.1 Classification and nomenclature

Botanical classification allocates plants into divisions which have certain gross features in common; these are successively subdivided on the basis of similarity into smaller orders, then families, genera and eventually species.

The botanical names of timber species are those of the trees from which the timber is produced. It is important to use the proper species name if the specifier or user wants to get exactly what he is looking for because different species of the same genus can produce timbers with quite different features, e.g. sapele with its pronounced stripe figure and utile which is quite plain, are different species of the genus *Entandrophragma* (i.e. *E. cylindricum* and *E. utile*) respectively.

Trade or common names can be misleading since the same name can be used for different species in different countries or they may differ according to tradition in regions of a country. BS EN 13556: 2003 *Round and sawn timber. Nominclature of timbers used in Europe* indicates the botanical species and preferred commercial, or standard names. It also indicates where a commercial name is botanically inappropriate.

15.2.2 Softwood and hardwood

The commercial division of timbers into hardwoods and softwoods has evolved from long traditions when the timber trade was dealing with a limited range of species. Today, however, this division bears no relation to the softness or hardness of the timber. Softwoods are produced from coniferous or cone bearing trees which have needle-like leaves and are mostly evergreen e.g. pines and yew. Hardwoods are produced from broad-leaved trees which produce seeds contained in an enclosed case or ovary, for example an acorn or walnut.

15.2.3 Structure

All living organisms are composed of cells. In the growing tree different cell tissues perform different functions; some store and convey food manufactured in the leaves, some convey liquids while others provide strength and elasticity to the tree to withstand storms and gales, and to support a heavy crown of branches. Most of the conducting and supporting cell tissue is vertically arranged in the tree as can be seen from the grain direction of timber. Food storage and conduction across the tree is largely a function of narrow ribbons of weak cells, known as rays, which run across the grain.

15.2.4 Sapwood, heartwood and pith

The trunks of many trees show a darker coloured area in the centre; this is the heartwood. Its function is almost entirely to provide the mechanical support of the tree. Around the outside of the heartwood is a ring of often lighter coloured sapwood which conducts water from the roots to the leaves. Not all trees show a difference in colour between the sapwood and heartwood, but both exist in all mature trees. The cells undergo chemical and physical changes in their conversion from sapwood into heartwood. Food or waste materials may be converted into complex organic substances resulting in colour changes; outgrowths from the cell walls, called tyloses, may occur, blocking the conducting cells so that wood that was permeable as sapwood may become impermeable as heartwood.

Sapwood should be regarded as having low resistance to fungal or insect attack regardless of species; the resistance of heartwood varies considerably depending on the species. Trees with richly coloured heartwood can provide highly decorative timber. Many of the softwoods used for structural purposes, e.g. European redwood, will today nearly always contain a significant amount of sapwood as the trees now harvested are small in diameter and contain a higher proportion of sapwood.

15.3 Moisture content

15.3.1 Moisture in timber

Wood will always contain water, and 'green' wood as felled is saturated, with the cell cavities full of water. Moisture content (mc) is expressed as a percentage of the oven dry weight of wood, thus green timber may have a moisture content greater than 100%. As the wood dries out, water leaves the cell voids until the moisture content of the wood reaches about 28%. This does not produce significant shrinkage. Below this value water, which is bound in the cell walls, starts to be removed and the wood will shrink as it dries. Similarly expansion occurs as the moisture content increases up to this level—the 'fibre saturation point'.

$$mc\ (\%) = \frac{\text{weight of wet wood} - \text{weight of dried wood} \times 100}{\text{weight of dried wood}}$$

Wood loses moisture until the amount of water in the wood balances the amount of water in the air around it, i.e. the wood reaches equilibrium. The equilibrium moisture content will vary as the humidity of the surrounding air changes. Wood which is thoroughly air dried under cover in the UK will attain a moisture content of around 18%. To achieve moisture contents lower than this, kiln drying will be necessary. Correct specification is particularly important if the timber is to be used in heated buildings. Timber below about 20% moisture content is also sufficiently dry to prevent fungal decay occurring.

15.3.2 Drying

The process of commercial timber drying involves evaporating moisture from the surface of the wood. This is usually carried out by heating the timber in a kiln although timber can also be air dried either outside or preferably in a properly designed drying shed. The surface layers dry by evaporation and the deep-seated moisture gradually moves towards the surface, setting up a moisture gradient in the piece—drier towards the outside and wetter towards the centre. Timber drying is a relatively slow process which may take several days to several months, depending on the species of timber, its thickness and the drying facilities used. Drying is more rapid from the ends of boards or logs, unless they have been treated with an end grain sealing treatment. Timber which is dried in a kiln is usually given a conditioning treatment at the end of the process to even out the distribution of moisture in the pieces.

15.3.3 Measuring moisture content

There are two commonly used methods of measuring moisture content in wood. One, using moisture meters, is a non-destructive test which is based on the fact that the electrical properties of timber vary with its moisture content. This method is particularly appropriate for routine checking (e.g. quality control) and site use. The second method is destructive in that it involves taking the piece of timber (or a sample from it), weighing it to determine the mass of the wood plus water, drying it completely to obtain the dry weight, and using the formula given earlier to calculate the moisture content. This method gives values that are more accurate than those provided by electric moisture meters.

Most common meters work on the principle that as the moisture content of a piece of timber increases, its electrical resistance decreases. Thus timber with a low moisture content has a high resistance. Conductivity is usually measured between two or more pin or blade-like electrodes which are pushed or hammered into the timber. Further information is given in the TRADA Wood Information Sheet: 'Moisture meters for wood'.

The main advantage of a moisture meter is that it gives instant readings which, although they may not be highly accurate, can be repeated many times to give an overall picture of moisture content and its distribution. The portability of the instrument is also a major advantage and moisture meters have gained wide acceptance in determining moisture content during the processing, storage, transportation, installation and in-service checking of timber and timber products.

Temperature and species corrections should be applied to improve the accuracy of the results. Even so, the values will not be completely accurate because the relationship between conductivity and moisture content is not precise. The usual standard expected of such meters is that most readings will be within 2% moisture content of the true level between about 8 and 25%. Some meters have scales which suggest they can be used outside this range but such readings should be viewed as indicative only, unless proved by calibration against oven drying.

The readings given by moisture meters are influenced by the presence of salts in the wood, such as those from waterborne preservatives, flame retardant treatments and contamination by sea water. Such salts increase the conductivity of the wood, particularly at higher moisture contents, and give a falsely high reading for moisture content. The magnitude of the effect is variable and it is not usually possible to correct the readings. Therefore when the presence of such salts is suspected, moisture readings which are higher than expected should be treated with suspicion. When using moisture meters on panel products, be aware that panels are not consistent between brands. Correction values are available for some board products from some meter manufacturers. For others, the moisture content readings obtained with resist-ance-type meters should not be treated with confidence. The effect with some varieties of exterior plywood can be particularly severe. In extreme cases, meters may indicate moisture values approximately double those obtained using the oven dry method. If high accuracy is required, advice should be sought from the board or meter manufacturer. A small calibration device can be used to verify the correct operation of most types of resistance meters. These are available from meter manufacturers and independent suppliers.

15.3.4 Moisture content of timber in service

The moisture content of wood changes in response to the temperature and humidity of its surroundings i.e. it is a hygroscopic material. In constant conditions of temperature and humidity, the timber will, in theory at least, eventually reach a constant moisture content—the so-called equilibrium moisture content (emc) for those conditions. In practice, such stability of conditions rarely occurs and therefore a true equilibrium moisture content is never reached. The response of timber to changes in temperature and humidity is quite slow and tends to average out minor fluctuations

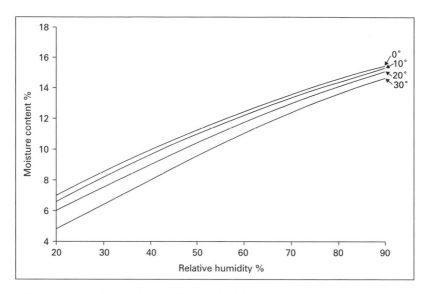

Figure 15.1 Averaged emc values derived from recent research on 20 species
(temperatures in °C).

in conditions such as 24-hour variations in central heating. The outer layers of the timber respond more rapidly to changes than the inner sections of a piece.

Protective or decorative coatings, such as paints, varnishes and exterior wood finishes slow down the response to a degree roughly related to the thickness of the coating. They will not prevent the moisture content of the timber from changing. In general terms, changes in moisture content of timber in buildings are measurable on a seasonal basis, rather than in terms of days or weeks.

The actual emc value varies slightly between different species of timber and also depends on whether the wood had to gain or lose moisture to reach the equilibrium level. Wood-based boards, such as plywood, chipboard and fibre building board often have lower emc values than the timbers from which they were made.

Figure 15.1 shows averaged emc values of 20 species of timber widely used in the UK over a temperature range of 0–10 °C and a humidity range of 35 to 75% RH. Table 15.1 gives approximate emc ranges for timber in a range of different service environments.

15.3.5 Shrinkage

Figure 15.2 shows generalised swelling and shrinkage characteristics of wood. The characteristics are stable above fibre saturation point and are shown as a linear change in dimension as the moisture content reduces below this level. This is a very simple approximation of a complicated picture but is quite adequate for a 'rule of thumb' assessment. Actual shrinkage and movement values vary between species and can be influenced by mechanical restraint. For most practical purposes the following assumptions should suffice:

- timber does not shrink or swell significantly along the grain
- shrinkage starts as the timber dries below about 30% moisture content

Table 15.1 Equilibrium moisture content for timber in a range of different service environments.

Service environment	Moisture content %
Structural timber	
External exposed	14–25 + (depending on weather conditions)
External covered	14–20 (depending on weather conditions)
Internal (unheated and heated)	6–12 (depending on heating regime)
Joinery	
External joinery	12–19
Internal joinery (unheated)	12–16
Internal joinery (heated to 12–21 °C)	9–13
Internal joinery (heated to above 21 °C)	6–10
Internal floors (from BS 8201)	
Unheated	15–19
Intermittent heating	10–14
Continuous heating	9–11
Underfloor heating	6–8

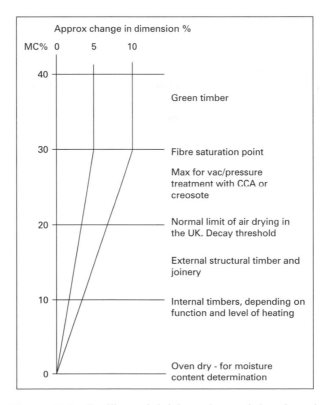

Figure 15.2 Swelling and shrinkage characteristics of wood.

- it shrinks and moves almost twice as much across the width of a flat sawn board (i.e. in a tangential direction) as it does across a quarter sawn board (i.e. radially)
- tangential shrinkage or swelling can be estimated as roughly 1% for every 3% change in moisture content below 30%; radial shrinkage is about half this.

If timber is put into service at a moisture content higher than that which it is likely to reach in time (i.e. higher than the likely emc), two interrelated problems can occur. One is shrinkage and the other is distortion. For example, if a piece of timber at 25% mc is put into a 16% emc environment it can be expected to shrink 3% tangentially and 1.5% radially. This may not seem much on a 25 × 25 mm batten but it represents 13 mm on a 600 mm deep laminated beam and 230 mm on a 10 m wide floor. Whilst it is unrealistic to apply high precision to matters involving moisture content, a severe mismatch between the moisture content of supply, storage or installation and the timber's eventual emc will often lead to problems in service.

Distortion is caused by the difference in shrinkage in a tangential direction compared with that radially, coupled with the fact that the grain of a piece of timber rarely runs true. Thus large changes in moisture content below fibre saturation point can result in the bowing or twisting of studs or the cupping of floor or cladding boards. Careful design to accommodate anticipated movement, coupled with sensible moisture content specification will avoid such problems.

15.3.6 Movement

As explained above, a stable emc is never reached in practice. The change in dimension exhibited by timber after its initial drying shrinkage is termed movement. This varies between species. Table 15.2 indicates the movement values of a number of common species. The classes are based on the sum of the tangential and radial movements corresponding to a change in humidity conditions from 90% to 60% relative humidity at 20°C.

Table 15.2   Movement values of some common species.

Small			
abura	aformosia	afzelia	agba
andiroba	cedar, S. American	Douglas fir	gedu nohor
guarea	hemlock, western	idigbo	iroko
jelutong	lauan	limba	mahogany, African
mahogany, American	makore	mengkulang	meranti
merbau	missanda	muhuhu	muninga
obeche	rosewood	sepetir	Sitka spruce
teak	utile	western red cedar	
Medium			
ash	birch	danta	elm, English
jarrah	keruing	mansonia	maple, rock
niangon	oak, American red	oak, European	omu
opepe	parana pine	redwood, European	sapele
sycamore	whitewood, European		
Large			
beech, European	karri	ramin	

- Small: under 3.0%
- Medium: 3.0–4.5%
- Large: over 4.5%

Where movement tolerances are critical, a timber with small movement characteristics should be considered.

15.4 Choice of species

Timber combines a number of properties that make it a uniquely attractive material for a wide range of purposes. Colours range from pale cream through browns to black; some are red, grey, green; some are plain, some are highly figured and others striped. The variety is matched by the range of properties available—from light and soft materials to hard, strong and heavy.

Timber selection needs to be based firstly on its suitability for the job. The specifier must identify the hazards likely to be encountered during the life of the product or component under consideration and must define the life expected. For structural purposes, strength and mechanical properties are likely to be the predominant factor. Natural durability or ease of preservative treatment is important in the choice of timbers to be used externally or in situations where high moisture contents may prevail. For interior situations, factors such as hardness, wear resistance and working qualities may be more significant.

The initial selection criteria will define a number of species that are potentially suitable for a particular purpose. The choice can then be refined on the basis of other factors, such as colour and texture, cost and availability.

The range of timbers available in the UK is wide and varies from time to time with changes in the political and economic situation, both here and in the supplying countries. The availability of a particular species in the sizes required should be checked before design and specification. Table 15.3 lists a number of different timber species and gives their common uses and density along with a basic comment on material properties. Further information is available in the TRADA Wood Information Sheet: 'Timbers—their properties and uses' and in the TRADA booklets *Timbers of the World*.

15.5 Size

Timber is an internationally traded material and the UK has traditionally imported timber from a wide range of countries, as well as producing some of its own. Although timber structures and non-structural applications can be designed using material of a wide variety of sizes, the sawmills producing timber do not know the end use for the material at the time it is cut. They therefore produce a range of, customary sizes. Specifying timber in these sizes is the most economic method, both in terms of material usage and cost.

Specifying timber sizes can be more complex than specifying the sizes of other structural materials because the dimensions of a piece of wood vary depending upon its moisture content. The terminology associated with the specification of timber sizes can cause confusion to those who are not used to the trade.

For further information the reader is referred to TRADA Wood Information Sheet. 'Softwood sizes—European standards' and the relevant European Standards listed in the reference section.

15.6 Strength grading

Strength grading provides a means of assessing the strength of a piece of timber. This can be carried out visually or by machine. Changes to the standards governing structural timber design, most significantly, the introduction of European strength classes, have resulted in changes to strength graded timber. Another significant change is in the name—what was known in the UK as 'stress grading' is now 'strength grading'.

Visual strength grading rules define the size, type and number of strength-reducing characteristics allowed in each grade. Strength-reducing characteristics include natural features such as knots, wane and slope of grain, plus splits and shakes which may have developed as a result of drying. The grader assesses each piece and stamps it with the appropriate mark. Since the strength of timber is species related, the same grade in different species will have different strengths. This results in a very wide range of combinations of species and grade. However, to simplify design, species and grade combinations of similar strength are grouped together into strength classes. The strength classes are defined in BS EN 338 *Structural Timber*. Strength classes and are described in the TRADA Wood Information Sheet: 'European Strength classes and strength grading'.

Machine grading is based on the relationship between strength and stiffness. The machine grades each piece and stamps it with the appropriate mark. An additional visual assessment takes account of strength-reducing characteristics not automatically sensed by the machine.

The use of timber in construction in the UK is governed by the requirements of the building regulations. The regulations and their associated documents for all areas refer to BS 5268 The structural use of timber Part 2 Code of Practice for permissible stress design, materials and workmanship and to BS DD ENV 1995–1–1 Eurocode 5: Design of timber structures: Part 1.1 'General rules and rules for buildings as evidence of compliance with the regulations for general structural timber'.

Both BS 5268 Part 2 and Eurocode 5 refer to timber strength graded to the standards listed below:

- BS EN 518: *Structural Timber—Grading—Requirements for Visual Grading Standards*. Note: this standard defines the requirements for visual grading rules to be acceptable in Europe, it does not define grading rules itself.
- BS 4978: *Visual Strength Grading of Softwood*. This standard defines the visual grading rules used in the UK for strength grading softwoods. The rules comply with the requirements of BS EN 518.
- BS EN 519: *Structural Timber—Grading—Requirements for Machine Strength Graded Timber and Grading Machines*. This defines the requirements for grading machines and details the additional visual assessment required for machine strength graded timber.
- Economic Commission for Europe (ECE) *Sawn Timber. Recommended Standard for Strength Grading of Coniferous Sawn Timber*. This standard is little used in the UK.

Table 15.3 Table of timbers.

Species H: Hardwood S: softwood	Treated	Interior joinery	■	Flooring	Construction	Average density kg/m3	Normal uses
Afzelia	H	✓	✓	✓	✓	✓	830	Stable, durable, exudes yellow dye when damp
Ash, American/European	H		✓	✓		✓	710	Bending—good, resists shock loads
Beech, European	H		✓	✓	✓	✓	720	Bending and turning—excellent, hard wearing
Birch	H			✓			670	Bending—excellent
Cedar, S. American	H	✓	✓	✓			470	Stable, fragrant odour, density very variable
Cherry, American	H		✓	✓			580	Fine and straight grain
Chestnut, sweet	H		✓	✓	✓	✓*	560	Use non-ferrous fixings in exposed conditions
Douglas Fir	S	✓P	✓			✓*	530	Avoid damp contact with unprotected iron
Ekki	H					✓	1070	Very durable structural timber
Elm, European	H		✓	✓		✓	560	High resistance to splitting
Greenheart	H					✓*	1040	Very durable structural timber
Hemlock, western	S	✓P	✓	✓		✓*	500	Uniform in colour and texture
Idigbo	H	✓	✓	✓		✓	560	Stable, liable to iron-stain in damp conditions
Iroko	H	✓	✓		✓	✓*	660	Very resistant to decay, high strength
Jarrah	H				✓	✓*	820	Very resistant to decay, used in marine work
Kapur	H				✓	✓*	770	Pin worm holes limit usage
Karri	H				✓	✓*	900	Durable structural timber
Kerung	S	✓P	✓		✓	✓*	740	Tends to exude resin and split
Larch, European	S	✓	✓	✓	✓	✓	550	Cladding, decking, may be used in windows
Mahogany, African	H	✓	✓	✓	✓		530	Long established red wood
Mahogany, American	H		✓	✓	✓	✓	560	Stable and easily worked

Species	Type				Density	Characteristics
Maple, rock	H				740	Hardwearing, turns well
Meranti, dark red	H	✓P	✓		710	Uniform grain
Meranti, light red	H	✓P	✓		550	Uniform grain
Merbau	H	✓	✓	✓*	830	Stable, liable to iron-stain in damp conditions
Nyatoh	H	✓P	✓		720	Extensively used for garden furniture
Oak, American red	H		✓	✓	790	Bending—excellent
Oak, American white	H		✓	✓	770	Appearance similar to European oak
Oak, European	H	✓	✓	✓*	720	Hardwearing, bends well, liable to iron-stain
Obeche	H		✓		390	Stable
Opepe	H	✓	✓	✓*	750	Very resistant to decay
Parana pine	S	✓		✓*	550	Attractive colour variations
Pitch pine	S	✓		✓*	670	Very distinctive appearance
Ramin	H	✓	✓		670	Easily machined, used for mouldings
Redwood, European	S	✓	✓	✓*	510	Excellent for general joinery and construction
Rosewood	H		✓		870	Highly decorative
Rubberwood, Hevea	H	✓	✓		560	Used for furniture and manufactured goods
Sapele	H	✓			640	Pronounced stripe when quarter sawn
Scots pine (UK grown)	S		✓	✓*	510	Same species as European redwood
Spruce, Sitka	S		✓	✓	450	Appearance similar to whitewood
Sycamore	H	✓			630	Turning excellent
Teak	H	✓	✓	✓*	660	Stable, very resistant to decay and chemicals
Tulipwood 2	H	✓	✓		510	Furniture framing and interior joinery
Utile	H		✓	✓	660	Good general purpose hardwood
Virola, light	H	✓	✓		530	Easily worked
Walnut, African	H		✓		560	Decorative
Walnut, American	H	✓			660	Highly decorative
Walnut, European	H		✓		670	Highly decorative
Western red cedar	S	✓P	✓	✓*	390	Stable, fragrant odour, may corrode metals
Whitewood, European	S	✓P	✓	✓*	470	Excellent for general joinery and construction
Yew	S		✓		670	Variegated appearance

[1] The density quoted is for European ash, American ash is 560–670.

[2] Note the British Standard recommended name for tulipwood (Liriodendron tulipifera) is American yellow poplar

- BS 5756: *Specification for Visual Strength Grading of Hardwood*. This Standard applies to temperate and tropical hardwoods.

Timber imported from Canada and the USA may be graded to their national standards. Timber strength graded to the following standards is accepted by BS 5268 Part 2 and by Eurocode 5.

- National grading rules for dimension lumber. NLGA, Canada, 1994.
- National grading rules for softwood dimension lumber. NGRDL, USA, 1975.
- North American export standard for machine stress rated lumber, 1987, and supplement No. 1, 1992.

All structural timber used in the UK should be strength graded and marked under the supervision of a certification body approved by the UK Timber Grading Committee. The approved certification bodies are listed in a leaflet Approved Certification Bodies for the supply of strength graded timber, published by the Timber Trade Federation. The certification bodies oversee the training and certification of graders and the operation of grading machines. They also monitor the quality of grading carried out by the companies under their control.

The certification body or grading agency logo or mark forms part of the grade stamp which must appear on all strength graded timber.

15.6.1 Visual grading

BS 4978

BS 4978 lays down the rules for visual strength grading in the UK. These comply with the European requirements set out in BS EN 518. All timber graded to BS 4978 must be marked so that specifiers, inspectors and site staff are able to identify immediately the grade and moisture content level of the timber supplied.

Two principal grades are defined: GS—General structural and SS—Special Structural, graded at two levels of moisture content: dry and wet.

- *Dry graded timber* is assessed when the batch of timber has an average moisture content of 20% or less, with no reading exceeding 24%. Dry graded timber should be used for Service Classes 1 and 2. Timber over 100 mm target thickness is exempted from this requirement.
- *Wet graded timber* is graded at a moisture content above 20%. Timber specified for use in contact with water or for use in Service Class 3 defined in BS 5268–2.

Marking

Each piece of visually strength-graded softwood will be stamped indelibly on one face with:

- the grade (the strength class may also be marked)
- the species or species combination
- information to identify the company responsible for the grading
- the certification body
- the British Standard number i.e. BS 4978

Figure 15.3   Example of a visual strength grading mark.

- if dry graded timber, the word 'DRY' or if kiln dried, 'DRY' or 'KD'
- if wet graded timber, the word 'WET'.

See Fig. 15.3 for an example of a strength grading mark.

The Standard makes allowance for marking to be omitted for aesthetic reasons. In these circumstances a certificate of compliance can be issued for each parcel of a single grade of softwood.

Sizes

BS 4978 allows a minimum cross sectional area of 2000 mm² and a minimum thickness of 20 mm for strength graded softwood. Permissible deviations and processing reductions for constructional timber are given in BS EN 336. These are outlined in the TRADA Wood Information Sheet: 'Softwood sizes—European Standards'.

BS 5756

BS 5756 defines five strength grades:

- **HS** grade for tropical timber.
- **TH1** and **TH2** for general structural temperate hardwoods (oak and sweet chestnut)—for timber of a cross-sectional area less than 20 000 mm² and a thickness less than 100 mm. TH1 is a higher grade than TH2.
- **THA** and **THB** for heavy structural hardwoods (oak and sweet chestnut) for timber of a cross-sectional area of 20 000 mm² or more and a thickness of 100 mm or more. THA is a higher grade than THB.

Marking

Hardwoods strength graded in accordance with BS 5756 should be marked with:

- the species and grade (the strength class may also be marked)
- information to identify the company responsible for the grading
- the certification body
- the British Standard number i.e. BS 5756
- if dry graded timber, the word 'DRY' or if kiln dried, 'DRY' or 'KD'
- if wet graded timber, the word 'WET'.

15.6.2 Machine grading

BS EN 519

Most machines operating in the UK operate by applying a defined load to each piece of timber as it passes through the machine. The resulting deflection indicates the grade which is then marked on the piece. The machine settings need to be strictly assessed and controlled to maintain consistency of grading.

Softwood strength graded by machine in the UK or in Europe to BS EN 519 will normally be graded to the strength limits laid down for the nine softwood strength classes defined in BS EN 338. These are shown in the TRADA Wood Information Sheet: 'European strength classes and strength grading'. BS EN 519 does not define grades or strength classes for timber. Hence it is possible for special grades to be developed and defined for particular purposes. BS 5268–2 has recognised a special grade developed for the manufacture of trussed rafters, designated TR 26.

Each piece of machine strength graded timber will be marked with:

- the grade and/or strength class
- the species or species combination
- the Standard number i.e. BS EN 519
- information to identify the company and the machine responsible for the grading.

A National Annex to BS EN 519 for the UK defines 'Dry machine graded timber' and 'Wet machine graded timber' in the same way as for visually graded timber in BS 4978 (see above). Timber graded to BS EN 519 for use in the UK will also be marked with:

- the name or mark of the certification body
- for dry graded timber, the word 'DRY'; if kiln dried, the letters 'KD' may be used
- for wet graded timber, the word 'WET'.

Like BS 4978, BS EN 519 allows marking to be omitted in exceptional circumstances for aesthetic reasons and replaced by a certificate of compliance or each parcel of timber of a single class.

Sizes

BS EN 519 does not stipulate a minimum size but requires timber to meet the requirements of BS EN 336. This includes a National Annex which gives sizes of structural softwood customary in the UK, see Wood Information Sheet: 'Softwood sizes—European Standards'.

Timber machine strength graded in North America is likely to have been graded in an output-controlled machine. In this case, sufficient load is applied to each piece of timber to induce a defined deflection. The load required indicates the grade, which is marked on the piece. The consistency and quality of grading is checked by measuring the strength of timber specimens from the daily output of the machine.

15.7 Structural use of timber

BS 5268 The structural use of timber is the UK Code of Practice covering the major structural uses of timber, particularly in building. In total it consists of seven parts; Part 2 covers general structural timber.

BS 5268–2 brings together design stresses, design methods and some general guidance on the structural use of timber. It covers a wide range of softwoods, hardwoods, plywood, tempered hardboard, structural chipboard and jointing devices, and provides design information for solid timber, glulam and built-up composite structures such as box and 'I' beams.

Tables 15.4 and 15.5 which are shown below give the species and grade combinations and details of the strength classes that they satisfy for a range of commonly used hardwood and softwood species.

15.8 Storage

Bad handling and poorly organised storage of timber and wood-based products are major causes of wastage on building sites. The results of a study by the Building Research Establishment showed that for the main building materials the average wastage was in the order of 10%, with instances as high as 20% for some materials.

Table 15.4 Common softwood species and grade combinations that satisfy the requirements for strength classes C14–C30 when graded in accordance with BS 4978.

Species (source) Grades to satisfy strength class

	C14	C16	C18	C22	C24	C27	C30
British pine	GS			SS			
British spruce	GS		SS				
Caribbean pitch pine			GS			SS	
Chilean pine							
Douglas fir (Britain)	GS		SS				
Douglas fir—Larch (Canada, USA)		GS			SS		
Hem-fir (Canada, USA)		GS			SS		
Larch (Britain)		GS			SS		
Parana pine (imported)		GS			SS		
Redwood (imported)		GS			SS		
Sitka spruce (Canada)	GS		SS				
Spruce-pine-fir (Canada, USA)		GS			SS		
Southern pine			GS		SS		
Western red cedar (imported)	GS		SS				
Western white woods (USA)	GS		SS				
Whitewood (imported)		GS			SS		

Table 15.5 Hardwood species and grade combination which satisfy the requirements for strength classes D30–D70 when graded in accordance with BS 5756.

Species	Grade to satisfy strength class				
	D30	**D40**	**D50**	**D60**	**D70**
Balau					HS
Ekki				HS	
Greenheart					HS
Iroko		HS			
Jarrah		HS			
Kapur				HS	
Karri			HS		
Kempas				HS	
Keruing			HS		
Merbau			HS		
Oak	TH1, THB	THA			
Opepe			HS		
Teak		HS			

By observing a few simple rules about the correct specification of moisture content and by careful site storage 'drying out' problems such as shrinkage, distortion and staining can be averted. In situations involving high-class joinery or other prestige work, proper care takes on special significance.

15.8.1 Carcassing timber

When there is no suitable building available, spread a fine granular material such as sand or ashes on a well-drained space in the open and stack the timber on bearers to keep it off the ground. Bearers should be arranged so that the timber will lie flat, otherwise warping can result. The stack must be covered to keep off the rain, but in such a way as to allow free circulation of air and the ability to dry, thereby overcoming condensation problems under the covering. The covering is also essential to provide protection against direct sunlight.

15.8.2 Structural components

It is essential to store structural components, such as trussed rafters, timber frame components and ply-box beams properly. They should be stored above ground on levelled trestles or other suitable bearers, and generally laid horizontally. Like carcassing timbers, structural components should be covered by a tarpaulin to protect them from rain and sun. While providing this protection, attention should also be given to ensuring a good circulation of air around the components.

15.8.3   Trussed rafters

Where trusses are laid flat, bearers should be placed to give level support at close centres, sufficient to prevent long-term deformation of all truss members. If subsequent bearers are placed at different heights, they should be vertically in line with those underneath.

Alternatively, if trusses are stored in the upright position, stacking should be carried out against a firm and safe support. They should be supported at the positions where the wall plate would normally occur and at such a height as to ensure that the rafter overhang clears the ground. Such requirements should be clearly stated in the relevant specification.

Because of the variation in delivery times it is common practice to order trusses before commencement of the contract. This can result in trusses being available on site some considerable time before they are needed during which time they could suffer from the weather. TRADA strongly recommends that they should be sheeted over, but in such a way as to allow the circulation of air. Further information on the care of trussed rafters can be found in BS 5268–3. Timber should always be handled with care. Split ends and damaged edges and corners are caused mostly by carelessness. Use an extra line when hoisting roof members to guide them past projections.

15.8.4   Joinery

External joinery may generally be treated in a manner similar to that for structural components with support being chosen carefully to avoid warp or twist due to unnatural loading. Priming offers little protection against the uptake of moisture. Should the horns or any other sections of primed timber be cut off, the exposed bare timber should be reprimed. Similarly, in the case of timber that has treated with preservative, all cut ends borings, notchings etc. should be liberally swabbed with preservative. Ideally internal joinery and flooring should be kept in a heated dry store to maintain the correct moisture content. However, a garage completed earlier could be used as a convenient storage area, providing it is well ventilated. The stored components should not rest directly on fresh concrete and the material should be close piled and fully wrapped. Good ventilation is essential. Factory finished components such as kitchen units and fitments may be treated in a manner similar to that for internal joinery and flooring. Additionally, however, particular care is necessary to avoid damage, especially with completed units.

15.8.5   Board materials

Boards of all types must be kept flat and dry. Store as internal joinery and ideally boards should be 'sticked' to allow air to circulate.

15.9   Building programme

- Build to a programme that protects timber and joinery from the wet. Do not allow floor joists and roof timbers etc. to stand exposed to the elements for longer than necessary.

- Glazing before fixing floorings, linings etc. minimises the possibility of timbers being soaked and therefore reduces excessive shrinkage and twist.
- Plan deliveries of joinery to coincide with the progress of work so as to avoid prolonged site storage. Prepare storage in advance.
- Check quality, specifications and moisture content of the timber, and the standard and finish of the joinery at the time of delivery. This is the time to raise any points with the supplier.
- Ensure that proper protection has been afforded to the material in transit.
- Use proper mechanical handling equipment whenever possible. Do not damage the product, especially edges, corners and wrappings, which should be maintained as long as possible. Instruct and train handlers and always supervise off loading.
- Timber can pick up moisture from wet trades so ensure as much time as possible for drying out the building before introducing kiln-dried timber components.
- Heating should be introduced as soon as possible but care should be taken to ensure that only gentle heat is used at first.
- Ventilation at all stages of construction is vital.

15.10 Deterioration of timber

15.10.1 Timber, fungi and insect pests

Timber is composed of wood cells (fibres) with hollow centres. The cell walls comprise cellulose cemented together with lignin, a complex organic material constituting some 20% by weight of dry wood. The ratio of wood to air; the shape, variety and arrangement of the constituent fibres, and the small proportion of loosely bound 'extractives' vary to give timbers which differ widely (e.g. balsa to lignum vitae), mainly in terms of their physical and mechanical properties. The chemical constitution of various woods varies little except for the 'extractives' which confer the specific properties of colour, odour and resistance to bio-deterioration by fungi and insects.

Sugars and starch are found in parts of the living tree; they can persist in the log or converted timber for several years under the right conditions, and they can influence the susceptibility of timber to degradation by fungi and insects. Fungi can cause staining, decay and weakening; insects disfigure the timber or render it unserviceable by boring holes in it or consuming it. Fungi require the timber to have a moisture content of at least 20% if they are to develop and cause damage. Some insects on the other hand can attack 'dry' wood. In many situations of use, the moisture content of timber not in contact with the ground can be kept below the danger threshold for fungal attack by correct design and detailing to minimise wetting and to ensure continued weathertightness and adequate ventilation. Such measures do not, however, necessarily confer immunity to insect damage. It is for this reason that there is a role in wood preservation for formulations with insecticidal but not fungicidal properties. This is particularly the case in regions where termites are indigenous. Some timber species are resistant to fungal and insect attack. Typically, only the inner, or heartwood, zone contains the extractives which give resistant properties to the wood. It follows that the outer 'sapwood' is essentially unprotected from fungi and insects if the conditions for attack are satisfactory. This is important to bear in mind when the 'natural durability' of a timber is quoted. This is usually assessed by comparing the performance of wood stakes in ground contact against a reference species of known durability, and relates only to the heartwood.

The sapwood of most species is in the lowest group of the five-category durability classification. (The lowest group (Class 5) is considered 'not durable' whereas class 1 is 'very durable').

15.10.2   Fungi

The spores of wood-decaying or staining fungi are so widespread that avoiding attack by sanitary measures is impossible. Long service life in an environment suitable for fungal development depends on high natural durability or effective wood preservation with fungicidal chemicals. The degree of preservative protection required to prevent infection from spores is less than that needed to prevent colonisation from an existing fungal infection from adjacent wood or the soil.

Sap-stain fungi and moulds

Various fungi cause deep-seated stain (sap stain or blue-stain) or superficial discoloration (e.g. mould) of damp timber and wood products. Both types feed on the cell contents and stored food reserves and are therefore normally confined to sapwood. They do not cause loss of strength but can reduce the value of timber or even render it unsaleable by spoiling its appearance. Even if arrested in its early stages by reducing the timber moisture content, sap-stain or mould attack can revive if moisture content increases and cause disruption of paint or varnish films in service. A moisture content greater than 25% is required for active sap-stain development, whereas mould growth can continue down to about 20%, or even lower if high relative humidity persists. Sap-stain is of greatest consequence with softwoods and light coloured tropical hardwoods such as obeche, ramin, celtis, jelutong and pterygota, all species where even the heartwood does not contain sufficient extractive to provide protection. Debarked logs are most susceptible but infection can enter via log ends, branch stubs, damaged bark or bark beetle holes. Converted timber is also susceptible prior to drying.

Wet rot fungi

These fungi cause decay of wet timber and are subdivided into white and brown rots. The former destroy both the cellulose and lignin components of the timber whilst the brown rots attack the cellulose and leave the lignin as a brown residue. Numerous species of fungi are involved worldwide, varying in their vigour, timber preferences and susceptibility to control by wood preservatives. The first line of control is to keep the timber moisture content below 20%. In certain end uses, such as poles, posts and sleepers, this is impossible and reliance must be placed on the use of naturally durable species or wood preservatives. In other situations, such measures are a back-up defence or form of insurance against the failure of design and maintenance measures. Wet rot fungi cause loss of strength and eventually complete disintegration. Their rate of development depends on temperature, moisture content, oxygen availability and the degree of resistance of the timber.

Dry rot fungi

This title is a misnomer in that the brown rot fungi, which cause dry rot requires timber at a high (20%) moisture content. The distinctive feature of the dry rot is that it is able to translocate water over distance and form their own wet zone in previously dry timber under conditions of high relative humidity with poor ventilation. They are typically inhabitants of humid, unventilated spaces in buildings and once established, are persistent and require specific remedial treatment.

15.10.3 Insects

Worldwide, the most economically important insect pests of wood products are the various forms and species of termites. A wide range of beetle species whose larvae tunnel in wood for food and protection, are less important, though they are pests under some circumstances.

Termites

Termites are essentially a problem of tropical and subtropical regions but they also are found as a limited hazard to wood in service in some temperate countries, e.g. France, Japan, Korea, Germany. Their distribution is limited by low temperature due to geographical location and altitude. There are two main types which attack timber in service; dry-wood and subterranean termites. Of the 1800 or so species so far described, 10% have been recorded as causing damage in buildings, and 53 species are serious pests in this regard, 10 being dry-wood types.

Dry-wood termites live entirely inside the timber on which they are feeding, often hollowing large timbers but leaving a thin shell of outer wood for protection. Attack, once begun, takes place largely within the timber and may be well advanced before being recognised. Prevention must take the form of using resistant or chemically treated timber since building design features are ineffective against flying insects. Eradication is usually accomplished by a fumigation treatment but such treatment does not confer long-term protection which must be provided by *in situ* application of insecticides.

Subterranean termites live in nests in mounds or cavities in the ground. They do not produce frass as do the dry-wood termites and possess a different form of colony organisation. The nest sites may be hundreds of metres from the attack but movement is always within protective tunnels or runways constructed by the foraging workers from earth and chewed wood. Control of subterranean termite damage can be by the use of preservative treatment and design features to prevent entry from below ground. Species differ considerably in their distribution and voracity.

Beetles

The most important beetle pests of wood and wood-based products, particularly in tropical and subtropical regions, belong to the *Lyctidae* family e.g. *Lyctus Minthea*. The larvae form tunnels in the sapwood of solid timber or veneers of susceptible hardwood species. Susceptibility depends on two main factors: firstly the pores of the timber must be large enough in diameter to allow egg laying. This critical size varies with the species, but for *Lyctus brunneus* (a European species), is about 0.8 mm. Secondly, it must have a high starch content for, unlike most wood-boring beetles, these insects feed on the starch stored in the sapwood which can be completely disintegrated if conditions are favourable. Light-coloured, large-pored timbers, such as obeche and ramin are most susceptible. Softwoods, small-pored hardwoods and heartwood are not attacked. *Lyctus* beetles can cause damage to susceptible hardwoods in the UK often in local epidemics attributable to poor timber yard hygiene.

This group is also responsible for the only significant insect attack of panel products (plywood) in the UK.

Lyctid beetles are often known colloquially as 'powder post beetles', a name they share with the mainly tropical *Bostrychid* beetles which similarly produce powdery frass. Adult *Bostrychid* beetles bore short tunnels in which to lay eggs and so are not dependent on large-pored timber species. Attack is more readily detected early and protective measures can be put in hand. Logs and dried timber can be infested in the tropics. Both the adult beetles and larvae feed on the starch content of the sapwood of hardwoods. Imported timber infested with *Bostrychid* larvae is occasionally found in the UK but the attack quickly dies out and cannot spread.

Ambrosia beetles are small and infest freshly felled logs (or standing trees, if sickly or moribund). Attack by different species can occur in temperate or tropical regions, both softwoods and hardwoods being vulnerable. More than 1000 species have been named. The damage, caused by the tunnelling of the adult beetles is variously known as pinhole, pinworm or shothole and is characterised by small, circular holes with a dark lining and halo of stained timber. This colouration is caused by moulds which grow on the insides of the tunnels and on which the larvae and adult beetles feed. A high (over 35%) moisture content is required for this fungal 'ambrosia' to grow so that infestation dies out when the timber is dried. Attack of freshly felled logs, however, can be very rapid—within a few hours, unless protective treatment is carried out. Ambrosia beetle attack can penetrate to the centre of a tree and is not restricted to the sapwood. Somewhat unusually, it is the adult beetles that do the boring, creating circular holes 0.5–3 mm diameter according to the species of beetle, the tunnels being across the grain of the wood. The attack does not persist in dried timber and there is no danger of spread or reinfestation in the UK.

Longhorn beetles are widely distributed in tropical and temperate regions. They attack green i.e. wet, freshly felled and partially dried timber. Their name comes from the long antennae. Adults of some tropical species are 75 mm in length. Thus the oval tunnels which the larvae create can be large and destructive and so timber infested with this insect is rarely seen in the UK. Attack by forest longhorns can be prevented by prompt removal of bark from the felled logs but this process encourages Ambrosia beetle attack.

The house longhorn, *Hylotrupes bajulus* differs in its lifestyle from the forest longhorns described above, in that it infests and can reinfest dry sapwood of softwoods in service. It is a widely distributed species found in North and South America, South Africa and Europe. Damage is caused by the larvae which burrow into the timber from an egg laid on the surface. The larval stage can last up to 11 years and since a fully-grown larva can be 30 mm long, destruction of the timber can be extensive. Exit holes are oval and conspicuous. This is in contrast to the larval entry holes and the larval tunnelling which, although it may be very extensive and near the surface, is always behind a thin layer of undamaged wood.

In the UK the house longhorn is restricted by climatic factors to an area to the south-west of London where it is controlled by special provisions in the Building Regulations which require preservative treatment of roof timbers where damage by these beetles can lead to catastrophic failure.

Wood wasps are pests of sickly standing trees or freshly felled logs of coniferous trees. Eggs are laid through the bark and into the wood where the larvae that hatch from them, tunnel into sapwood and heartwood alike, pupate and eventually emerge as adults through circular flight holes. The fear of a wood wasp epidemic is a major factor behind the strict quarantine regulations imposed by Australia on imported wood products. Although three species of wood wasp exist in the UK they are not a problem in practice.

Various members of the beetle family *Anobiidae; Anobium* (woodworm or furniture beetle), *Xestobium* (death-watch), *Ernobium*, are significant pests of seasoned timber in service. The larvae use the timber as a food source and cyclic reinfestation can eventually lead to structural failure of building components and furniture. The various species have different preferences for timbers, but they are able to infest dry timber of a wide variety of species. In many situations, preservative treatment with a contact insecticidal formulation is the only practical preventative.

In tropical and subtropical countries many other insects infest timber in its various stages of utilisation, causing degradation by structural weakening or marred appearance. Many are of local or sporadic occurrence, so that only the most widespread and economically significant types have been introduced here. Fortunately, the variety and vigour of insect pests of timber are much lower in temperate regions than in warmer parts of the world.

15.11 Wood preservatives

15.11.1 Types of preservative

There are three basic types of material, which are suitable for the exterior treatment of wood: preservatives, paints, and pigmented exterior wood stains.

There are four main wood preservative types: tar oils, water borne, organic solvent and micro-emulsion. Wood preservatives (except for tar oils) are not usually designed or expected to perform, as exterior finishes. Their primary function is to penetrate and be retained in a 'shell' or envelope of the outer few millimetres or to a depth of several centimetres of timber to protect against stain, fungal decay, mould growth and insect attack. Surface treatments such as paints and exterior wood finishes will not alone confer protection so that susceptible timbers must usually be given an adequate preservative treatment prior to the application of the finish. Guidance on preservative treatments is given in BS 8417. *Preservation of Timber—Recommendations*.

Tar oils

One of the most important tar oil preservatives is creosote, which has been used extensively for transmission poles, fencing, sleepers and external timbers for over a century. When properly applied by pressure impregnation, or hot and cold tank steeping, it gives excellent durability and can be used in ground contact. Generally, very little maintenance is required but periodic surface applications may be necessary to renew the colour. More regular re-coating will be necessary if the original treatment was applied by brushing or cold dip. The smell and stickiness of creosote often rules out its use in many places and creosoted timber is not usually satisfactory for subsequent painting unless a sealer is used. It is however, oily and this has the effect of slowing down weathering. The use of creosote has now been restricted primarily to industrial end-use and traditional creosote is now no longer available over the counter for DIY applications. Creosote should not be used where there is a risk of 'frequent skin contact' so its use has been withdrawn for items such as garden fencing, playground equipment or for parks and gardens.

Water-borne preservatives

Most water-borne preservatives are applied by vacuum pressure impregnation. After impregnation, the traditional copper chromium arsenic types most frequently encountered become fixed and insoluble in the wood, giving high durability to external timber. With certain hardwoods there is little or no colour change, whereas some softwoods become greenish in appearance, although this tones down on weathering. There are CCA treatments available which are coloured to produce a more decorative effect. No maintenance is required but treated timber will become grey and dirty with prolonged exposure unless subsequently stained or painted. When CCA treatments are used as pretreatments for some exterior wood stains, they prolong the interval between necessary maintenance work. They have also been shown to retard the rate of natural weathering of unprotected timber as well as preventing decay. Water-borne preservatives may include a range of formulations used as alternatives to organic solvent based products.

These require the protection of a finish to prevent long-term leaching. Preservative formulations such as boron (disodium octaborate tetrahydrate) may exhibit salt efflorescence on dark-coloured, vapour-permeable finishes. Whilst this may be a nuisance, the performance of the coating will not be impaired.

Recent EU legislation has restricted the use of CCA water-based preservatives which contain arsenic and chromium. Recent developments with copper-based wood preservatives include the use of systems which replace the chromium and the arsenic with an organic biocide. These are accepted for use where the traditional CCA type preservatives are no longer permitted, i.e. for domestic situations such as decking, floor joists, and timber to be used in the garden or where it is likely to come in to 'frequent contact' with people.

Organic solvent-borne preservatives

Organic solvent preservatives can be applied by dip, double-vacuum or vacuum pressure impregnation treatments. They are sometimes applied by brush, but the degree of protection afforded is minimal. Some form of finish, either paint or pigmented wood-stain is essential for satisfactory exterior performance. Preservative treated wood should always be allowed to dry before a finish is applied. Organic solvent-based wood preservatives are available with added water repellents, which is advantageous for exterior use, but may interfere with the film-forming properties of applied coatings.

Micro-emulsion preservatives

In more recent times increasing legislation on the use and reduction in emissions of VOCs (Volatile Organic Compounds) had led to the development of water-based non-metal based formulations or emulsions for use as wood preservatives. A micro-emulsion can be defined as 'a clear, thermodynamically stable dispersion of two immiscible liquids'. In the micro-emulsion, water and hydrocarbon form an oil-in-water emulsion or alternatively water-in-oil emulsion. The active ingredients, organic biocides, and the carrier, water, are mixed with surfactants. These act as emulsifying agents and disperse the biocide.

15.11.2 Preparation of timber

In order to treat wood effectively with a preservative it should have a moisture content below f.s.p, (fibre saturation point) as it is impossible to effectively treat green timber other than by sap displacement or diffusion. If using organic solvents the timber should be dried to a moisture content appropriate to its end use. The timber should also be free from dirt, bark and surface finishes, which will affect the penetration of the preservative. Ideally, species with the same permeability should be treated together and for external joinery items it is advisable to treat the individual pieces prior to assembly, i.e. treat in stick form. Packs of timber and planed items should be separated by sticks or laths in order to allow greater access for the preservative solution between and around the timber. Some timber is incised prior to treatment which facilitates deeper penetration for items such as railway sleepers and telegraph poles.

15.11.3 Methods of application

There are various methods of application of wood preservatives and these include brushing, spraying, dipping, diffusion, steeping, immersion, and pressure treatment methods. One of the most effective commercial methods used in the UK is double vacuum pressure treatment. This method can be used for the application of micro-emulsion and organic solvent wood preservatives and provides effective long-term protection for timber in high hazard situations.

Double vacuum treatments are carried out in large steel pressure vessels where the wood preservative is applied to the timber under pressure, after an initial vacuum has been drawn, forcing the preservative deep into the timber. A final vacuum is then drawn after the preservative has been drained back to the storage tank. This assists with removing excess preservative from the surface of the timber.

Wood preservatives can also be applied by hand i.e. brushing or spraying but these methods do not provide effective protection for high hazard applications such as fencing and decking.

15.11.4 Commonly encountered problems

One of the main problems with preservative-treated wood is that it must be fully dried prior to use and before any finishes can be applied. This can lead to extended delivery times and delays to work being carried out on site. Precautions must also be taken when handling preservative treated timber whilst still damp. Also some wood preservatives are incompatible with finishes and adhesives. They can also have a corrosive effect on some metals, so it is important to check the compatibility of fixings with the manufacturers of the preservatives used.

Preservative-treated wood must always be disposed of carefully, as the waste (e.g. sawdust and off-cuts) may be considered hazardous.

15.11.5 Checking preservative treatments

Timber can be tested to see if a wood preservative is either present or if the treatment has been carried out effectively and in accordance to a particular specification. The amount of preservative present within a sample of wood can be assessed by chemical analysis, and will provide clarification if the retention achieved is adequate to provide the desired service life required. There are also test-kits available which can be used by operatives to carry out spot tests for the presence of wood-preserving chemicals.

15.12   Modified wood products

In the past five years a number of modified wood products have entered the UK construction market. There are a range of products including chemically modified and thermally modified wood. These products normally take low durability timbers (e.g. European softwoods) and the treatments improve their resistance to decay (durability). A number of other properties may also be altered. Modified wood has improved dimensional stability (e.g. reduced movement) which means it has improved performance for joinery applications. The durabilities and applications of individual products should be checked with manufacturers.

Further reading

Standards and Codes of Practice

The TRADA web site contains information on the subject of wood preservatives.
   www.askTRADA.co.uk
BS 4978: 1996 *Specification for Visual Strength Grading of Softwood*.
BS 5268: 2002 *The Structural Use of Timber*.
Note: This Standard is in several parts dealing with such topics as design; workmanship; trussed rafters; preservative treatment; timber frame walls; domestic floor joists; joists for flat roofs; ceiling joists; domestic rafters; and buildings other than dwellings.
BS 5756: 1997 *Specification for Visual Strength Grading of Hardwood*.
BS 8417: 2003 *Preservation of Timber—Recommendations*.
BS EN 518: 1995 *Structural Timber. Grading requirements for visual strength grading standards*.
BS EN 519: 1995 *Structural Timber. Grading requirements for machine strength graded timber and grading machines*.
ENV 1995 Eurocode 5: *Design of Timber Structures*.
EN 338: 1995 *Structural Timber-strength Classes*.
EN 912 *Timber Fasteners—Specifications for Connectors for Timber* CEN Brussels.

Other related texts

TRADA produce many publications on this subject. For further information on the use of timber in construction go to www.askTRADA.co.uk.

Chapter 16 EXTERIOR CLADDINGS AND INTERNAL FINISHES

Brian Barnes

Basic requirements for best practice (exterior claddings)

- Check overlaps
- Check sealants and fastenings
- Use only trained installers
- Check continuity of insulation
- Work to manufacturer's instructions

Basic requirements for best practice (internal finishes)

- Control the internal environment
- Check minimum drying times for all materials
- Follow manufacturer's procedures
- Discuss accuracy and tolerances with all concerned
- Manage interfaces effectively

16.1 Scope

Although this section deals mainly with good site (i.e. construction) practice, reference is also made to theory and design issues so that those responsible for managing, supervising and inspecting the cladding operations will understand the basic design requirements for satisfactory work. A further reason for this is an increasing trend for the contractor to be responsible for 'design and construction' and, hence, in the eyes of the client, the contractor is then also responsible for design. The introduction of the Private Finance Initiative (PFI) can put the further responsibilities of life cycle costing and maintenance onto the contractor.

The scope of internal and finishing trades is too wide for it to be covered thoroughly in this publication. Broad guidance is provided with regard to problems related to the general building environment and a number of items that are known to be repetitive are covered in some detail.

The selected advice that follows is the result of analysing repetitive and costly mistakes witnessed by the compiler and his working colleagues.

16.2   Introduction (exterior claddings)

Please note that the recommendations contained in Figures 10.1 to 10.10 in Chapter 10 are also relevant to this section of this chapter.

Those managing any cladding on site need to have a sound appreciation of the performance requirements of each type of cladding. This enables supervision and inspections to be focussed onto critical aspects.

Arguably the most important functions of any cladding are:

- structural stability
- durability
- resistance to weather and air infiltration
- thermal performance
- resistance to fire
- accommodation of movement
- acoustic performance
- acceptable appearance.

The actual list will depend on the particular requirements of the contract. For example, any building near an airport will, almost certainly, have an enhanced acoustic requirement and a bonded warehouse will require a high degree of security.

This list of the client's actual requirements, with the performance level clearly quantified, should precede the writing of any specification. The same list becomes a management tool during the procurement period as a checklist for use with specialist suppliers and subcontractors. During the installation of the cladding the same list can be used to make quality control checklists that pick up essential points for inspections.

Continuing advances in structural design methods, materials and construction techniques have resulted in some structural frames that are less rigid than in the past. The demand for the greatest possible useable floor area forces the vertical cladding into the narrowest zone possible but still requiring the cladding to perform fully in all respects. The cladding of walls and roof is also affected by a search for lighter or thinner materials. This also means the introduction of methods of attachment that may be more vulnerable to the rigours of service conditions than traditional design and construction.

16.3   General principles

A useful guide is BS 8200, Design of non-loadbearing external vertical enclosures of buildings. Please note a pre-condition to this Standard, namely: *it is intended to be used by designers as a comprehensive check list, incorporating useful guidance, and not for inclusion in specifications or other contract documents.* Such a precondition should have appeared in the Foreword and the reason that few know about it is because it was added as part of Clause 0 in Section 1. Notwithstanding this, most construction professionals will find this 70-page Standard a great help.

16.3.1 Structural stability

The cladding must be capable of withstanding all live and dead loads that it will be subjected to. This includes any temporary loading conditions. A number of recent buildings have been designed to withstand blast impact.

Even where the cladding is structurally adequate, loadings will lead to deflections that will themselves be subject to limits. Often it is the reduction effect on performance requirements that create such limits. Another deflection limit can be imposed by visual acceptance.

In many cases this issue emerges during or after construction. With any untried or unproven form of cladding construction there is a good case for examination of deflections and performance by testing a mock-up during the procurement period.

Supports and restraints to the cladding will play an essential part in the structural stability. The design, manufacture and installation of supports and restraints should be considered at the same time as the cladding. Some of the matters that need to be taken into account are:

- The fixing should be corrosion resistant. Once in position it is usually inaccessible and therefore impossible to replace. This includes the problem of bimetallic corrosion.
- The metal should be non-staining especially if the cladding material is vulnerable to discoloration.
- The metal should be sufficiently strong to allow it to support all the loads applied to it (including any temporary loading conditions).
- The metal should, where possible, be of a standard material size (special sections are costly).
- The workability of the material should be good while retaining its strength characteristics.
- The fixing must meet fire regulations (may need fire protection).
- The cost should be acceptable although integrity and safety should be of prime importance.
- The fixing should be capable of catering for the specified tolerances in the frame and in the cladding.
- The fixing may have to be designed to cater for movement.
- The fixing should not permit bimetallic corrosion to occur and separating membranes may be necessary.
- If any fixing needs to be inspectable after cladding erection, this may affect the design.
- The fixings should be straightforward to install (and should be fitted in accordance with an agreed procedure).

(The above list by courtesy of the Technical Department of Harris & Edgar.)

16.3.2 Durability

There is no clear agreement on the definitions or of terminology. Manufacturers often refer to life expectancy. Some clients require a specified 'maintenance-free life' or *period to the first major maintenance*.

Clause 27 in BS 8200 uses the term *design life*. As European Standards are becoming more 'performance based', this term provides us with a better starting point. If the performance criteria are complete and clearly quantified, then the end of the *design life* is reached when the cladding no longer *performs* as specified.

It is important to identify the life expectancies of all components in the cladding and also any shortening of that life by incompatible interfaces or other detrimental influences. One important influence is that the life expectancy of a loaded component is shorter than its life in an unloaded state. Durability can also be affected by adjacent materials and environmental pollution.

The next stage is to examine the components or interfaces that have the shorter lives. The accessibility of these aspects to be inspected, maintained or replaced will be a determining factor in the design life of the cladding. One example would be where the intersection of two gaskets at a corner requires the use of an applied sealant. Because it cannot easily be inspected or replaced, the life of the sealant should be at least the life of the components it joins. The design life of the cladding could therefore be determined by the life of the least durable component.

16.3.3 Resistance to weather

The main elements of any cladding are generally resistant to weather. Most problems appear where joints need to be made or details affect the continuity of the main materials.

Careful design, detailing and workmanship are necessary for interfaces, jointing, openings, penetrations, copings, cappings, cills, damp proof courses and flashings. All of these should meet the same performance requirements, i.e. be as weathertight and durable as the main element. Rain penetration at details can cause damage to other materials.

When storm conditions of wind and rain strike any building, the effects are generally most severe at details. It is also appropriate to remember that storms hitting taller buildings can direct rain upwards. The rule of thumb is that wind and rain can move upwards for the uppermost one third of the building.

16.3.4 Thermal performance

With changing attitudes to energy conservation, there are continually increasing requirements for thermal performances from new and refurbished buildings. This has a number of effects on the design of the cladding. The insulation should be protected from any moisture unless it is designed to be in a wet zone and then it should be more carefully chosen. In all cladding, cold bridges should be prevented or minimised.

Closely connected with the thermal performance is minimising the moisture, in the form of vapour, from inside the building. Vapour control layers are needed in many forms of cladding particularly where insulation layers could be affected by such moisture.

16.3.5 Resistance to fire

Refer also to Building Regulations.

The designer should check with national and other authorities with statutory powers any mandatory requirements for fire. Checks should also be made with the Loss Prevention Council and the owner's and tenant's insurers. These can be onerous and expensive. The rapid spread of fire in some recent examples has led to a re-examination by some authorities of cavity fire barriers. Some aspects that need to be considered are:

- the stability of the envelope during fire
- the limit of the spread of fire to or from adjacent buildings or parts of buildings
- the limit of the spread of fire over the external surface
- cavities that limit the spread of flame or hot gases
- that added components should not become detached or, if they can fall, the hazard should be assessed
- that the cladding materials should not form a significant part of the total combustible material
- the internal surface, which should have a low surface spread of flame classification
- the possible emission of toxic gases from cladding materials • openings and routes, which should be provided as a means of safe escape • other openings that should be provided for venting heat and smoke
- the cladding which should protect escape routes.

16.3.6 Accommodation of movement

All buildings are subject to cyclic and progressive movements and provisions should be made to accommodate the cumulative effect. The cladding, which is attached to the frame, will be subject to different causes of movement than the frame. If these act in opposite directions the degree of relative movement can be high. Relative movement can be caused by:

- differential settlement
- ground heave
- loading (dead and live)
  - steel frame, elastic deformation
  - wind
  - earthquakes
  - component loading
  - beam or slab deflection
- release of loading
- changes of moisture content or drying
- concrete frame creep or shrinkage
- thermal changes (insulation can increase the temperature range)
- chemical action including corrosion
- freeze–thaw action
- loss of volatiles can lead to shrinkage
- impact or explosion
- vibration.

Relative movement can be either reversible or irreversible and there are several ways in which movement takes place:

- transverse, i.e. opening and closing (as in a butt joint)
- shearing, this can be in two separate directions
- torsion
- bending.

Components or joints may have to, concurrently, accommodate more than one type of movement.

16.3.7   Provision for movement

Failure to provide for movement leads to stressing or overstressing of components, fixings or adjacent elements. Deformation or fracture can be the result.

The principal function of a movement joint is to permit the calculated movement to take place but at the same time maintaining the specified (or minimum) performance (see Fig. 16.1).

A movement joint also:

- can induce a 'crack' at a controlled location
- should prevent stresses or damage occurring during movement
- permits large panels to be erected and jointed
- accommodates induced and inherent deviations
- closes an otherwise open joint
- makes what would be a crack or a gap visually acceptable.

See Fig. 16.2 for a summary of the effect of external forces on panels.

Some of the factors that should be taken into account in movement joint design are:

- the need for a movement joint
- the causes of movement (major ones identified particularly)
- how movement affects sealant (transverse, shear etc.)
- the spacing of joints
- the extent of each movement, quantified in all directions
- the characteristics of movement (slow, cyclical etc.)
- the shape of joint
- exposure and degradation risk
- the specified performance at the joint position
- maintenance-free period or design life required
- the substrate material's properties (including need for primer)
- aesthetic requirements
- the sealant's properties
- maintenance
- construction accuracy and tolerances (i.e. induced deviations)
- the ambient temperature at the sealing stage
- local environmental conditions.

16.3.8   Appearance

The difficulty of specifying an acceptable appearance stems from the difficulty of putting the requirements into words. If it cannot be clearly described and quantified,

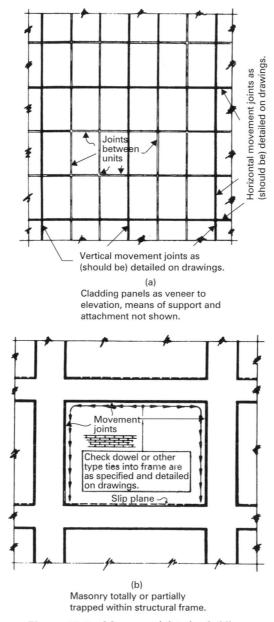

Joints
between
units

Horizontal movement joints as
(should be) detailed on drawings.

Vertical movement joints as
(should be) detailed on drawings.

(a)
Cladding panels as veneer to
elevation, means of support and
attachment not shown.

Movement
joints

Check dowel or other
type ties into frame are
as specified and detailed
on drawings.

Slip plane

(b)
Masonry totally or partially
trapped within structural frame.

Figure 16.1 Movement joints in cladding.

it cannot be measured and verified. On many sites the appearance becomes the main issue, frequently resulting in less attention being given to the other performance requirements.

Acceptable visual alignment is one aspect that could be quantified but generally is not. British Standards have side-stepped this issue perhaps because large-scale

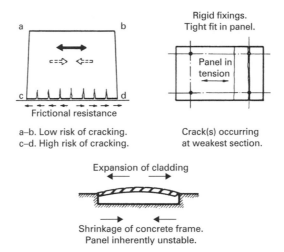

Figure 16.2 External forces on panel.

coloured photographs have never been part of the established black on white printed format.

One possibility is to select the cladding of a similar building and specify this as the standard required for appearance. This has its difficulties and so, for the foreseeable future, we will continue to rely on a post-tender mock-up and the difficulties of reaching mutual agreement.

BS 8200 provides a useful introduction into the design of external enclosures of buildings.

16.4   Recommended procedures

16.4.1   Preliminary appraisal

As soon as is practical, check the information relevant to the cladding for completeness, for clear performance levels, for compliance with Statutory Requirements, British, European or International Standards, or for compliance with any other authoritative standard of good practice.

Establish the appropriate quality management procedures starting with the checking of specification and design matters. Clear all missing, ambiguous or doubtful issues before ordering materials or negotiating with prospective specialist subcontractors.

Ensure that all working drawings and details are complete before work begins. Check that all materials are available in the quantity, to the quality and within the right timescale before placing and confirming the order.

The accuracy required in installing the cladding should be established at this stage. The permissible deviations in the cladding will probably be related to performance limits or to visual requirements. It is highly unlikely that the steel or concrete frame can or will be erected to the same degree of accuracy.

Obtain physical samples of all supports and restraints that will be used and analyse their adjustability in each of three directions. The maximum adjustability will indicate the out-of-tolerance that is therefore acceptable in the frame. **The decision must be taken at this stage either to construct the frame to the analysed accuracy or to redesign the supports and restraints to give them greater adjustability**.

Not taking this crucial decision (and constructing in hope) will inevitably lead to problems and delays that are unacceptable in today's fast-track construction.

16.4.2 Frame construction and survey

BS 5606 is most frequently referred to when problems occur. It has a different philosophy to earlier versions of this same standard and not all construction professionals have familiarised themselves with these recommendations.

One interpretation of the standard follows:

- Ensure that dimensional accuracy of the steel or concrete frame is realistic, necessary, has been specified and been fully understood.
- Ensure that the dimensional accuracy of the curtain walling specified has been understood.
- Ensure that any incompatibility between the two bullet points above has been recognised, and
  - that the design of the frame, cladding, fixing and jointing can accommodate such incompatibility
  - that clear details are drawn, particularly of the fixings, and specified to overcome the variability in practice on site.
- Construct the frame to the required accuracy with adequate intermediate checking plus remedial action where necessary.
- Survey the frame as far as possible ahead of cladding erection (possibly with both the frame subcontractor and the cladding subcontractor present) and preserve the survey records.
- Ensure that the responsibility for each activity has been allocated and accepted.

Surveys and checks of the building frame during construction are of value in firstly identifying out-of-tolerance works (so that remedial action can be taken) and secondly in identifying where particular pre-planned solutions will be required to accommodate the within tolerance variabilities (see Fig. 16.3).

16.5 Prefabricated cladding units

During the procurement period all delivery, handling and distribution matters should be fully discussed. These will all be affected by the available access to and round site, by the craneage and lifting facilities available and by the form of scaffolding.

At this early stage the supports and restraints should be fully discussed and understood. Discussions should also include temporary storage of the units, the means of final adjustments to the units when they are in position and any temporary stabilisation measures. Some, particularly the heavier, units may not arrive on site in their intended vertical position. The means by which units are turned should also be established.

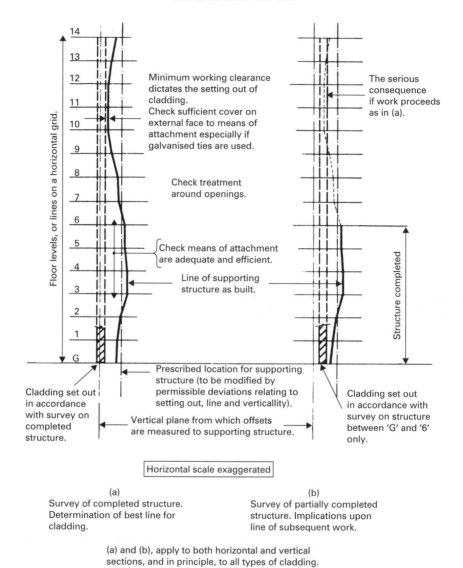

Figure 16.3   Survey of structure for select the setting out line for cladding.

Detailed discussions are needed regarding site joints between units. This will include the means of achieving acceptable seals and the accommodation of relative movement. The same discussions should include accuracy and tolerances for the frame, the units and their supports.

Examine all units before acceptance, preferably before unloading. This should be done with each consignment to check compliance with the specification. The checking should include the accuracy of the delivered units. Establish quality checklists for this purpose. Units that are damaged, distorted or non matching in colour should be rejected at the point of delivery. Panels should not be erected that will have to be removed and replaced at a later date.

BS 8297 provides useful information on the installation of pre-cast concrete cladding.

16.6 Masonry cladding to reinforced concrete frames

See also Chapter 10.

16.6.1 Materials

Bricks, blocks, natural stone, reconstituted stone, sand cement and additives all need to comply with the specification. Ask for verification, in the form of test results, before the first deliveries are made. On going test data may be needed for future deliveries if that is part of the quality management procedures. If test results are not available, consider instigating independent tests. This is done regularly with concrete cubes but many other materials are accepted 'on trust'.

The means of supporting these cladding materials should have already been established. Metal components such as cavity ties, shelf angles and bed-joint reinforcements should all be of adequate corrosion resistance. Stainless steel is the most common metal but there are other non-ferrous metals that are used in some cases.

Establish which British Standards these materials should comply with. BS 5628 is frequently quoted. This, in turn, provides cross-references to other relevant Standards. Your company quality procedures or the contract specification may require a higher or lower standard than the British Standard(s). British Standards only claim to represent a level of good practice. This can mean that the level recommended in the British Standard is a minimum. If you are required to carry out work to any lower specification, ensure that clear written instructions have been received.

16.6.2 Construction details

For reasons of economy and weight, today's masonry claddings are relatively unstable. They rely heavily on the structural frame combined with designed supports and restraints for their stability and durability. Masonry claddings are also affected by the other components that need to be incorporated particularly DPCs, trays and flashings.

Site managers and bricklayers should understand the other functional requirement of the wall such as appearance, resistance to weather, thermal performance and accommodation of movement.

Masonry can be supported on concrete footings, suspended slabs or concrete nibs but stainless steel brackets or shelf angles are more frequently used, particularly on multi-storey buildings. Using shelf angles gives a greater facility for accurate construction by accommodating deviations in the supporting structure (see Fig. 16.4).

The most recently published chart of permissible deviations in a British Standard is Table 2 in BS 8000–3. The compiler of this chapter feels strongly that masonry built in accordance with this Table would rarely achieve the accuracy normally required. Each site should examine and establish its own accuracy requirements. Clause 3.1.2 in that same standard provides the means for more stringent requirements.

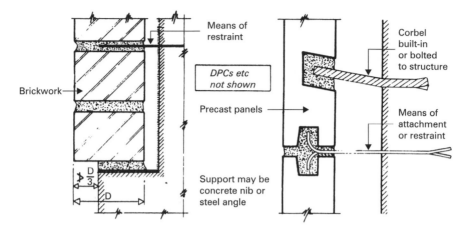

Figure 16.4   Attachment and support of brickwork and stone cladding.

See Fig. 16.5 for typical failures of brick panels, and Fig. 16.6 for failures at supporting nibs.

The alignment of the steel angle is critical. Masonry requires a minimum of a two-thirds bearing on that (or any support). The angle must be adequate to accommodate torsion and bending between its supports. The length of angle between supports will be part of its design. Each steel angle will also be designed to incorporate shims, normally only at fixing positions, but there will be a maximum thickness of shims for each angle. This should be established and not exceeded. Shims should always reach to the 'heel' of the support. If the structure is out of tolerance, standard brackets may have to be replaced by special ones.

Supports, and some restraints, often require drilling into concrete. Cast-in sockets or slots are generally cheaper and can be more efficient but such provisions have frequently proved to be out of position when cladding activities begin.

Where anchors are to be installed by drilling into concrete the approval of the structural engineer is needed. The installation should only be carried out by a fully trained operative giving special attention to:

• avoiding reinforcement
• maintaining the correct hole diameter during drilling
• cleaning out the hole by blowing before inserting the anchor
• tightening the anchor to a pre-determined torque value and locking in position to prevent subsequent loosening
• the re-drilling of concrete and steel angles on account of faulty holes should be agreed with the structural engineer
• encapsulated resin anchors, which are particularly sensitive to operator error
• precise installation and checking procedures, which should be implemented whatever type of anchor is used.

Whilst a BS exists for the plates from which the angles are formed there is no BS for the profiles formed from them. However, standard rolled sections may be used

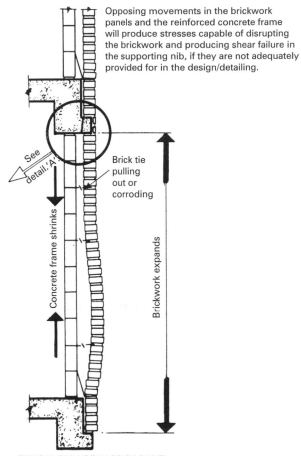

Opposing movements in the brickwork panels and the reinforced concrete frame will produce stresses capable of disrupting the brickwork and producing shear failure in the supporting nib, if they are not adequately provided for in the design/detailing.

See detail 'A'

Brick tie pulling out or corroding

Concrete frame shrinks

Brickwork expands

TYPICAL FAILURE IN BRICK PANEL

Figure 16.5 Typical condition for failure of brickwork panel.

provided allowance is made for the root radius. It is, therefore, essential to establish a clear specification when obtaining quotations or ordering.

If the engineer has not specified permissible deviations, then state on the order the following instructions:

- Cross section—the leg lengths of the angle shall be within ±2 mm of those specified
- Squareness—the internal angle of the support angles shall be 90° + 0° − 2°
- Length—the length of a member shall be within 4 mm of that specified
- Straightness—for any member required to be straight, its deviation from the straight line, between any two corresponding points on the member, L apart, shall not exceed 1 mm or $L/600$ whichever is the greater.

Stainless steel angles can be obtained either as rolled profiles or prefabricated by folding a flat plate. The rolled profiles that give a near-square corner between the legs of the angles are generally only available as special rollings in relatively large

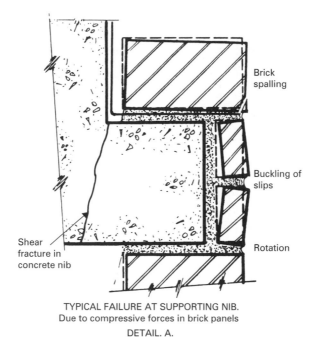

TYPICAL FAILURE AT SUPPORTING NIB.
Due to compressive forces in brick panels
DETAIL. A.

Figure 16.6   Typical failure at supporting nib.

quantities although some smaller profiles may be more readily available from continental sources.

There are no British Standards giving bending tolerances for folded plate stainless steel angles. Consequently, if the supplier does not exercise good quality control, then a 90 angle between the horizontal and vertical legs will not be achieved, and, with positive and negative angle deviations, this can produce difficulties if the horizontal legs of adjacent angle lengths are not aligned.

Profiles of folded plate angles are not a direct substitute for a rolled profile of the same overall dimensions and thickness. The reason for this is that the folded plate has a fairly large root radius resulting in less depth of the upstand angle in contact with the concrete. The net effect of this substitution would be to increase significantly the load in the fixing bolt and increase the deflection in the angle when loaded.

See Fig. 16.7 for diagrams of concrete and steel supporting toes, and Fig. 16.8 for details of steel angle supports to brick cladding.

16.6.3   Use of clay brick slips

During the 1970s and early 1980s greater use was made of concrete nibs to support the brickwork at upper levels. To hide the vertical edge of the concrete nib, clay brick slips were used. The results of differential movements between concrete and clay bricks resulted in many of the slip bricks cracking and being dislodged. Adhesion between clip bricks and support was frequently found to be poor. Cushioning designs were used with some success until the use of stainless steel angles became more widespread.

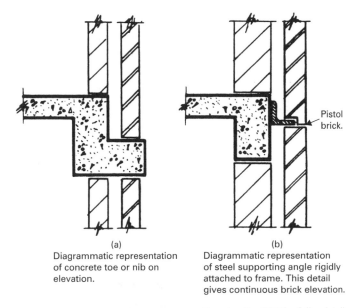

(a)
Diagrammatic representation
of concrete toe or nib on
elevation.

(b)
Diagrammatic representation
of steel supporting angle rigidly
attached to frame. This detail
gives continuous brick elevation.

See also Fig. 16.8 for fuller details.

Figure 16.7 Concrete and steel supporting toes.

This reference is included partly for information if remedial works to older buildings are needed, and partly as a warning in case the lessons learned get forgotten. An alternative to using slip bricks is the use of the specially shaped or cut pistol brick. An IStructE report (*Aspects of Cladding*) recommends that slip brick adhesion should be enhanced with mechanical fixings.

See Fig. 16.9 for details of concrete nibs masked by clay brick slips.

16.6.4 Concrete bricks and blocks, natural or reconstituted stone

Generally all building stone should be laid with its natural bed normal to the load. Natural bed is defined as the plane at which the sedimentary material was deposited. In cornices and copings, depending on the strength of the stone, the projecting stone is frequently edge-bedded. Quoin stones should be set on their natural bed to prevent the quoin end becoming face-bedded. Any variation to natural bedding should be clearly specified.

- The direction of the natural bed on archstones should be laid normal to the curve.
- All joints, horizontal and vertical, should be completely filled with mortar.
- Cavities should be kept clear of mortar.
- At openings, check and maintain accuracy and check the adequacy of the frame fixings.
- Any temporary timber wedges (if allowed) should be soaked before use so that they shrink out, and by so doing, do not damage the arris of the stone, above or below, or both, when removing. Do not use materials that could stain or corrode the stone.

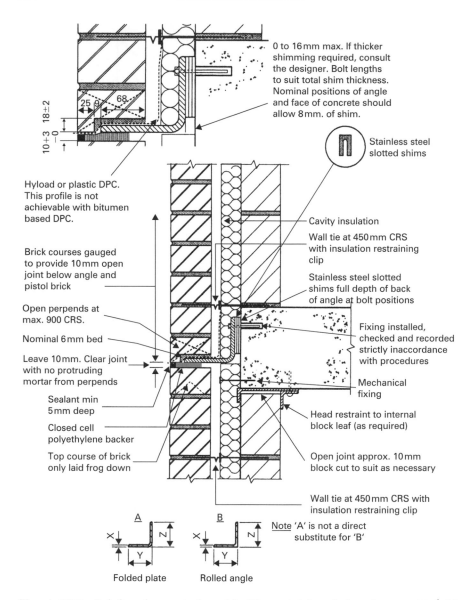

Figure 16.8 Reinforced concrete-framed buildings: stainless steel angle support to brick cladding.

- Stone surfaces should be wetted before the bed is laid. On no account should the bed be laid while water is lying upon the stone.
- The use of stone cladding as permanent formwork to in situ concrete is not recommended. Should its use be instructed, all units should be adequately supported and the joints between the units should be kept clear of cement grout. Should movement of the stone from its building line occur during or after the concrete has

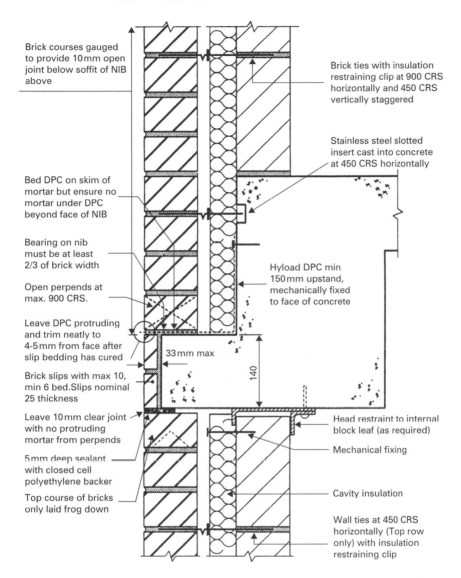

Brick courses gauged to provide 10mm open joint below soffit of NIB above

Brick ties with insulation restraining clip at 900 CRS horizontally and 450 CRS vertically staggered

Stainless steel slotted insert cast into concrete at 450 CRS horizontally

Bed DPC on skim of mortar but ensure no mortar under DPC beyond face of NIB

Bearing on nib must be at least 2/3 of brick width

Open perpends at max. 900 CRS.

Hyload DPC min 150mm upstand, mechanically fixed to face of concrete

Leave DPC protruding and trim neatly to 4-5mm from face after slip bedding has cured

33mm max

140

Brick slips with max 10, min 6 bed.Slips nominal 25 thickness

Leave 10mm clear joint with no protruding mortar from perpends

Head restraint to internal block leaf (as required)

Mechanical fixing

5mm deep sealant with closed cell polyethylene backer

Top course of bricks only laid frog down

Cavity insulation

Wall ties at 450 CRS horizontally (Top row only) with insulation restraining clip

Figure 16.9 Reinforced concrete-framed buildings: cladding supported by continuous concrete nib masked by clay brick slips. NB: current best practice suggests that clay brick slips should be fixed by a suitable adhesive but also assisted by mechanical anchorages.

been placed, no amount of pushing or knocking will get the stone back into its original position without first removing the concrete backing before it sets.
• If the cladding is backed with brick or blockwork (cavity or solid), the two skins should be built with each other as far as practically possible. This can frequently introduce the need of supervising two trades working closely in conjunction with each other.

- The bedding mortar for cills and thresholds should be solid at each end but the central part of the bed should be left open to prevent fracture in case of differential movement. Point later with matching mortar.
- All edges to be jointed or sealed should be cleaned and kept clean.
- Strut free-standing elements to avoid wind damage. Generally limit to two courses in a working day. (Without strutting, limit height of fresh masonry to 1 m.)
- Protect new masonry from the weather (rain or frost). All projecting and vulnerable areas of finished stonework should be substantially protected against physical damage.
- Particular care should be taken with scaffold staining on finished facework. The innermost board should be replaced with a dry, clean board during rain and at night and weekends.

Useful information on the above may be found in BS 5628 and BS 8298.

16.7　Masonry cladding to structural steel frames

See Fig. 16.10 and Fig. 16.11 for diagrams of typical masonry cladding to steel-framed buildings.

Before starting work ensure that the specification, drawings and working details are complete and that they all comply with current best practice.

The recommendations for this section are very similar to those in Section 16.6 above.

The accuracy of many steel frames has frequently caused problems when the cladding is fitted. Greater attention is needed to the design and adjustability of the supports and restraints.

The frame movement characteristics should be fully understood and incorporated into the pattern of movement joints. This is particularly important in the case of lighter frames.

The pattern of wall ties should be established taking into account the need for both structural stability and for accommodation of movement.

Much of the modern cladding of this type involves the use of very thin (*circa* 50 mm) stone slabs fixed to a backing frame. It is suggested that expert advice is sought from CWCT when constructing this type of cladding.

16.8　Profiled metal panel roofing and cladding

16.8.1　General

Profiled metal has become the cheapest solution for roofing and cladding in recent years, especially for industrial buildings. Today's life cycle costing will determine if it is the most economical alternative.

The basic construction consists of outer and inner metal sheets separated with a sandwich of insulation. The insulation should have a vapour control layer (VCL) on its inner surface to minimise the moisture-carrying vapour leading to condensation within the insulation. Some roof panels require the addition of a breather membrane on the outside of the insulation to minimise condensation (and some leaks) falling into the insulation. The base material is generally steel.

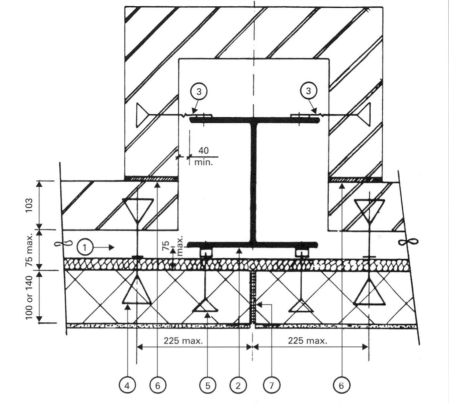

Notes:
1. 50 mm minimum clear cavity.
2. Main column.
3. Flexible tie, as described in Note 5, must have suitable drip to avoid passage of moisture across cavity coming into contact with steel column.
4. Double triangular wall ties, for flexibility, to be provided each side of joint. Note 75 mm maximum cavity width.
5. Flexible frame ties either bolted to flange or fixed using self drilling and tapping screw. Above 3 storeys use slotted ties with channels to accommodate vertical differential movement.
6. Brickwork expansion joint. Joints in internal leaf to be staggered from those in external leaf.
7. Contraction joint in blockwork. Consider incorporating bed joint reinforcement above and below openings to prevent cracking at reduced section due to longitudinal contraction of wall.

Figure 16.10 Typical masonry cladding to a steel-framed building: external pier without sway (taken from Brick Cladding to Steel-Framed Buildings, Brick Development Association).

This can be built up on site or can be brought to site as a *composite* panel in which the insulation is bonded to both sheets to avoid site assembly and the need for a separate vapour control layer and a breather membrane.

The steel is produced in coil form. Resistance to corrosion is achieved by a coating of zinc or of zinc/aluminium *galvanising*, which in turn is protected by a coating. These protected coils are passed through profiling machines to give a wide range of shapes. These have particular structural properties in which the deeper profiles make the sheet

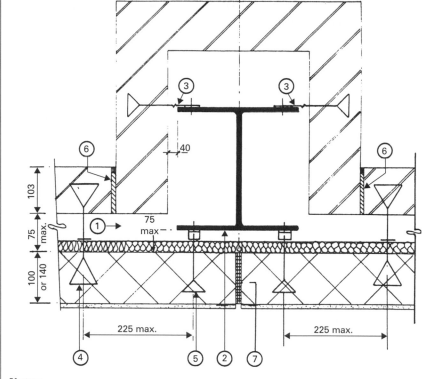

Notes:

1. 50 mm minimum clear cavity.
2. Main column.
3. Flexible tie, as described in Note 5, must have suitable drip to avoid passage of moisture across cavity coming into contact with steel column.
4. Double triangular wall ties, for flexibility, to be provided each side of joit. Note 75 mm maximum cavity width.
5. Flexible frame ties either bolted to flange or fixed using self drilling and tapping screw. Above 3 storeys use slotted ties with channels to accommodate vertical differential movement.
6. Brickwork expansion joint. Joints in internal leaf to be staggered from those in external leaf.
7. Contraction joint in blockwork. Consider incorporating bed joint reinforcement above and below openings to prevent cracking at reduced section due to longitudinal contraction of wall.

Figure 16.11   Typical masonry cladding to a steel-framed building: external pier with sway (taken from Brick Cladding to Steel-Framed Buildings, Brick Development Association).

more appropriate for roof use and the medium-depth profiles would be more suitable for vertical cladding. The flattest profiles are those used as inner liner sheets that generally need less corrosion resistance due to their less demanding final environment.

16.8.2 Purpose

The purpose of any system is to create a selective filter between the outer and the inner environments. The performance specification needs to be carefully drawn up with clear levels for each performance requirement. Check particularly the requirements for durability, for air infiltration limits, for thermal and acoustic levels and for fire requirements.

16.8.3 Design

Designs of metal panel systems are normally carried out by specialist material suppliers or by the subcontractors. Manufacturers produce extensive load/span tables together with standard construction details. The Metal Cladding & Roofing Manufacturer's Association (MCRMA) produce an excellent series of design booklets and The National Federation of Roofing Contractors (NFRC) produce an installation guide, which includes much useful information on design, matters. There is no single British Standard covering this aspect of construction.

The designer should first establish all the environmental conditions that the building must withstand. Prevailing weather conditions will affect the roof slope, fixing details and choice of sealing methods. Exposed and coastal sites have different requirements to inner city, sheltered sites. Some marine conditions will require a more careful choice of sheet and coating and similar problems can be experienced in industrial areas with gaseous pollution. Buildings with high internal humidity or corrosive elements such as swimming pools, kitchens or industrial processes will require extra attention to the inner sheet and its protection.

Sheet colour is another factor to be considered. Generally, lighter sheets are affected less by thermal movements, which can affect the durability of the fixings. The location and particular facing direction of any sheeting has an effect on the guarantees given on the coating. The coating manufacturers readily provide this information.

16.8.4 Materials

Sheeting

Steel sheeting should be protected from damage (see also Section 16.8.6). This may occur due to handling and installation or from environmental hazards. For the outer sheets, the normal minimum recommendation for wall sheets is 0.5 mm and for roof sheets 0.7 mm. The galvanised protection should not be less than 275 g/m². The two most commonly specified coatings are PVDF (previously known as PVF2) and PVC Plastisol. PVDF is nominally 27 microns thick and is mostly used for internal surfaces. It is available in metallic colours, which is one reason why it may be specified. Plastisol is nominally 200 microns thick and is provided with a leatherette finish. It is the most commonly used finish on metal panels in the UK. New coatings will continue to be developed.

Liner sheets can also be coated with polyester paints but these may not have adequate lifespans for many buildings.

Rooflights

The provision of light for building users in the most economical manner has resulted in the practice of providing natural light into the building by incorporating translucent sheets. A more expensive solution is to incorporate pre-assembled rooflights.

The rooflight material should be carefully chosen. There is no ideal solution. The first requirement is normally for wind uplift followed closely by safety for roof and maintenance operatives (against the risk of falling).

The Statutory Requirements for rooflights and fire must also be complied with. The rooflight can provide an inexpensive alternative to smoke venting in the event of a fire due to its low melting point. The light transmission properties of rooflight materials reduce with age and this deterioration should be recognised in the specification and choice.

Most rooflight materials have thermal movement characteristics greater than the metal panels (PVC is notable for this) and special precautions should be taken to cater for relative movement.

Insulation

A wide variety of materials is available to choose from. Composite panels arrive with the insulation bonded to the inner and outer sheets, having been injected there during the manufacturing process. The most convenient for the built-up, on-site installation is in the form of rolls that are laid tight fitting between the structural members. It is important that no gaps are left and that the insulation is kept dry.

Thermographic surveys are becoming more widely used and cold bridges caused for any reason are then immediately evident. The cost of rectifying such problems is out of proportion to getting it right in the first place.

Fasteners and fixings

In metal panel roofing and cladding the NFRC recommend that 'fastener' is the term normally used to describe the mechanical device by which any sheet is secured to a frame or other component. The term 'fixing' is normally accepted to mean the result achieved by the use of a fastener.

Primary fasteners are those that are needed for structural performance. This would include the fasteners that are used to attach the sheets to purlins or to 'Z' spacers. Secondary fasteners are not relied on to contribute to structural performance but may be used to transfer loads from sheet to sheet. Examples of the latter would be side-lap fastenings and flashing fastenings.

The fastenings require careful selection followed by careful installation. Workmanship errors are more common than material failure. Examples of such errors are:

• not using the correct fastening
• not working to the manufacturer's recommendations
• non-perpendicular drilling
• incorrect tightening
• failure to pull the sheets together.

Primary fastenings are usually fixed using a power tool to drill a self-tapping screw into the material beneath. Correct tightening compresses the neoprene sealing washer between the sheet and the steel washer of the fastening. Some sealing washers are designed as load indicators in so far as when the compressible washer reaches the correct diameter the fixing is weathertight. The same compressible washer also indicates, from its compressed symmetry, that the fastening has been installed squarely. Plastic colour matching caps are available mainly for visual reasons but they offer some protection to the head of the fastening.

Sealants

The most commonly used sealant is a non-hardening, permanently flexible, blend of butyl and PIB (polyisobutylene) rubbers. It is provided in roll of a predetermined cross-section with a tear-off backing strip enabling it to be positioned without stretching of the lower surface. The backing strip is removed prior to positioning the overlapping sheet. This strip sealant has good adhesive properties provided the temperature and humidity are within the manufacturer's recommended range. The sealant can also withstand limited shear movements of the adjacent sheets. All laps of sealants should be properly made and for shallow pitches the NFRC recommend a double strip of sealant strip.

Flashings, cappings, corners

Ancillary sheeting items should be made of the same material as the external sheeting. A better visual finish is achieved if such flashing and cappings are made from a heavier sheet, as a crisper line is then the result. Where roof cappings or linear flashings abut the profiled sheet the resultant gap should be filled with polyethylene or neoprene profiled fillers. These have adhesives on one or both contact surfaces but are designed to be just in compression when the sheet to flashing fixing is complete.

16.8.5 Handling and storage

Material handling and storage should be in accordance with the manufacturer's recommendations. Sites should make provision for storage with due regard to access for installation, space required, security and cleanliness in order that the cladding materials may retain their quality prior to installation. Relatively minor damage to the end or sides of sheets can lead to poor seals. Wet insulation will cause subsequent problems and rooflight materials require greater care. All ancillary items should be retained in secure storage.

16.8.6 Problems

By far the most common problem with metal panels is water penetration particularly in roofing. Wind-driven rain can enter through any poorly made laps and joints or through any poorly installed fixings. This particular problem is most frequent at any detail such as penetrations, flashings, change of direction or interfaces with other elements.

In many cases when defects are discovered they are found to be due to poor workmanship. The cause can also be poor design, where details might be missed or a combination of poor design and careless workmanship. The discovery of the actual cause of any leak is a long and costly process because the offending drip may occur some way from the actual leak. When leaks are investigated condensation should be looked for first. The latter gives all the impressions of a leak but is frequently a design problem. When condensation has been eliminated or cured the search for leaks should resume.

Remedial works to metal panel leaks are expensive (partly the cost of access), disruptive and reputation-damaging. It is therefore essential that the design is thoroughly checked and the installation is rigorously supervised by both the specialist and the lead contractor.

Material problems are not so common but they can include damage to sheets and coatings and premature corrosion at site-cut edges. Overloading, sometimes by site operatives, can lead to unacceptable deflections.

Metal panels are not maintenance free. Clients should be presented with comprehensive maintenance manuals to enable them to extend the life of the system before re-coating or major maintenance is required.

16.8.7 Workmanship and supervision

Delivery sequence

The sequence should reflect, as far as possible, the erection programme. Large loads arriving at one time may need to be stacked in such a way that the first items needed are accessible. Regular coordination with the suppliers is vital.

Off-loading

All off-loading proposals should be checked with the cladding specialist. Cooperation with the source of the deliveries can lead to information regarding safe handling. All lifting devices should be designed to avoid damage to the units and for maximum ease of use commensurate with safety. All erection gear must be tested to ensure compliance with the requirements of the appropriate Factory Acts and Regulations.

When off-loading packs of cladding by fork-lift, take care to avoid damage to the sheets on the underside of the packs. Where possible, packs should be off-loaded with wide (100 mm minimum), soft slings e.g. nylon, but in no instance should chains or the like be used. Avoid handling sheets unnecessarily. Sheets should be lifted, never dragged off the stacks or packs. The installer should be made responsible for off-loading.

Careful consideration should be given to the maximum sizes of sheets that can be safely and satisfactorily dealt with.

Storage

The storage areas should be accessible for delivery and removal. They should be clear of site traffic and of working areas. They should be clear, clean and capable of incorporating appropriate protection. For valuable items they should be secure.

If the materials cannot be kept under permanent cover, a light frame and waterproof cover should be provided. Leave space between the cover and the materials to allow air to circulate. Inspect regularly to ensure that moisture has not penetrated the stacks.

Time on site

As a general rule, the shorter the time that any materials are on site the lower the risk of damage or theft. Where this is unavoidable there is a greater need for security and probably periodic inspections for water or any other contamination.

Protection

Discussions regarding the protection of any materials should involve the supplier and the specialist. Each phase should be considered to agree on what is necessary and to provide it as economically as possible. At this stage it is the protection during delivery, off-loading, storage and after erection that should be discussed, together with agreement on responsibility.

Identify those subsequent trades that could damage the completed installation and decide on isolation or local protection, particularly at points of access.

Distribution

The distribution routes from the storage area to the installation point should be fully planned. The heaviest and the longest or largest units are the obvious ones to consider most thoroughly. This will reveal the need for any loading platforms or horizontal distribution difficulties.

Survey

Care should be taken to ensure that the frame is constructed to the specified accuracy. The frame must be surveyed as it is erected, ahead of cladding erection. A joint survey is generally the most effective. An early warning system can help to maintain progress by deciding on appropriate measures. Consideration should be given to the possibility of a few specially designed and fabricated fixings for the occasional out-of-tolerance position.

Method Statement

Establish a clear erection procedure, preferably at the design stage. The method statement should be discussed and agreed prior to any work on site. This should act as the reference document. Create flexibility wherever possible and make contingency plans for repetitive difficulties.

Working drawings

The working drawings should be sufficiently detailed to allow proper monitoring of the installation.

Markings

All components should be marked and numbered to facilitate distribution and subsequent erection.

Access

If scaffolding is provided it should reflect the access positions needed by the cladding installers. It should be designed for all temporary loadings and for any adaptations. The location of any scaffold ties should be planned and checked for practicality. All scaffolding and access provisions must be safe and correspond to current safety regulations.

Quality and inspection

Arrange for constant supervision and regular inspections of all erection work. All quality controls should be agreed, understood and carried out. The qualifications, experience and training of each gang of installers should be checked before they start work and the first day or two's work will require careful inspection.

Interfaces

All interfaces between cladding and other elements should be carefully supervised. The frame to cladding and the cladding to roofing interfaces are crucial examples. This activity should also include internal packages. This will involve the need for setting out by different subcontractors to be coordinated.

Installation details

With horizontal cladding, the appearance of laps can be improved by locating the main fasteners towards the end of the overlapping sheet (but no nearer than 50 mm) and by locating the sealant behind the fastener. The direction of laying should be towards the direction of the prevailing wind.

Check that the minimum specified end and side laps are achieved and check whether or not it is sealed with a mastic, sealant or sealing strip. Corrugated or profiled sheets cannot overlap and nest without gaps.

The temperature at which sheeting is installed is important; for any materials with high coefficients of expansion, it may be desirable not to fix at any temperature below 10 °C.

Generally, holes should be drilled and all swarf removed before fixings and washers are positioned.

Care should be taken to ensure that all fastenings and fixings are correct and are properly installed. Regular checking should be carried out. Care should be taken to prevent distortion of the profile.

If the designed fixing cannot be practically installed do not permit a make-shift compromise to be used. Make certain that the new solution takes into account all factors considered in the original design.

The appearance of the cladding is enhanced by a high standard of workmanship during fixing and by careful alignment of fastenings.

Sealants should be carefully installed. A high level of workmanship is necessary for the sealant to act in accordance with its design properties. The installation should be in accordance with the manufacturer's recommendations. Ensure that the literature is on site.

Carry out frequent and regular checks on the sealants during and after application. The efficiency of the bonding to adjacent materials requires particular attention.

Damage to panels during construction, especially at edges and corners, can make the formation of a weatherproof seal virtually impossible.

Penetrations of the metal panels will require additional design attention, the highest level of workmanship and extra supervision.

If the cladding has an applied finish, check whether a precise degree of gloss or colour matching is difficult to achieve. If it is, ensure that all sheets for one particular contract are obtained from one batch. Some finishes also exhibit directionality, and this must be taken into account during erection.

16.9   Fully supported, membrane-type, flat roofing

16.9.1   General

Most of this chapter is devoted to vertical forms of cladding with the exception of the section immediately above, which included metal panel roofing. Of the other types of roofs that are available, flat roofs are perceived to have more problems than pitched roofs. This short section is included to offer an outline of some of the problems that could be encountered. The design of a truly flat roof is unlikely. The roofing industry has agreed on the following terminology:

- flat roof—up to 10 slope
- low-pitched roof—10 to 20 slope
- pitched roofs—above 20 slope.

The membranes for full-supported roofs may include:

- built-up bituminous felt
- mastic asphalt
- single-ply membranes (up to seven different types)
- lead
- copper
- zinc
- liquid applied membranes structural waterproofing systems.

This section will be applicable more to the first three mentioned in the list above.

16.9.2   Purpose

The main functions of a flat roof can include:

- Keep rain, snow and wind out
- Resist wind pressure and wind suction/uplift
- Meet the required design life or maintenance-free period
- Meet the aesthetic requirements
- Provide the required airtightness
- Keep the interior warm or cool, as appropriate
- Keep sound in or keep sound out
- Provide a means for hanging or creating the ceiling
- Provide a vapour check (minimise condensation in roof space and insulation)
- Meet the statutory fire regulations
- Support its own weight and dead loads
- Support the designed snow loading and live loads
- Keep people out (security)
- Surface and roof to accept defined roof traffic
- Provide provision for plant and service distributions
- Accept service penetrations, rooflights and other appropriate details
- Provide falls for drainage and include outlets for rainwater
- Be maintainable and, if appropriate, replaceable.

Other requirements might be imposed by the client and local authorities.

16.9.3   Design

Once the main functions have been identified, the performance requirements should be quantified. The design and detailing can follow the choice of a roof type:

- warm deck roof
- cold deck roof
- inverted warm deck roof/upside-down roof/protected membrane

and the choice of a membrane, the most common of which are:

- built-up bituminous felt (can be two or three layer)
- mastic asphalt
- single-ply membrane.

Each of these membranes, and others that could be chosen, have advantages and disadvantages, which the skilful specifier takes into consideration. A well-prepared specification will lead to the choice of the most appropriate type of roof and type of membrane and result in fewer problems. A poor specification will enable inappropriate roofs and membranes to be considered.

Planning permission may be easier to obtain if the client considers the use of a roof garden as an amenity. In this particular case, because of the soil above it, and the watering necessary, the membrane requires to be 100% perfect before adding soil or landscaping starts. The choice of membrane and joints should take into account biological conditions and root growth.

16.9.4   Materials

The limits of this chapter prevent a full analysis and comparison of all possibilities. As one example, there are seven distinct materials that are all classified as single-ply membranes:

- PVC   Polyvinylchloride
- CPE   Chlorinated Polyethylene
- CSM   Chlorosulphonated Polyethylene
- VET   Polyvinylchloride blended with Ethylene Vinyl Acetate
- PIB   Polyisobutylene
- FPO   Flexible Polyolefin
- EPDM   Ethylene Propylene Diene Monomer.

All seven have advantages and disadvantages, of which some designers might not always be aware. This list indicates the difficulties that might be experienced by a specifier trying to choose the best. The information regarding advantages (and perhaps a little information on the disadvantages) of the membranes can be found from the following sources:

- Built-up bituminous felt   Association of British Felt Manufacturers
- Mastic asphalt   Mastic Asphalt Council
- Single-ply membranes   Single Ply Roofing Association
- Lead   Lead Sheet Association
- Copper   Copper Development Association

• Liquid applied   International Liquid Applied Membrane Association

There are other materials to be incorporated in the roof construction that should also be taken into consideration. There is a wide choice of insulating materials that can be affected by fire or environmental conditions. Insulating materials can include:

• rigid urethane foam
• extruded polystyrene
• expanded polystyrene
• phenolic foam
• cellular glass
• mineral wool (rigid)
• perlite/fibre
• woodfibre
• cork
• foamed concrete.

Take additional care, in both design and installation, in and around details (upstands, edges, changes of direction, flashings etc.). This also applies to service penetrations, holding-down fixings, roof lights etc. In these particular aspects late instructions/changes can jeopardise the weathertightness of the roof.

Check the adequacy of the specified material (particularly for the degree of exposure) and possible reactions with adjacent materials.

Check that all local, differential and structural movements have been identified. Then check that they have been accommodated (or minimised) in the design. Include interface and deck movement.

Check that during installation all sheet laps are adequate and thoroughly jointed. Prevent any damage to the membrane after laying.

Avoid trapping construction or other moisture when the roofing is laid. Check that the condensation risk has been analysed. Install the vapour control layer with great care to prevent punctures.

Check that the adhesion and fixings are adequate for any temporary working conditions and for the permanent situation.

Check that the minimum falls and cross-falls have been specified and achieved. Ensure there is no ponding.

Check that the rainwater discharge outlets are adequate, accessible for maintenance, installed low enough and that the membrane is adequately connected to them.

16.10   Internal finishes

16.10.1   General

The building environment

Control of the internal environment is important for the satisfactory installation of all internal finishes.

Many internal finishes are either sensitive to moisture or require drying conditions to permit evaporation of moisture or solvents in a setting or hardening process.

Many finishes should only be applied within a given temperature range. Some are also sensitive to dust either during application or drying and a few are sensitive to sunlight.

Large quantities of water are often used, even in buildings with mainly dry methods of construction. Rain and atmospheric moisture can add to the total amounts of water that needs to evaporate. A 1000 m², 200 mm thick concrete floor slab may have to lose 5 tonnes of water to reach its equilibrium moisture content.

Drying out depends on the rate and effectiveness at which the external envelope can be made weathertight, on weather conditions and on through draft and air movement. Another important factor is the rate at which moisture naturally leaves any material. A relatively slow, even rate of drying generally results in the least damage to materials in the form of drying shrinkage, cracking and distortion due to differential drying through the thickness of the element. This is very difficult to achieve particularly with today's trend of fast-track construction.

Accelerated drying often has to be used. Heaters, especially those using gas or liquid fuels, are frequently too fierce. In addition, the process produces very high moisture levels in the enclosed space. If this moisture is not adequately vented it can cause problems with condensation. If heaters cannot be used continuously, on a 24-hour basis, they should be turned off for at least two hours before the building is to be shut up for the night to allow dispersal of the warm humid air which would otherwise cool and cause condensation. This could mean 2.00 or 3.00 p.m. in winter.

Accelerated drying using dehumidifiers is more satisfactory and generally to be preferred. In this case, it is essential to seal the area being dried from the ingress of external air. Adequate attendance is needed to dispose of the water collected by the dehumidifiers.

Specialist advice will probably need to be sought on both natural drying times and on accelerated drying times without producing problems. The choice should then be made either to incorporate these times into the construction programme or, if the drying time is unacceptable, look for alternative solutions.

Measurement of moisture in floors

The most common request for measurement of moisture content of materials is made by floor layers who generally require screeds and concrete to have achieved a moisture content of not more than 5%, by weight, before sheet or floor tiling is laid. This measurement can be by electric moisture meter or by determination of the relative humidity of air held in close contact with the floor, for example, by using a hair hygrometer under a sheet of polythene.

Details of available methods can be found in British Standards and in BRE Publications.

Measurement of moisture in wood (See also Chapter 15)

The moisture content of wood, on delivery, during or after storage or after installation can readily be determined using an electrical moisture meter. Hammer probes are available for use with timbers of larger dimensions where surface readings may be misleading.

Before the installation of exceptionally high-quality, low-moisture-content joinery, the relative humidity of the air in particular areas of a building may need to be monitored over a period of weeks using recording hygrographs. The stage at which the area is dry enough to receive the joinery is then accurately determined.

Dust

It is easy to overlook the time-honoured specification clause requiring surfaces to be clean, dry and sound from grease. Dust is the most common contaminant and the least easily controlled. It is not unusual to use a stiff broom to convert surface dust into airborne dust, which then settles on every available surface. Industrial vacuum cleaners may be the best solution in many situations.

Sunlight

A few materials, and in particular hardwoods, are subject to major colour changes on exposure to sunlight. The changes may be bleaching or, sometimes, darkening depending on the variety of wood. It is sensible to protect stored hardwood joinery, doors, skirtings, and mouldings etc. from direct exposure to sunlight.

Materials for internal finishes

The range and variety of such materials makes it impossible to comment on all of them. The following points relate to materials that are particularly sensitive to moisture or that feature frequently in failure investigations.

Wood, paper and many fibre-based materials readily pick up atmospheric moisture. When damp they are very prone to support growth of disfiguring moulds. Blockboard, laminated boards and some plywoods are particularly sensitive and, on drying out, develop cracking of the outer veneers or surfaces that makes subsequent decoration difficult and sometimes impossible.

In damp conditions, plasterboard and related materials can bow or sag after fixing and do not tighten up adequately on subsequently drying out.

Many insulating materials can absorb very large amounts of water. The materials can sometimes be sandwiched between other materials that can reduce the rate of drying so that they remain wet for months and so give rise to subsequent problems often in the form of condensation.

Suspended ceiling tiles are frequently affected by moisture and warp very easily.

Many fire-protection materials contain salts that are water soluble or even hygroscopic. In some cases manufacturers give clear warnings for storage and installation. With other materials their sensitivity may not be so apparent or the warnings so clear.

Fire retardant paints and particularly clear lacquers can perform very poorly unless applied in really good, dry conditions and manufacturers usually disclaim responsibility for defects that might arise. Correction can be expensive, for example having to strip a clear lacquer that has dried cloudy.

Carpets and fabrics are examples where manufacturers do not anticipate that their moisture-sensitive product might be taken into a partially complete and still damp building. The manufacturers assume that the required conditions are known and produced.

The main requirement is to seek the manufacturer's advice or not to fix unless dry conditions are achieved.

16.10.2   Floor screeds

This is arguably the most common area for problems among internal finishings. The principal reference documents are BS 8203 and BS 8404–1, 2. The notes that follow are based on evidence of repetitive problems.

Bay sizes

Bay sizes control the total length of joints within a given area. The smaller the bays the longer the length of joints with correspondingly greater risks of curling, hollowness and crumbling at those edge joints. Small bay sizes make very little difference to the development of minor shrinkage cracking or crazing. Current practice is to lay as large an area as possible. Minor shrinkage cracking, which should not be automatically judged as a defect, can be filled in with a floor-levelling compound if necessary.

Cement content

It is rare for a screed to fail solely because of low cement content. Note however that balled up cement can thereby reduce the effective cement content so that the mix is too weak for it to be fit for its intended purpose. In cases of dispute, requests for the determination of the cement content of a set screed should be resisted. There is evidence to show that the results of such tests are of doubtful accuracy and are of little assistance in judging whether or not a screed is satisfactory.

Mixing

The cement should be evenly distributed through the mix to provide consistent strength. Free-fall mixers have been shown permit some balling up of the cement and their use should be avoided wherever possible. Delivered mixes or more effective site mixers are now more commonly used.

Compaction

The most common fault is to create a thin well-trowelled finish over a weak, crumbly, open-textured underlayer. Compaction of large areas by trowel or even hand tamper is not a practical proposition and rarely gives an acceptable result. Consider rolling, laying in two truly monolithic layers or use a plate vibrator. Even with mechanical aids, special care is needed to achieve compaction at joints and at the perimeter.

Curing

This is essential to permit the development of the designed strength. Seven days' curing should be the absolute minimum. Note that over-dry mixes may have barely enough water in them, when placed, to hydrate the cement. Poor curing produces the most severe shrinkage cracking and is the main cause for curling at edges and joints with the associated hollowness and risk of cracking under load. Polythene sheet, when not adequately fixed or weighted down, can create a wind tunnel and accelerate rapid and uneven drying rather than curing.

Impact testing

An impact test developed by the BRE can be used both to assess screeds that are failing in use and also to test newly laid screeds to check that they are sound. BS 8203 gives guidance on specification and the method of test.

16.10.3 Plastering

This section should be read in conjunction with Section 16.10.1.

Supervision and inspections

- During the procurement period check the specification for the plastering or rendering type and mix. Check also the specified finished accuracy and the required or minimum thicknesses.
- Discuss the level of working lighting required and its position to achieve the specified finish. It is useful to consider the permanent type and position of lighting that the finished work will be subjected to. Wall lighting in (long) corridors will dictate the need for a far higher quality of plaster finish than elsewhere.
- Contact the manufacturers for their Technical Guidelines, particularly for dry-lining.
- Check the substrate for accuracy, cleanliness and key.
- Check the need for provision for movement. Movement joints in brickwork or blockwork walls will require stop-joints in the plastering or subsequent tiling.
- Some cement-based renders craze in time. Check with the tile manufacturer for their recommended type, strength and thickness of backing.
- Check for completeness of the preceding trades. Incompleteness can lead to return visits, damage or unacceptable patches in the finished work.
- When work starts, ensure that continual checks are made on the materials and the mix. Also check regularly that the required finish and accuracy is being achieved. Check the widths and plumb of reveals and narrow features, and the widths of finished corridors.
- Allow the specified or recommended drying times.
- Insist on regular clearing of surplus materials, waste and rubbish.

16.10.4 Joinery

This section should be read in conjunction with Section 16.10.1.

Supervision and inspections

- During the procurement period, check on the delivery periods for all items. Some specialist ironmongery items (e.g. 'suited' locks) require a long lead-time. Early delivery will generally increase the loss by theft. Late delivery affects following trades and handovers.

- Decide whether ordering set lengths of certain items like skirtings might minimise wastage and cost.
- Careful storage in dry conditions is essential for all joinery items. Identify the permissible range of moisture contents and check before fitting if there is any doubt. Ironmongery requires the most secure storage you can arrange. Just-in-time is the best policy.
- Joinery items require either protection or extreme care. Relatively small amounts of damage can result in work having to be replaced.
- The more specialised the joinery work, the greater the need for site measurements to be made by the specialist.
- Cheap joinery and cheap ironmongery frequently *fail*.
- The most *damage-sensitive* joinery should be fitted as late as possible.
- New gangs on joinery should be checked early and until they *prove* themselves.
- Glance at all doors and frames as you pass. Parallel edges speak for themselves and twisting can be spotted early.
- Close to handover, fitted joinery may need protection against cleansing agents.

16.10.5   Paints and coatings

This section should be read in conjunction with Section 16.10.1.

Supervision and inspections

- Ensure that the manufacturer's recommendations are on site. Extract relevant check lists for site inspections. Accept the principle of a 'paint system' so that a full appreciation is given to preparation and primers.
- Decide early on what tests will be carried out, particularly checks on film thicknesses.
- Before continuing, check the quality of each substrate. Allow for all preparations including any pre-treatment, e.g. knots and shakes, and porous areas, in good time.
- Consider the manufacturer's recommended temperature and humidity limits. Decide where space heating is needed. Minimise access to other trades through any area at this stage of finishes. Examine any source of dust that could affect the work. Spray paints can drift.
- Extract all recommended drying times. Include the times between coats.
- Any snagging that can be done before these finishes saves time later (fewer return visits and patchy work).
- Agree on protection of fittings, flooring, etc. as an essential part of the method statement.
- Where masking tape is needed, insist on its use. Time the removal of that tape carefully.
- Wallpapers need to be ordered in good time, imported materials particularly. Be wary of a client's last minute changes. Wallpapers for any area should come from a single batch. Any full height feature could provide a change point for rolls from

one batch to another and internal corners can often be used (do not use external corners).

- In some uses coatings may need to have the ability to breathe. This property should be checked.
- Where appropriate check for flammability.

Further reading

Exterior claddings

Standards and Codes of Practice

BS 1494–1: 1964 *Fixings for Sheet, Roof and Wall Coverings.*
BS 4868: 1972 *Profiled Aluminium Sheet for Building.*
BS 5250: 1995 *Control of Condensation in Buildings.*
BS 5427–1: 1996 *Design of Profiled Sheet for Roof and Wall Cladding on Buildings.*
BS 5606: 1990 *Accuracy in Buildings.*
BS 8000–7: 1990 *Workmanship on Building Sites. Code for Practice for Glazing.*
BS 8200: 1985 *Design of Non-loadbearing External Vertical Enclosures of Buildings.*
BS 8203: 1996 *Installation of Resilient Floor Coverings.*
BS 8204–1, –2: 1987 *Screeds, Bases and in-situ Flooring.*
BS 8297: 1995 *Code of Practice for Design and Installation of Precast Concrete Cladding.*

Other related texts

Manual of Good Practice in Sealant Application: CIRIA, London, CIRIA Report SP6, 1976.
A Suggested Design Procedure for Accuracy in Building: CIRIA, London, CIRIA Report No. TN 113, 1983.
Water Leakage Through Façades: ASTM, Philadelphia USA, ASTM Report STP 1314, 1998.
Modern Stone Cladding: Design and Installation of Exterior Dimension Stone Systems: M. D. Lewis, Philadelphia USA, ASTM, 1995. *Masonry: Design, Construction: Problems and Repairs*: ASTM, Philadelphia: ASTM Report No. STP 1180, 1993. *Durability of Building Sealants*: A. T. Wolf ed., Paris, Rilem Report 21, 1991. *Aspects of Cladding*: IStructE, London SETO, 1995. *Off-site Fabrication*: A. G. F. Gibb, Latheronwheel UK, Whittles Publishing, 1999.

MCRMA publications—technical papers

A series available from the Metal Cladding & Roofing Manufacturer's Association Ltd, 118 Mere Farm Road, Noctorum, Birkenhead, Mersyside. L43 9TT (Tel. 0151 652 3846).

NFRC publication

Profiled Sheet Metal Roofing & Cladding (3rd ed.): available from the National Federation of Roofing Contractors, 24 Weymouth St, London W1N 3FA (Tel. 020 71436 0387).
Note: The results of an extensive cladding research programme conducted by Taylor Woodrow and Loughborough University is available via *www.cwct.co.uk*. This is in the form of two CDs (Cladding buildability and Cladd: ISS) and deals with tolerances and interfaces.

Internal finishes

Standards and Codes of Practice

General

ENV 1996: Eurocode 6: *Design of Masonry Structures.*

Screeds and floor finishes

BS 8000–9: 1989 *Workmanship on Building Sites: Code of Practice for Cement/Sand Floor Screeds and Concrete Floor Toppings.*
BS 8203: 1996 *Code for Practice for Installation of Resilient Floor Coverings.*
BS 8204–1: 1987 *Code for Practice for Concrete Bases and Screeds to Receive in situ Flooring.*
BS 8204–2: 1987 *In situ Floorings. Code of Practice for Concrete Wearing Surfaces.*

Plastering and rendering

BS 5492: 1990 *Code of Practice for Internal Plastering.*
BS 8000–10: 1989 *Workmanship on Building Sites: Code of Practice for Plastering and Rendering.*

Partitioning

BS 5234–1, 2: 1992 *Partitions (including Matching Linings).*
BS 8000–8: 1994 *Workmanship on Building Sites: Code of Practice for Plasterboard Partitions and Dry Linings.*

Joinery and ironmongery

BS 1186 *Timber for and Workmanship in Joinery*: Part 1: 1991 *Specification for Timber (Withdrawn)*; Part 2: 1988 *Specification for workmanship.*
BS 8000–5: 1990 *Workmanship on Building Sites: Code of Practice for Carpentry, Joinery and General Fixings.*

Tiling

BS 5385–1: 1995 *Code of Practice for the Design and Installation of Internal Ceramic Wall Tiling and Mosaics in Normal Conditions.*
BS 8000–11: 1995 *Workmanship on Building Sites: Code of Practice for Wall and Floor Tiling.*

Suspended ceilings

BS 8290–3: 1991 *Suspended Ceilings. Code of Practice for Installation and Maintenance.*

Paints and coatings

BS 6150: 1991 *Code of Practice for Painting of Buildings.*
BS 8000–12: 1989 *Workmanship on Building Sites. Code of Practice for Decorative Wall Coverings and Painting.*

Chapter 17 GLASS

David Doran

Never accept unmarked glass when safety glass has been specified and make sure the contractor treats it with the loving care it deserves!

–Roy Kinnear, Sandberg, 1996

Basic requirements for best practice

- Keep site storage to a minimum consistent with contract programme

- Check thickness and type of glass before installation

- Remember that satisfactory performance is dependent on both the glazing and its support

- Where possible insist on trial pre-contract assemblies to iron out possible problems with buildability

---

17.1 Introduction

Glass is a brittle material which gives very little warning of impending failure. Most glass is sodium-lime silicate. It is composed of a combination of oxides the principal constituent of which is silicon dioxide–sand mixed with other oxides in approximately the following proportions:

Silicon dioxide 69–74%
Calcium oxide 5–12%
Sodium oxide 12–16%
Magnesium oxide 0–6%
Aluminium oxide 0–3%

In construction glass is usually found in sheet or block form. The material is commonly used as a cladding but may be used structurally in floors, facades, staircases and canopies. In block form it is found in non load-bearing (and lightly load-bearing) internal walling and pavement lights.

Outside the scope of this chapter, it is also used as reinforcement in glass-fibre reinforced concrete (GRC) glass-fibre reinforced plastics (GRP), mirrors, shop fittings and also in glass-wool used for insulation purposes.

17.2 Sheet Glass

There are many types of sheet glass including:

- Annealed glass—the main method for producing glass is by the float process–accounting for perhaps 90% of manufactured flat glass–in which molten glass (at about 1500°C) is poured over a flat bed of tin and allowed to cool. The cooling process (annealing) ensures that the finished product is virtually stress free. The width of glass so produced is usually limited to about 3 m. For normal purposes glass is manufactured in the thickness range 2–25 mm although 0.5 mm thickness is available to the electronics industry. The fracture mode is usually a breakdown into large shards.

Flat glass that has been produced by the float process may, if required, be re-heated to about 1000°C (sufficient to soften but not melt the glass) and shaped to suit architectural requirements.

In order to deal with more exacting situations, alternatives exist which include:

- Thermally toughened (or tempered) glass—glass produced by the float process is then subjected to temperatures of 650/700°C and then quenched. This process raises the strength of the material but also affects the failure mode in which a panel will shatter into numerous small pieces. (see also Section 7.5 on defects.)
- Heat-treated glass—produced in a manner similar to that of thermally toughened glass but with additional control so that residual stresses are less than for toughened glass. As a consequence this type tends, on fracture, to break into shards. Heat-treated glass also reduces the incidence of nickel sulphide inclusions (see also below) and is less prone to optical distortion than toughened glass.
- Laminated glass—formed of two or more layers of glass with various bonded inter-layers. Most types of glass may be subjected to this procedure. Bullet proof, blast resistant and fire resistant properties may also be achieved by the use of laminated glass.
- Other treatments of glass
  - Surface coatings to reduce emmisivity, to filter out or reduce unwanted types of light or to reduce solar gain in hot weather.
  - In hermetically sealed units incorporating separate panes of glass with an air gap (up to 20 mm) to produce thermally efficient double or triple glazing. As a further refinement Argon, an inert gas, may be introduced into the cavities of double glazing to enhance the thermal properties.
  - A range of tinted glasses are available for both aesthetic and thermal enhancement.
  - Fire resistant properties may be enhanced by the use of wired glass or laminates with a sandwich of intumescent material between outer panes. In general, borosilicate glass exhibits lower expansion characteristics than soda–lime silicate glass.

Figure 17.1   Structural glass: Glasgow Medical School – roof support beam, general view.

Figure 17.2   Structural glass: Glasgow Medical School – roof support beam, detail.

- Recent developments have seen the emergence of self cleaning glasses which incorporate 15 micron thick titanium dioxide photo-catalytic coatings. This allows rainwater to spread evenly thus uniformly washing the surface.
- Decorative finishes can be produced by texturing the surface of the glass.
- Toughened glass may be used where safety is paramount. In such situations it is a legal requirement that glass panels be marked accordingly (known as manifestation marking). It should be noted that where glass performs the role of a barrier or where it protects a drop in level it must be safety glass in accordance with Part N of the Building Regulations. In the latter case it is also necessary to provide containment after breakage.

17.3   Glass blocks

These may be hollow translucent brick-sized units for internal walling or solid blocks for use as pavement lights. The walling blocks are usually pattern-moulded on one or two faces. They may be fire resistant and are usually non-combustible and, being hollow, have good heat insulation properties. Mortar for jointing purposes usually consist of 2 parts Portland cement; 1 part lime and 8 parts sand. As an alternative specially manufactured compounds are available.

17.4   Best Site practice

- Damage to glazing after installation can be minimised by well defined inspection and maintenance procedures.
- Before any work is carried out the contractor must check with the designer whether glass is being used structurally. For example, it is common current practice to stiffen glass facades with glass fins glued to the glass face with silicone type adhesives. It is also critical that external glass panels are of sufficient thickness to safely resist wind loading.
- Check that glass thickness complies with construction drawings and specifications.
- Early signs of problems in completed work will be apparent from displaced glazing bars, distorted gaskets, poorly formed seals and wind/rain penetration.
- Distorted reflections in glass panels may suggest that poor quality glass has been used and should be replaced.
- Where glass has to be drilled to accommodate fixing bolts it is essential to adhere strictly minimum edge distances and these must be carefully checked before proceeding with installation. For best results drilling should be carried out on flat bed drilling equipment. It is essential that bolts do not make direct contact with the glass but are adequately restrained within washer assemblies. In this sense bolts act in a manner similar to friction grip bolts used in structural steelwork.
- Cut edges of glass will normally be rough and require some treatment such as polishing before use in a site assembly.
- The contractor should be fully aware of the arrangements in place to accommodate structural and thermal movements and that these are fully understood and adhered to.

- Strict adherence to tolerances must be observed. Too large and a pane may not fit within the frame: too small and the assembly may be prone to leakage.
- It must be remembered that a good result will only be produced by attention to both the glazing, its **framing and support.**
- For refurbishment or repair work to domestic properties, the correct putty or putty substitute should be used to achieve the best results. Putty is a compound of whiting (chalk) and linseed oil. Do not apply putty when there is any visible moisture on the surface to be treated. In normal conditions a period of 48 hours is required for putty to harden before painting. As an alternative to putty, specially manufactured compounds may be used.
- Note that alkalis from chemical paint strippers may etch the glass.
- Particular care must be taken where operatives are fixing, repairing or removing glass from roofing assemblies including roof-lights. In appropriate cases independent means of access must be provided to avoid direct operative contact with the glass. *Fragile roof* signs should be displayed at roof access points.
- Self cleansing glass **must** be installed with the coated face exposed to the weather.
- Temporary storage on site should be kept to a minimum consistent with meeting programme requirements. Storage areas must be kept free from risk of falling objects, abrasives and chemical contaminants. It is essential that glass in storage is maintained in a position that does not apply stress to the glass as this may severely damage or stress the material beyond the level required in service. The prudent use of separators may be required to achieve this. Coated glass is particularly sensitive to storage damage.
- Operatives handling glass (including loading and off-loading transport) should be provided with appropriate protective clothing.
- On major curtain walling contracts there are great benefits to be gained from testing pre-contract mock-ups. In addition to checking for wind and water-tightness the process provides a useful check on buildability. The author is aware of situations where designers have used components from *two* different proprietary systems. A pre-contract test has shown the use of these in *one* assembly to be incompatible.

17.5 Defects

Like many construction materials, glass may be subject to some defects which include:

- **Nickel sulphide inclusion.** These came to light in the late 1980s. They are minute inclusions formed within the glass when nickel [usually from the sand] and sulphur combine. These inclusions expand with time and exposure to the light and may exert sufficient stress on the glass to trigger failure. Heat soaking in the glass manufacture will usually reduce or eliminate nickel sulphide inclusions.
- **Weld spatter.** (Sometimes called weld *splatter.*) May occur due to small specks of hot metal [possibly from local welding or metal grinding operations] on the glass surface. These specks may embed themselves in the glass and their consequent expansion and corrosion due to heat and moisture may shatter the glass.

- **Edge shelling.** This may occur when the edges of glass are damaged, for example, in poor packaging and transport to site. This damage together with other accumulating stresses due to wind loading, cramping from glazing bars etc may induce a critical point at which failure may occur.
- **Thermal shock.** Failures have occurred due to sudden bursts of heating in office premises after holiday periods or periods of very cold weather. This is especially so in highly insulated buildings with very powerful air conditioning plant. It is understood that glass treated with solar films are particularly sensitive to this type of problem. Examination of such failures has shown the origin of the breakage to be at an otherwise innocuous minor defect in the glass, where stresses remained subcritical until the superimposition of the thermal shock.
- **Fatigue loading.** Glass, unlike most other materials, has no inherent ability to absorb long-term movements. When subjected to external edge loads, stresses increase dramatically at the point where the load is applied. If the edge of a pane is subjected to point loading (for example because a shim is accidently left in place) other cyclic and long-term building movements are transmitted to the glass and failure may occur due to fatigue.
- **Scratch damage.** The surface of glass will usually be scratched and very severe scratching may cause failure. However such severity will usually produce complaints on aesthetic grounds before failure is likely to occur. Scratches deeper than 400 μm and detectable by a fingernail test will usually cause failure.

Acknowledgements

In writing this chapter I have drawn heavily on the work of Roy Kinnear formerly of Sandberg, Dr Arthur Lyons and various colleagues at ARUP (Bob Cather, Edwin Stokes and others). I would also like to pay tribute to Bill Black (Drivers Jonas) who reviewed the draft and made a number of helpful suggestions.

Further reading

Standards and Codes of Practice

BS 952-1: 1995 *Glass for glazing. Classification.*
BS 952-2: 1980 *Glass for glazing. Terminology for work on glass.*
BS 8000-7: 1990 *Workmanship on building sites. Code of practice for glazing.*
BS 5051-1: 1988 *Bullet-resisting glazing.* [Now superseded by BS EN 1063:2000.]
BS 5357: 1995 *Code of Practice for installation of security glazing.*
BS 5544: 1978 *Specification for anti-bandit glazing (glazing resistant to manual attack).*
BS 6262:1982 *Code of practice for glazing for buildings.*
BS EN 572: 2004 *Glass in building-basic soda lime silicate glass products.* Parts 1–9.
prEN 12725: 1997 *Glass in building. Glass block walls.*

Other related text

Structural use of glass in buildings: IstructE, London, 1999.
Materials for architects and builders: A. Lyons, Butterworth-Heinemann Oxford, 2007.

Construction materials reference book: D. K. Doran ed., Butterworth-Heinemann. Oxford, 1992. (Second edition in preparation.)

Approved UK Building Regulations: HMSO, London, 2008.

Priorities for health and safety in the glass industries: Information sheet No. 1, HSE Information Services, Suffolk, 2002.

Centre for Window and Cladding Technology (CWCT) documentation includes:

- TN1 *Air leakage through windows and glazed cladding systems*
- TN11 *Glass types*
- TN12 *Specification for hermetically sealed units*
- TN13 *Glass breakage*
- TN18 *Gaskets*
- TN35 *Assessing the appearance of glass*
- TN41 *Site testing for water-tightness*
- TN58 *Repairs to glass*

Note: Other information is available from The Glass and Glazing Federation (G&GF), London.

Chapter 18 BUILDING ENGINEERING SERVICES

John Armstrong

Basic requirements for best practice

- Be clear about 'who does what'
- Ensure the quality control procedure is adequate
- Understand Health and Safety responsibilities of all parties
- Maintain site cleanliness at all times
- Monitor the approval process of materials and equipment
- Promising early availability of an area may create problems
- Commissioning and testing must be properly specified and managed
- Identify and categorise defects and manage correction

18.1 Introduction

18.1.1 General

Building engineering services covers a very wide range. Each service can have many subdivisions, for example sprinkler systems can range from a simple inlet pipe from the local water supply authority and connected to a valve set serving a sprinkler distribution system, to a complex arrangement of a separate storage tank, electric or diesel pumps, many valve sets and extensive distribution systems such as 'in rack' used for high-piled storage.

The working environment is the *end product* of any services installation and the client judges our total effort on the project on how the services perform. If the occupants are uncomfortable by being too hot or too cold or if the lighting levels are poor or the toilets do not flush properly then, in the client's eyes, the main contractor will have failed.

18.1.2　Standard symbols and notations

Although there are generally used symbols, check each project separately as some may be used out of context.

18.2　Types of contract

18.2.1　Design and build

The detailed design, supply and installation work is usually carried out by a specialist subcontractor. The performance specification and types of system are usually determined by the main contractor's designer, who appraises the subcontractor's working drawings prior to installation. This information is given also to other members of the design team, engineers, architects and surveyors for their work. Main contractor performance specifications do not usually identify equipment and plant by manufacturer or reference number.

Where the main contractor produces a basic design to be developed further by a subcontractor, the information and documentation will be more detailed. In the case of a full design by the main contractor, all relevant details will be provided for the subcontractor. The main contractor designer may be an in-house group or an outside consultant appointment.

18.2.2　Main contractor build only

All design is carried out by independent consultants or designers.

18.3　Checking out the proposed building services subcontractors

Prospective subcontractors need to be identified and prequalified. Post-tender, standing instructions or procedures of the main contractor should be used, where available, to verify the capabilities of subcontractors. Ensure minutes and agreements reached off site, e.g. in regional offices, are available on site and understood by both the main contractor and subcontractor's staff.

18.4　Subcontractor requirements

Services subcontract tenders often have qualifications and 'builder's attendances'. Identify these while the tender is being analysed. Be clear about 'who does what' and 'who provides what' under the terms of the contract. Watch out for the following, often omitted on purpose by the subcontractor:

- Off-loading heavy plant and equipment
- Off-loading materials
- Distribution of materials and plant to floor entry or installation point

- Scaffolding
- Offices, storage, workshops and compounds
- Hard standing, fencing and gates
- Access roads
- Sufficient power available on site?
- Payment of electrical power and telephone costs
- Site transport?
- Provision of fuel and power for commissioning and testing
- Rubbish clearance
- Drilling holes for pipework and ductwork support rods and fixings in general
- Access gantries (ladders, platforms, trench covers etc.)
- Approved welders (including testing of their competence)
- Specified requirements for non-destructive testing (NDT) e.g. radiography of welds or special insurance tests
- Painting.

18.5   Separate services subcontractors

The major problem here is coordination. Be aware of who is responsible for coordination and how it is to be carried out. Watch out for problems with different contractors in the same space:

- Above false ceilings, with the ceiling itself, location of pipes, ducts, cables, luminaires, sprinkler heads, smoke detectors etc.
- Vertical shafts
- Below false floors, e.g. different services crossing in a restricted space
- Who provides electrical services to the HVAC equipment and their controls?
- Interconnections between one discipline and another, such as fire alarm panels picking up signals from HVAC equipment
- Final electrical connections to equipment such as:
- roller shutter doors
- client's equipment
- Final connections to kitchen equipment.

Once a contract has been let, hold regular site meetings to monitor progress and to take corrective action when necessary. All subcontractors involved must be present.

18.6   Procurement schedules

These should be used for major M&E plant and equipment, ductwork manufacture, prefabricated pipework, large cables, switchboards, generators, etc. Ideally updated on a monthly basis showing:

- That the manufacturer has been approved
- That the manufacturer has received an order and the date placed

- The proposed date of certified drawings to be issued and approved
- The manufacturing timescale
- The date that manufacture started
- The earliest date available on site
- Dates of works audit inspections by:
  - designer
  - main contractor
  - client
  - insurers
- The programmed date for delivery to site
- Remarks column to show: – on site
  - in store
  - any problems.

18.7 Factory visits

These should be made when the delivery date of equipment is on a critical path. Letting a manufacturer know that the main contractor is intending to visit the works, with the subcontractor, should ensure the equipment is actually being made and not just an order on paper. Where necessary, conformity to specification should be checked on such visits.

18.8 Method statements

These should be part of the contractual requirements.

A contractor should always be asked to provide method statements where activities are sublet or considered to be complex or awkward to perform. The subcon-tractor's quality plans should contain a list of method statements he proposes to prepare or procure from his specialist subcontractors, c.g.

Basic construction

- Mechanical services
- Electrical services
- Public health engineering
- Lifts and escalators
- Sprinklers, hose reels, dry fire riser.

Include pressure testing of air and water systems and NDT.

Special construction

- Embedded floor coils
- Lightning protection
- Diesel generators.

Systems preparation (HVAC)

- Flushing e.g. heating, chilled water pipework (specialist)
- Chemical cleaning of chilled water pipework (specialist)
- Chlorination of domestic water services, (specialist).

Commissioning

- Air/water balancing.

Building management systems (BMS) and controls

- Motor control centres and control outstations, procurement, including software development and off-site testing
- Field wiring MCCs
- Control testing
- BMS testing
- points, outstations, supervisory centres
- Instructing clients staff.

Note: Method statements must be approved before work starts.

18.9 Quality control

18.9.1 Introduction

Before the subcontractor starts manufacture at works or on site, follow any standing instruction or procedures to ensure that work complies with the specification and drawings. This will minimise the accumulation of defects at the end of the contract, prevent uncorrected defects being covered up, avoid lowering of standards and save the time spent in correcting them at a later date.

Ensure that the subcontractor's project quality plan includes supervision and recorded constant monitoring, so the main contractor can check that the control procedure is adequate and maintained.

18.9.2 Workmanship

The services subcontractor and his specialist sub-trades should as part of their project quality plans and method statements produce checklists for inspections covering:

Mechanical

- Welding standards for pipework etc.
- Jointing standards and intervals on pipes and ducts
- Adequate supports on pipes—intervals, strength, appearance
- Adequate supports on ductwork—intervals, location (adjacent to walls, equipment, dampers)

- Valves, dampers and plant, the right way round for air or liquid flow
- Insulation integrity, especially vapour seals
- Chilled water/air pipe or duct bracket inserts to prevent condensation
- Services installed to drawings
- Materials and plant to specification
- Valves and dampers installed in correct locations
- Access for removable items of plant or parts of plant, e.g. can the strainer cage be removed?
- Correctly applied anti-vibration measures
- Provision of sleeves on pipes passing through walls
- Protection of open ends of ducts and pipes (no protection = blockage = delay and extra costs)
- Labelling of valves, controls, equipment
- Provision of statutory notices.

Electrical

- Services installed to the current edition of the IEE regulations
- Services installed to drawings
- Materials and equipment specification
- No cables to be installed when the cable or ambient temperature is at or below freezing point
- Depth of external cables and distance from other cables and services
- Provision of cable markers to external cables
- Cable identification/correct use of coloured cables
- Labelling of equipment
- Provision of circuit schedules mounted in distribution boards
- Provision of statutory notices
- Fire alarm pushes not painted out, they must be *red* in colour
- Earthing of metal stud partitions, raised floor pedestals etc., especially in buildings with information technology (IT).

The lists are not exhaustive and cover key items and aspects.

18.10 Health and safety and statutory authorities

See also Chapter 19.

18.10.1 Health and safety

Understand the health and safety responsibilities of the main contractor and yourself. No services subcontractor or any specialist sub-trades should commence work on site before the approval by the main contractor to their contract health and safety plan.

18.10.2   Statutory approvals

These are obtained by the design team, but check that drawings issued to site have been approved by the appropriate authority. Various authorities will visit the site to check on the installations. Clients may also have their own specialists.

18.10.3   District Surveyor

See also Section 18.10.5.

*   Fire damper locations in ventilation ducts
*   Fire resistant cladding to ventilation ducts, pipework and electrical cable ways
*   Smoke pressurisation tests.

   Construction staff should confirm that the building structure and fabric as designed, can be constructed to the required level of air tightness, e.g. staircases, lift lobbies, builders' work ventilation ducts.

18.10.4   Building Inspector

*   Emergency lighting—to British Standards.
*   In-slab and above-slab drainage—adherence to drawings and specification; air tests will be witnessed on the drainage installation
*   Toilet ventilation—adherence to bye laws; air volumes to design levels
*   Smoke clearance systems
*   Sprinkler systems
*   Car park venting systems
*   Pressure testing of building envelope (where required under Part L2A of building regulations).

18.10.5   Fire Officer

*   Audibility tests for fire alarm sounders
*   Emergency lift operation
*   Location of fire alarm pushes
*   Location of fire alarm sounders
*   Smoke/heat detector; fire alarm operation
*   Location of hose reels
*   Location of wet and dry riser landing valves
*   Location of fire brigade inlet points and control rooms
*   Smoke pressurisation tests
*   Emergency lighting.

18.10.6   Water supply company

*   Rising mains, cold down services, overflows etc.
*   Access to valves
*   Adherence to byelaws
*   Installation methods
*   Materials

- Prevention of freezing
- Equipment and materials to current Water Research Council (WRc) approved lists.

18.10.7   Gas supply company

- Adherence to regulations
- Completion of gas safety certificate for the installation.

18.10.8   Electricity supply company

- Adherence to regulations
- Completion and inspection certificates filled in as prescribed in the current Institution of Electrical Engineers (IEE) Regulations for electrical installations.

18.10.9   Insurance company

The insurance company is usually that of the employer/tenant.

- Sprinkler installations—adherence to either the current edition of the Loss Prevention Council (LPC) rules or the relevant rules for American employers. Sprinkler installations are checked on completion. It is not unusual to have alterations requested by the insurance company before acceptance. This relates to seeing the job in the flesh as opposed to on drawings.
- Lifts-tests witnessed on completion. Tests are usually to the appropriate British Standard (but see also Section 18.10.5). Make sure that the lift manufacturer and installer is told who wants to witness the tests, otherwise he will charge if a re-test is needed.

18.11   Drawings

In addition to tender and contract drawings usually produced by the design team, the services subcontractor must produce working drawings. Obtain from the subcontractor the list of drawings he proposes to produce with a resourced programme for their release to the approval circuit. Get the designers to review the adequacy of the intended drawing list, monitor production against programme.

Keep on site a drawing register showing the latest revisions of *For Construction* issue drawings. In addition to the usual sketches, general arrangements, details and schedules, there are:

- Functional diagrams or schematics: always produced by the designer, showing where the plant pipework and ductwork valves and dampers, controls and instruments are all located.
- Single line diagrams: showing electrical equipment location and interface.
- Coordinated services: these should be produced to show the physical relationship of service to service, service to structure and service to fabric, preferably with

dimensions. Coordinated drawings are essential where services are congested, exposed to view in membranes such as false floors and ceilings or surface mounted. Individual services present degrees of difficulty to resolving coordination problems due to:
- size e.g. ductwork, large cable bending radii
- priority location e.g. falls of rainwater, soil and waste pipes from outlets
- risk, e.g. keep water pipes away from electrical switchboards. Beware the large scale General Arrangement drawings with *all* services shown on and called coordination drawings. This they are certainly not. Coordination drawings produced by CAD must be generated from software with three-dimensional (3D) capability.
- Manufacturers' certified drawings: produced by plant and equipment manufacturers and are only made available once an order is placed. Until certified drawings have been received by the subcontractor, builders' work drawings and the detailed plant room layouts cannot be completed. These, in turn, affect other drawings such as the ductwork layouts by the sub-subcontractor. Do not use manufacturers' catalogue details. Invariably they are wrong for your particular case.
- Builder's work: identify plant bases, holding-down bolt holes, holes through floors and walls, ducts, shafts etc., these are usually produced by the subcontractor. Holes through structural members are generally shown on the relevant structural drawings.

18.12   Programme

The single line labelled *M&E Services* on a bar chart indicates a complete lack of understanding of the M&E involvement.

The services subcontractor must be involved in producing a detailed programme as soon as an order or letter of intent has been placed, to be integrated into the overall programme. These must be broken down in enough detail for the various activities to be 'dovetailed' into the main construction work:

- Separately identify services
- Breakdown each service, e.g. heating pipework, chilled water pipework, ductwork and if necessary sub-activities within the service
- Location of services—floor by floor, plant rooms, vertical shafts
- First and second fix activities must also be identified, especially where many services are involved, e.g. ductwork followed by grilles in a false ceiling tile, or sprinklers followed by heads in a false ceiling tile.
  Other activities to note are:
- Builders' work to be available
- Plant and equipment delivery dates
- Delivery dates for specials (can be on a very long delivery)
- Statutory services *on*, make sure whoever is responsible properly monitors this happening

- Services operating dates (identify each major service and plant separately)—it is critical to ensure power is into the outstation motor control centre before flushing, chemical cleaning and air system *blow through* (system preparation) can commence
- Pressure testing: separately identify service by service, floor by floor, shaft by shaft—do not apply insulation until testing is complete
- Cable continuity tests
- Earth resistance tests
- Specialist testing, e.g. radiography of welds and special insurance tests
- HVAC systems preparation is important, comprising for pipework, flushing and chemical cleaning, and pre-commissioning
- HVAC balancing (regulation of air/water)
- Connecting the controls
- Commissioning and testing (C&T): show each system, with start and completion dates.

Note: The final four points above can only be carried out in this order with limited scope for increasing resources to reduce time. These time periods being incompressible can become the area of contract overrun.

18.13   Sequence of work

18.13.1   Continuity

One of the problems faced by services subcontractors is lack of continuity of the construction work ahead of their own work.

Most service installations have pipes, ducts or cables in long lengths. Parts of the construction work not built along any one service route can stop the subcontractor working or completing a particular service. The requirement for an area to be finished early should be researched thoroughly.

For an area to be provided with operational services early you must research thoroughly from where the services are generated. Promising an area early where some of the equipment is to be located on a yet to be completed floor will create problems. Buildings with a lot of services above false ceilings and below raised floors will need careful sequencing. Low-level services require protection before a raised floor is installed.

18.13.2   Access

Subcontractors require access to plant rooms that are rubbish free and weathertight. Plant rooms in basements and lower levels are rarely 'protected' by the floors above and can become waterlogged, requiring pump-out. These levels often contain electrical switchrooms with equipment that cannot tolerate damp conditions. Services shafts must not be used as unauthorised rubbish chutes.

Large plant items often require walls to be left down to gain access. Large heavy loads also need road access to the point of off-loading. Identify and sequence these and any other relevant activities.

18.13.3 Example of installation sequence

See also Figs 18.1 and 18.2.

1. Install electrical trunking and conduit.
2. Install ductwork/terminal point.
3. Install pipework: heating/chilled/sprinkler.
4. Pressure test pipework.
5. Pressure test HV ductwork.
6. Wire up lighting.
7. Insulate ductwork and pipework.
8. Install false ceiling grid.
9. Install light fitting boxes, HV grilles and diffusers.
10. Install tiles (ceiling becomes *locked*).
11. Fit lighting gear trays.
12. Wire up light fittings.
13. Install HV flexibles and drops to diffusers and insulate.
14. Install sprinkler drops.
15. Cut tile for sprinkler pipe.

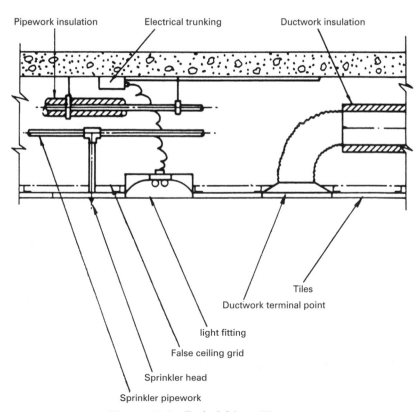

Figure 18.1 Typical false ceiling space.

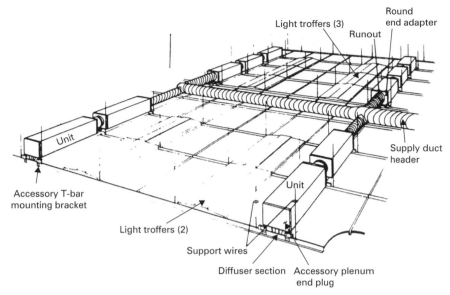

Figure 18.2 Typical false ceiling as seen from above.

16. Install tile.
17. Install sprinkler head.
18. Install lamps and diffusers.
19. Close ceiling.

18.14 Practical checks on site

- Adequate points for fixing drop rods for suspended services: additional steel may be required if this has been overlooked at the design stage. Fixing of primary brackets to steel beams before fire spray/cladding.
- Design changes: additional pipe runs or equipment may put an unacceptable load onto the steel work or a floor etc.
- Drawings: execute work only to drawings marked 'For Construction'. Always read the drawing thoroughly. Elements omitted can be very costly to put right. Always ask the designer for missing information. If a pipe size is missing, *do not guess*. You are bound to be wrong.
- Anchor/thrust blocks for underground water pipes: these must always be installed to approved detail drawings. Casually dumped concrete will not prevent pipework under pressure buckling and rising out of the trench. This is expensive to remedy.
- Access: it may be essential to provide access panels in plaster ceilings, vertical shaft enclosures for commissioning or future servicing, therefore the panel positions need to be known to ensure the space is not occupied by air diffusers, sprinklers, fire detectors or some other service equipment.

- Existing services should have been located and plotted, but:
  - Be observant—think!
  - What buildings were on the site previously?
  - Where were the buildings located?
  - Ask the local people. Are buildings to be demolished?
  - **Double check!**
- Electricity:
  - Substations located inside buildings sometimes serve others as well. Check with the local electricity supply company before any demolition work proceeds.
  - Buried cables can be owned by the local electricity supply company or a national distribution company.
  - If buried cables are indicated, keep on digging carefully until you find them. They are often deeper than you think.
  - Enquire from voice and data cable providers what telecoms cables are on or near site. Contact Highways Department regarding any traffic control cables around site. Often these are located below pavements and can be damaged by hoardings.
- Gas and water mains operate at varying pressures. Damage to medium- and high-pressure mains which pass through a site could cause widespread disruption to users located well away from the site.
- Site access: heavy loads such as packaged boilers, transformers etc., require good access to the point of unloading if programme dates are to be met. Access will be necessary for large pieces of equipment to be brought into the building and to final positions. This will necessitate walls being left down in some cases.

Identify the sizes and weights of the equipment and prepare a method statement on how the equipment is moved into position. The most troublesome areas are staircases with applied finishes, which require closing to foot traffic. Remember that other trades require access at all times and early agreement on this is essential. If ignored, claims from subcontractors for excessive walking time may ensue.

In tall buildings the same problem can arise if lifting facilities are not available when required or are inadequate.

- Datum lines and levels: clearly identify and protect those areas where successive trades rely on them for setting out. Remember that any setting out is only as good as the reference points and levels that are provided. Bad setting out leads to costly remedial action, e.g. if walls, ducting and false ceiling grids in cellular spaces are not set out from the same datums (structural grids) ducting spigots and holes in ceiling for diffusers can be seriously misaligned.
- Adherence to drawings: ensure that 'Builder's Work' activities are carried out in accordance with the issued drawings. Preferably get the subcontractor to check the main contractor setting out. Areas to watch out for are:
  - inserts or pockets in plant bases
  - plant bases, especially if the pipework or ductwork is being prefabricated by the services subcontractor.

- cable entry points, especially long radius bends—Cables cannot be pulled if the bends are too sharp or tight.
- Tile and support grid dimensions must be known at the outset to ensure that the underside of luminaires and air terminal devices are positioned at the correct level. If changes occur, tell the other trades.
- Coordination: if the services have been coordinated the sequencing of the various activities can be identified, but site management must also take into account the main contractor element as well.
- Service shafts need access to install the various services, DO NOT make access difficult or impossible by getting the main contractor work out of the way and letting the subcontractor sort out his own problems.
- Trenches obviously come before the installation of the services but to limit the time any trench is open, liaise with the subcontractor and utility companies about the date they can actually start the work.
- Soils and wastes, when made of cast iron, require space for jointing. They are also governed by the fittings available. These services need to go in first.
- Who does what? Check the division of responsibility at trade interfaces, they are often specified for design rather than construction convenience.
  - Electrical feeds to HVAC plant and electrically operated doors, lifts etc., are often 'by others'.
  - Final electrical connections are often omitted. The electrical feed usually terminates in an isolator, leaving the cables from the isolator to the piece of equipment to someone else or no one!
  - Installation of recessed luminaires is often omitted by the services subcontractor leaving the job to the ceiling subcontractor.
  - Steelwork gantries for plant and equipment are omitted on the grounds that the services subcontractor does not design steelwork.
  - Access stairs, ladders and guard rails are generally 'builder's work', but it must be clarified at an early stage.
  - Steelwork for pipe supports should be included in the services subcontractor's tender. Vertical shafts create problems when a steel floor to a service shaft has to be installed at each level after the services have been completed.
  - Access doors and panels in walls, partitioning etc., must be such that the services equipment is also accessible.
- Earthing pits: soil conditions affect the location and depth of such pits. Monitor such work carefully as a lot of time and money can be wasted in obtaining the required resistivity levels.
- Mock-ups: these can be of great benefit especially where repetition occurs. Costs may not have been allowed by the subcontractor but even if they have not, mock-ups are worth investigating as potential savings can be made by all concerned.

Note: Cosmetic (aesthetic) mock-ups without all of the services installed may ultimately be proven unacceptable and unachievable.

- Electricity supply company: requirements for their incoming switch rooms vary one to another.

Watch out for grano floors to ± 2 mm level. Access doors to the supply company's own specification—but provided by the main contractor.

- Shafts and pits for lifts, scissor lifts and dock levellers: these must be built square and vertical. Tolerances should be say + 10 mm but should always be – nil. Shafts and pits too small of square or plumb will require remedial work before installation on the equipment can start.
- Lift shafts: all entrances have to be installed relative to the plumb of the shaft. *Fast-track* construction requires the entrances to be fitted before the front wall of the lift shaft can be constructed.

18.15   Weatherproofing

Services installations can be damaged, sometimes irreparably, if allowed to get wet. General dampness is also a problem with electrical work. Most vulnerable are control panels and switchboards—these are expensive to repair, especially if they have to be taken away for repairs.

Extensive damage can be done when wet screeding is laid, and water mixed with cement filters through to metalwork on the ceiling below, causing severe rusting or corrosion.

18.16   Accuracy of work

The first trade on site usually *greedy*, taking up some one else's allocated space. Ensure that dimensioned locations are maintained. The areas most at risk are vertical shafts, false ceilings and below false floors.

Often building and manufacturing tolerances do not equate. Specified tolerances must be clear and liaison at all stages of construction is desirable, e.g.:

(a) continuous runs of luminaires
(b) ceiling openings
(c) variations in ceiling flatness
(d) weight of equipment on the ceiling
(e) special tiles.

See also the notes on datum lines and levels in Section 18.14 above.

18.17   Fire barriers

The integrity of barriers must be maintained where services pass through.

18.17.1   Fire regulations

Due consideration must be given to the building regulations and local regulations.

Often some specific point occurs late in the contract, for example the ceiling may be 1/2 hr fire resistant but no information about this has been given to manufacturers of equipment located in or above the ceiling membrane or, alternatively, spread of flame limitations may apply.

Disastrous effects could result from irresponsible or unplanned placing of fibrous material over the backs of luminaires without prior consultation with the luminaire supplier.

18.17.2 Ductwork

Ductwork fire dampers create one of the problems in making good. If the ductwork installation is allowed to proceed without discussion on making good, it may become impossible in some cases. Ductwork should ideally be connected to one side of the fire damper, the making good completed, then the ductwork erected to the remaining side. This requires effective supervision (see Fig. 18.3).

18.17.3 Pipes

Most specifications require these to be sleeved. Making good creates similar problems as for ductwork. Sealing of sleeves and ducts must be detailed in the specification and drawings.

18.17.4 Cables

Proprietary materials, some in foam material, are available for sealing cable entries. Check with the designer on the method proposed and its acceptability to the local authority fire officer (Fig. 18.4).

18.18 Materials fit for purpose

Get subcontractors to schedule all materials and commodities specified as requiring sample approval. Monitor the approval process. This is usually covered in the specification but to avoid future criticism, draw to the specifier's attention any requirements not complying with current good practice.

- Manhole covers must be capable of withstanding traffic if located in car parks or anywhere a vehicle can obviously be driven. On one site, the wheel of a lorry jammed in a manhole causing extensive damage to the manhole and the lorry, all because the manhole cover was incorrectly purchased for 'light traffic'.
- External louvres must be sized to suit the application. Watch out for the shape, material and colour, and rain drainage, which must also be specified but may be by a different discipline.
- Ceiling grid: find out at an early stage if this is also acting as the main support for heating, ventilation and lighting equipment.

Also are the luminaires or air terminal devices intended to support other services? Such things as the weight of special tiles are often overlooked. Similarly, pressures on the side of the equipment can close the mouth opening thus preventing attachments from being inserted.

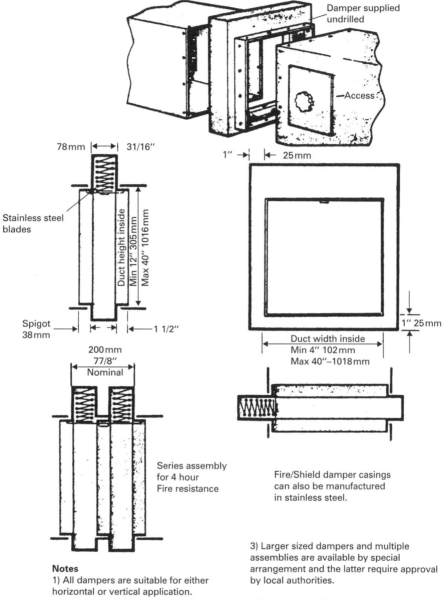

Figure 18.3 Typical proprietary fire damper details.

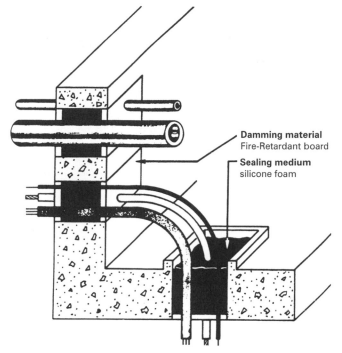

Damming material
Fire-Retardant board

Sealing medium
silicone foam

Figure 18.4 Diagrammatic details of typical floor and wall penetrations.

18.19 Cleanliness

- Open ends, especially of vertical ducts and pipes, drainage sockets and floor outlets must be temporarily capped once the installation work stops for whatever purpose. Baulks of timber, bricks etc. cause blockages that are difficult and expensive to locate once the building is nearing completion.
- Cement droppings on galvanised ductwork are expensive to remove later, apart from any chemical reactions that may have been started.
- Paint splashes are in the same category as cement droppings above.
- Trenches in building: once services are installed and insulated in trenches, it becomes difficult and expensive to remove accumulated rubbish. Trenches must have the covers placed as soon as the insulation has been completed. The covers will also prevent damage to the insulation.
- Plant rooms: subcontractors' rubbish is often the main problem. Ensure that there is adequate provision for it to be dumped before final removal.

Finishing trades such as plastering and painting can cause damage to installed self-finished equipment if that equipment is not adequately protected. Stove enamelled finishes covered in paint splashes are costly to put right.

- Lifts: all work must be complete and rubbish removed from the shaft and on top of and inside the lift car before any lift operation takes place.

- Protection must be provided to any plant, equipment or quality finishes. Contractual responsibilities and costs must be agreed early on in a contract as the effects of damage and its rectification can far outweigh the cost of protection.
- Site damage: a considerable amount of pre-delivery work can be done to keep this to a minimum. Include in early discussions with suppliers: packaging, storage, individual packing versus bulk packing, palletising, off-loading by crane, fork-lift truck or manhandling etc.

Any proposals relevant to the last four points should be covered in the subcontractor's quality plan.

18.20   Authorities

18.20.1   Connections

Check that the various statutory authorities have been asked for a supply and that it will be available when required. Regularly monitor their dates and resources *before* they are due to start.

Where existing supplies need to be turned off to enable new work to be connected, ensure that all the necessary procedures have been followed with the particular supply authority.

18.20.2   Road closures

Ensure that all necessary procedures have been followed including announcements in the local press.

18.20.3   Voice and data cables

Although wire ways and routes are provided, voice and data cables are often installed ignoring such provisions. Ensure that the installation work follows the prescribed routes.

18.21   Subcontractor control

18.21.1   Management and supervisors

Refer to any standing instructions or procedures and your project quality plans for the main contractor's responsibilities. Refer to and monitor the subcontractor against the controls provided in his quality plan.

18.21.2   Documentation

Ensure that the services subcontractors have copies of all the relevant documents on site, and at least three copies of each drawing (where appropriate) for the following uses:

- installation
- marking up progress
- noting alterations and exact locations for record purposes.

18.21.3 Materials

See also Section 18.18

Checking is made easier by having a copy of the orders placed by the sub-contractor.

Non delivery can be due to the subcontractor not providing sufficient information, not paying the supplier (may not even involve the contract in question), or the manufacturer being short of basic materials or labour. Check materials in stores on a regular basis—then shortages will be more noticeable.

18.22 Commissioning and testing (C&T)

18.22.1 General

Commissioning and testing should always be carried out by the subcontractor and his specialist sub-trades. On large complex projects commissioning management may be specified, or included as a management function by the main contractor. Commissioning management is the planning, organisation, coordination and control of all C&T activities. It requires broad knowledge and involves the preparation and then management of logical sequences, resourced networks and programmes. Only a few firms have the key personnel to do this adequately.

Commissioning is, therefore, a managed process taking installed engineering services through an unavoidable set of subroutines comprising, systems preparation, e.g. flushing and chemical cleaning of pipework, pre-commission checks, regulation (e.g. air and water quantities), connecting controls and on to environmental testing, system proving, and possibly fine tuning.

Commissioning must be properly specified. The subcontractor should provide method statements (including those of his specialists) in his quality plans (see Section. 18.8) demonstrating that the specified commissioning requirements are achievable in the contract period, their method statements should be provided early in the subcontract period, monitored and updated regularly.

Beware the designer specification that calls for extensive environmental testing, systems proving and fine tuning. These are rarely achievable in the contract periods of tight programmes. The level of what can be achieved in the contract must be established very early.

A well-specified commissioning process will refer not only to design criteria standards and codes, but the CIBSE and BSRIA technical memoranda and guides given in the reference, at the end of this chapter.

18.22.2 Commissioning

Commissioning is the advancement of an installation from static completion to full working order to specified requirements.

Each system has the following needs to be met:

- System cleanliness, internally and externally
- Correct installation and no damage—free movement of moving parts

- Visual checks to such items as air seals in ventilation systems (builders' work ducts, access doors etc.) to answer the question *Is the system complete?*
- Mechanical checks to such things as security of fixings, drive belt tension and alignment, correct lubricants, correct rotation, correct flow direction, correct selection of equipment to suit the design parameters
- Electrical checks include local isolation of equipment; no unshrouded live components in panels; all connections are tight; fuse ratings correct; all safety devices are operative and correctly set
- Setting to work and starting up plant items that form a system.
- Regulation—regulate air, liquid and gas circuits to ensure that the design quantities are delivered to the correct point (see also Section 18.8).

18.22.3 Works tests

These should be scheduled by subcontractors as 'hold points' and listed in their quality plans.

- Where the specification calls for works test certificates, make sure they are available before accepting delivery. It is too late to put it right when the equipment has arrived and has not been adequately tested to the specification. Do not accept equipment if the certificate is missing.
- Equipment likely to be affected: diesel generators, packaged boilers, chillers, large pump sets, fans, air handling units, pressure vessels, switchgear, busbars and transformers.
- Where such tests require to be witnessed by the client and main contractor, a fully descriptive statement on what tests will be carried out and how must be available for review *before* the tests take place.

18.22.4 Site testing

- Pipework is pressure tested in sections or as a whole. Insulation must not be applied or the pipes concealed until tests are completed, witnessed and signed as being acceptable.

  Air tests on rain-water pipes in habitable areas may be required by the client or local authority.

- Ductwork and pipework is pressure tested when designated high velocity.

- Electrical tests to current edition of IEE regulations are carried out during installation and on completion. They include:
  - protective short circuit current at the origin
  - earth fault loop impedance at the origin
  - continuity of ring final circuit conductor
  - continuity of protective conductors and equipotential bonding
  - earth electrode resistance
  - insulation resistance of the fixed installation
  - insulation resistance to earth of each item of equipment tested separately
  - polarity and position of single devices for protection and switching
  - earth fault loop impedance for operation of devices relied upon for earth fault protection

- operation of residual current operated or fault-voltage operated device for earth fault protection.
• Lifts, once the installation is completed, are tested according to the relevant British Standard.

Ensure that those persons requiring to witness tests, both at works and on site, are informed in advance of such tests being made.

18.22.5   System proving

In many cases, the evaluation or proving of system performance can only be tested when the following criteria have been met or experienced:

• maximum and minimum outside design temperatures and humidity
• the building fully occupied
• clients' machinery and equipment installed and working, and its effect on temperature, humidity, lighting and noise levels.

Systems test results have to be analysed and any design or installation faults or shortfalls investigated and corrected as fine tuning.

Similarly in practice, a full heating and cooling season has to be experienced before a system can be finally proven.

18.22.6   Main contractor involvement

Because C&T takes place once a project is nearing completion, be aware of the attendance work required by the subcontractor, such as the removal of false ceiling tiles for access.

Similarly because no one can guarantee C&T without problems, *things will sometimes go wrong* and it is at this point that the subcontractor needs all the cooperation and help he can get to rectify the problems.

18.22.7   Employer involvement

On many projects, employers have their own inspectors. Ensure that these are advised when C&T is programmed to start for each system.

18.23   Defect management

This is carried out once the subcontractor has offered the installation for acceptance. The subcontractor's quality plans should specifically cover his proposals for final inspections. Snagging is entirely separate from commissioning and testing. Look at each area in detail and record each item not meeting the specified requirements.

It is important to categorise snags e.g. DEF—defect; INC—incomplete; DAM—damage; VAR—variations. Be careful about creating a DES—design class. Deal with these *off the sheet*. Try to avoid the designers creating lists on their own. Go round with them. Once on a sheet, a debatable snag is difficult to remove.

Checking must be thorough. Standing in the entrance of a room and *looking around* is of no use at all.

- Cleanliness—look inside units, on top, behind and underneath
- Cleanliness of gutters and rainwater pipes—check for no obvious blockages
- Neatness and workmanlike appearance of finished work
- Absence of damage (i.e. it has all been put right)
- Evidence of vibration and noise
- Ensure that all controls are demonstrated at suitable time intervals, to make sure they operate as designed
- Does the plant and equipment start and stop as intended?
- Check on maintenance access
- Visual checks on lighting levels
- Subjective view of the environment—too hot/cold, stuffy, draughty?
- Record area by area, for example:

    - plant-rooms, separately
    - floor by floor—larger floors split into smaller areas
    - designated rooms separately
    - external areas.

Identify and categorise each defect separately floor by floor and room by room: an entry reading *straighten all light-switch plates* cannot be crossed off if there are six in number and only five done. It might take longer to write each item out but it identifies each one more easily, and is easier to monitor.

The separate *walk round* project handover inspection should be coordinated with the senior Main Contractor manager or agent on site. Any defects should be few, so they can easily be listed.

18.24   Operating and maintenance (O&M) manuals

O&M manuals, with *Record* or *As Fixed* drawings must be available in accordance with specified requirements when the project is handed over. Progress these items from the first site meeting.

Understand the specified requirements and get the subcontractor to schedule the documents and prepare a production programme for their delivery. Allow at least two (say four-week) approval cycles. This documentation is rarely all approved first time. The subcontractor's production proposals should be scheduled in his quality plan. **Note**: If the subcontractor cannot provide the building operators with the appropriate documents in time, he should run and maintain the installations until documentation is supplied.

18.25   Operating systems for the employer

Make sure that the necessary insurance cover has been effected before running plant and equipment for the benefit of the client.

Whilst running plant, follow maintenance procedures. The warranties applying to the equipment may be invalidated by a failure to do so. The subcontractor responsible for the installation should be employed to carry out the maintenance. Do not accept the installations from the subcontractor, but witness their handover.

18.26 Instructing client's staff

The employer's (client's) requirements are often not clear at the time the designer writes his specification. Unless clarified soon after the appointment of the subcontractor, and left until the end of the job, difficulties can arise. By the end of the job the designer and client will be clear in their needs. These will be more onerous than the subcontractor has priced for.

Early in the job, reach an understanding with client, designers and subcontractors. Get the subcontractor to produce a method and resource programme. This should be listed in his quality plan. Monitor the availability of both the client's and subcontractor's staff. A good specification would state the what, why, when, where, how, by whom and duration of training and instructions. It should cover:

- Familiarisation—e.g. a geographical walk round the plant, equipment and systems informing management and operatives
- Instructions—the specified level of information imparted to operatives and supervisors
- Training—a specified level of trade skill for staff and operatives e.g.:
  - plumbers trained up in water treatment
  - supervisors trained up to BMS engineer.

Further reading

Standards and Codes of Practice

BS 8313: 1997 *Code of Practice for Accommodation of Building Services in Ducts.*

Other relevant texts

BSRIA (Building Services Research and Information Association)

Commissioning HVAC Systems, Division of Responsibilities: TM 1/88.1, 2002.
Commissioning of VAV Systems in Buildings: AG 1/91, 1991.
Pre-Commission Cleaning of Water Systems: AG 1/01, 2001.
The Commissioning of Water Systems in Buildings: AG 2/89.3, 2002.
The Commissioning of Air Systems in Buildings: AG 3/89.3, 2002.
Commissioning of Pipework Systems—Design Considerations: AG 20/95, 1996.
Illustrated Guide to Mechanical and Electrical Building Services: ILL, 2005.
Guide to Legionellosis—Operation and Maintenance: AG 19, 2000.
Safe Thermal Imaging of Electrical Systems: AG 17/97, 1997.
Improving M&E Site Productivity: TN 14/97, 1997.
Building Energy Management Systems. The Basics: DLP 1, 1988.

Heating Controls in Large Spaces: TN 23/97, 1997.
Handover, O&M Manuals and Project Feedback Toolkit: BG 1, 2007.
Water Treatment for Building Services Engineers: AG 2/93, 2003.
Achieving Minimum Outdoor Air—Commissioning and Test Procedures: AG 17, 2000.
Photovoltaics in Buildings—Testing, Commissioning and Monitoring: S/P 2/00290 REP, 1998.

CIBSE (Chartered Institution of Building Services Engineers)

Air Distribution Systems: Commissioning Code A, 1996.
Boiler Plant: Commissioning Code B, 2002.
Automatic Controls: Commissioning Code C, 2001.
Refrigeration Systems: Commissioning Code R, 2002.
Water Distribution Systems: Commissioning Code W, 2003.
Commissioning Management: Commissioning Code M, 2003.
Testing Buildings for Air Leakage: TM23, 2000.
Lighting: Commissioning Code L, 2003.

Commissioning Specialists Association

Commissioning Engineers Compendium

Chapter 19   HEALTH AND SAFETY LAW

Malcolm James

This chapter will

- **Give an understanding of the context of the law**
- **Explain some legal definitions**
- **Describe the operational benefits that could be obtained by following the principles in the law**
- **Suggest that safety can be a commercial asset**
- **Give a précis of the more common aspects of the law dealing with the law on health and safety applicable to construction sites**

19.1   Introduction

19.1.1   General

There is great stress both in society and in industry concerning the need to improve health and safety at work. Persons, groups of persons and corporate bodies who have any sort of control over work processes are all required to take steps to identify the risks to the health and safety of both their employees and of anyone whom might be affected by their work activities during and before that work is started. They are then required to put measures in place that will prevent or control any significant or unacceptable risks.

This is regarded as such an important principle, both nationally and internationally, that there is an extensive amount of legislation, both as Acts of Parliament and in regulations drafted under the Acts, dealing with health and safety at work. As already mentioned these legal duties apply to everyone involved with work, whether corporate businesses or individuals including individuals or companies acting as agents on behalf of a business or who supply, design or manufacture equipment and substances for work. It is therefore fair to state that there is no one involved in work that is outside the legislation.

While there has been significant progress in the field of accident prevention over recent years, so much so that this progress has been cited as one of the greatest advances in construction (*New Civil Engineer*: 9/10/08), it is still a fact that there is a greater likelihood of a person suffering injury or ill health in the construction industry than in most other industries. (See Health and Safety Statistics published by the Health and Safety Commission.) Typically a worker in the construction industry is twice as likely to suffer some form of injury than those in all industries and four times as likely to suffer a fatal injury. The annual report published by the Health and Safety Commission contains details of accidents relating to severity, industry and activity. These figures amply demonstrate the nature of the risks involved in the construction industry and why there is such stress on trying to develop work processes aimed at achieving effective control of these risks. As a result there is a strong focus on the part of the enforcing authorities, the Health and Safety Commission, the Health and Safety Executive and local government, to attempt to control the health and safety risks in the construction industry through regular inspection backed by a series of regulations, approved codes of practice and guidance.

This chapter attempts to explain some of the more important aspects of the legal requirements in the Acts and regulations as simply as possible. However it is important to realise that although this should enable the reader to have a good appreciation of the subject, space prevents a comprehensive review of all this legislation and no review such as this one, can be a substitute for eventually reading both the documents themselves and the supporting approved codes of practice and guidance.

It has been the case, and possibly still remains so in some quarters, that these legal requirements and duties are seen as, or treated as, 'bolt on extras' that are imposed by government and must be endured or avoided where ever possible. In many of these types of situations the legal duties are thought of as increasing the costs of the work above those that are strictly necessary for its successful completion. There could even be some justification for this view when the regulations are poorly applied, and the legal requirements (or what is claimed to be the legal requirements) are used to demand layer after layer of safety provision The basis of this often being the perceived need to provide cover for the primary or principle levels of protection in case these prove to be inadequate and fail. The alternative is rather to properly assess the risk and decide, apply and enforce the most appropriate means of dealing with it. Such an 'add on' mentality will almost certainly mean that the cost of the work will be unnecessarily increased and it will be more difficult to carry it out effectively. To operate in this manner is a very inadequate way of fulfilling and dealing with the duties imposed by the health and safety legislation. It can also be counter productive as it does lead to persons taking hazardous short cuts in sheer frustration. To allow the principle of protecting persons from harm by applying safety measures in a sort of scatter gun approach is inefficient, ineffective and unprofessional.

For example, theoretically it should be possible to ensure that all the items in a temporary structure used, either for access or as a place of work, are of adequate strength for their purpose. As such it should not need additional 'redundant' members to make it safe. However to ensure that the basic structure is safe, does mean that those supervising, constructing and using the structure all play their part in making sure that the various members are present, properly in place and securely fixed. Failure to properly maintain any structure, particularly a temporary one, could easily lead to a catastrophic failure. However, including additional members as a safety precaution will make little difference if these too are not properly positioned and maintained. Therefore the important thing is for the basics to be both appropriate and correctly installed and maintained. In such a way the resultant provisions will be efficient, effective and provide adequate means of controlling any risks to the health and safety of those involved with or affected by the work.

This example illustrates that a sound understanding of the practicalities of the safety systems and equipment being used is required, and of how these may be used (and misused). Hence the essential need for those planning and supervising the work to be truly professional to be able to apply the requirements of the law on health and safety in an economic and effective manner so anyone who may be affected by the work is not exposed to unacceptable and unnecessary risks.

It is also clear that successful, efficient and effective systems of work are those that incorporate the requirement for health and safety in a seamless manner which blends the requirements for effective work stations, high levels of productivity and reduce waste, both human and material, to a minimum. This means that health and safety cannot be a 'bolt on extra', and must be approached with the same level of professionalism that is required in other, more 'traditional' forms of engineering.

Finally the law on health and safety has been the driving force behind the development of many of the best systems of work facilitating higher and more economic standards for doing such work. As a result these systems have been adopted world wide even in countries that have no legal requirements for such measures. Being part of this process of creating a working environment that effectively controls any risks to the health and safety of persons is as much a part of being a successful engineer as anything else.

To be able to organise work so that health and safety requirements are blended with efficient construction processes at a minimum cost requires a high level of competence and expertise. It also requires those involved with work not only to understand the means and equipment required as already explained but also the legal requirements for the work, so that these do not contradict each other as far as possible. Therefore this chapter is intended to provide a basis for the development of such a sound legal understanding and to be an aid to higher levels of professionalism.

19.1.2 Why comply?

Perhaps a major reason for seeking to comply with the law is the fear of prosecution, a fear that has probably been enhanced by the enactment of the Health and Safety Offences Act 2008. While this act does not create additional duties it does allow an increase in the level of fines in a magistrates court, increase the number of offences triable in either the higher or lower courts and allow the imposition of custodial sentences for up to 12 months. However complying with the legal requirements for health and safety to avoid prosecution or getting involved in civil litigation is only one reason for being concerned with health and safety matters, there are at least four others.

The first is a moral question in that we do live in a society where taking reasonable care of our neighbour is regarded as being a 'duty' on all of us. As such we should not require those who work for us, whether they are paid for this or not, to go into situations over which we might have some degree of control, where they run the risk of being hurt. We have the duty at least to shout a warning and if we do not then we carry some part of the responsibility for any accident that might happen.

Secondly there is an emotional issue. Few if any would relish the thought that they had caused or allowed someone to be injured. In cases where there has been a major serious accident, those even partly responsible can carry the burden for many years sometimes even to the end of their lives. Such a burden can be life crippling, perhaps leading to premature ageing and death. Further it is a very unpleasant experience to have to tell a relative that their son or husband has been badly injured while working at a site where you have some responsibility for the safe management of the site.

The third issue is one of efficiency. The incident rate for accidents and near misses of all kinds, whether reportable under the regulations or not, is a good indicator of the level of control exercised by everyone on site. Where the number of such incidents is high then this indicates that there is a lack of control not only over health and safety issues, but also over quality, productivity, waste control and environmental pollution. This means there will almost certainly be serious hidden financial implications both for the client and the contractor where there are poor standards of health and safety. This aspect is looked at in a little more detail below.

Lastly there is a question of professionalism. If a site is badly controlled it raises the question of the competence of those involved. Civil engineers frequently complain about their status and wages compared to other professions, but as members of an industry suffering a high accident record, (possibly indicating poor levels of management control as mentioned above) the responsibility for this surely lies with those who are supposed to be planning, managing and supervising the work. As such it may not be reasonable to expect the wider world to recognise that engineers are worthy of greater respect.

19.1.3   The costs of health and safety incidents

Returning to the matter of efficiency, the financial losses arising from poor health and safety management can be understood by considering the probable financial consequences of an accident. Although there are costs in ensuring acceptable levels of health and safety, there are also very substantial costs where these levels are not met. Being prosecuted for failure to comply with the law is probably only a small proportion of these costs. Studies carried out show that the incidental costs of an accident is very high and continues long after the accident has occurred and the site returned to normal as considered below.

The first major impact on costs following an accident comes from a fall in productivity. This is through the immediate stopping of work to deal with those injured and in cleaning the site. Almost certainly there will be further time losses as site staff discuss the accident between themselves. Even if each worker on the site only spends five minutes to talk about the accident that can still add up to a substantial sum. Then there will be the consequences from operating for a while without the injured worker and there could be further time lost if statements are required. Time could be lost through possible reductions in site morale and an unwillingness to do certain work. Other time losses will occur as senior management consider the implications of the accident, plan and implement revised work methods, deal with solicitors, (doubly so if the accident leads to a civil claim) or with an HSE inspector if they are considering whether to prosecute. Then there is also the likelihood that the company insurance premiums will increase. These and similar losses arising from a serious accident have been known to wipe out some two years of operating profit.

On top of this there could be cost implications from the damage done to a company's reputation which can take many years to overcome. Businesses have been closed down through their inability to gain sufficient work due to gaining a bad reputation after suffering a serious accident.

The costs of an accident as outlined above are only that which is carried by the contractor. However there are yet other costs such as those carried by the state in social security payments and hospital costs. There are also costs carried by the individuals and families involved through the loss of earnings or the loss of a loved one.

These are the direct financial losses caused by an accident but as has been already pointed out accidents are mostly the manifestations of poor control and incompetence and this is the cause of yet more hidden loses even when no accidents occur. These losses affect the quality of the final product and generate an increased need for maintenance of the structure which probably has to be carried for the life of the structure.

Striving to comply with the law, particularly with the its basic philosophy, in a manner that accurately reflects the risk and looks for solutions that are cost effective and improves working efficiency, is a powerful way of reducing costs to a real minimum and of showing high levels of professional competence. It is almost inconceivable that a construction contract could be really successful in every way, including economically, unless the planning for and control of the health and safety of all those persons who work at a site or who could be affected by it, was not of the highest order.

19.1.4 'Selling' safety, making the best of a good health and safety record

If the lack of safety results in significant financial losses as outlined above, then controlling the work so as to minimise losses and particularly those of human and material waste, is a cost benefit. This can therefore be an aspect that could be 'sold' to clients.

To do this there will clearly need to be proof of control over safety showing the effectiveness of management. This will enable a comparison to be made with the level of accidents in industry generally and indicate the level of a company's competence.

This information can then be offered to the client in a way that emphasises how the client himself can be seen to be carrying out his legal duties in appointing competent contractors, so avoiding the possibility of the client being drawn into any possible prosecution or litigation should there be an incident.

Finally if the link between the control over safety and the control over efficiency is accepted, then this will show that the client will get value for money.

19.1.5 Civil litigation

As well as that area of criminal law on health and safety that is the subject of this chapter, there is also a much wider area of litigation that deals with civil claims. This is that body of law where individuals or companies claim against others for the losses they may have suffered. Probably claims under the civil law are more numerous than proceedings under the criminal law and can be far more costly.

Often a claimant will cite an alleged failure under the criminal law, in effect a contravention of one or more requirements described in this chapter, as the source of his, or her loss. Clearly therefore if there has been a successful prosecution under the criminal law then it will generally be much more difficult to defend against an associated civil claim.

In addition a civil claimant could cite a number of persons or companies as being partly to blame for his/her loss and the court could award damages on the probability that the various defendants had some responsibility for this. Where someone has suffered a serious accident and could be perhaps crippled for life then the potential damages will be very high and a person or company being found only partially to blame could be faced with a very high bill. While it would be normal, especially for a company, for that company's insurer to pay these costs the knock on effect can be very expensive just through the inevitable higher insurance premiums that will continue for many years.

Experience shows that if a defendant can prove through good, clear records and possibly photographs taken at the time of the accident, that he had done all that he reasonably could, in effect ensured that the site was well run with good safety standards, then he would be far more likely to avoid being found at fault. To be able to do this however does require all those in the company, including the site engineer, to know and apply good methods of safety control that adequately deal with the risks as already outlined.

19.2 The nature of the criminal law on health and safety

The legal framework dealing with health and safety matters is very extensive and the legal requirements for health and safety in construction works can be set out in various Acts of Parliament, such as the Consumer Protection Act 1987, the Building Act of 1984 and the Control of Pollution Act 1974. In addition there are Acts that deal with specific work areas such as the Road Traffic Act 1991, the Offshore Safety Act 1992, The Energy Act 2004, the Gas Act 1995 and the Offices, Shops and Railway Premises Act 1963, to name but a few.

However the main Act dealing with health and safety of persons at work is the **Health and Safety at Work etc. Act 1974 (HSWA)** and it is this Act and the regulations made under it that are summarised in a general way in this chapter. As there are over 30 sets of regulations that could apply to construction work, it is quite impossible to deal with all of these in the space allowed. Therefore only those regulations listed below have been reviewed, either fairly completely or in part. This review should give a fairly comprehensive idea of the main and most generally applicable requirements for construction work particularly as the principles in many of the regulations reinforce each other.

- Health and Safety at Work etc. Act 1974 (HSWA)*
- Management of Health and Safety at Work Regulations 1999 (MHSW)
- Construction (Design and Management) Regulations. 2007 (CDM)
- Work at Height Regulations 2007 (WAH)
- Lifting Operations and Lifting Equipment Regulations 1998 (LOLER)
- Construction (Head Protection) Regulations 1989 (C(HP)R)
- Provision and Use of Work Equipment Regulations 1998 (PUWER)*
- Manual Handling Operations Regulations 1992 (MHO)
- Personal Protective Equipment at Work Regulations 1992 (PPE)
- Regulatory Reform (Fire Safety) Order 2005 (RR(FS)O)*
- Electricity at Work Regulations 1989 (EWR)
- Confined Spaces Regulations 1997 (CSR)
- Diving at Work Regulations 1997 (DWR)*
- The Ionising Radiations Regulations 1999 (IRR)*
- Control of Substances Hazardous to Health Regulations 2002 (COSHH)*
- Control of Vibration at Work Regulations 2005 (CVW)
- Control of Asbestos Regulations 2006 (CAR)
- Control of Asbestos in the air regulations
- Control of Lead at Work Regulations 2002 (CLW)
- Control of noise at Work Regulations 2005 (NWR)
- Reporting of Injuries, Diseases and Dangerous Occurrences Regulations 1995 (RIDDOR)
- Health and Safety (First Aid) Regulations 1981 (HS(FA))
- Health and Safety (Safety Signs and Signals) Regulations 1996 (HS(SSS))*

Note: The main requirements in the above regulations and those in HSWA are paraphrased in Sections 9.4–9.6 of this chapter dealing with the various duties on persons, the requirements for various tasks and the various documents required. The paraphrases are then linked to the specific regulations or sections in the Act so that if necessary the relevant part of the law can be studied in full.

Those marked * have only been reviewed in part where only those parts reviewed will be generally applicable to construction sites. However those so marked may be more relevant to specialised activities and a greater specialist knowledge of these marked regulations will be necessary.

As stated earlier, readers of this chapter must make themselves aware of any other specific requirements that may be applicable to their sites. Readers should also be aware that as the regulations have been authorised by Section 15 of HSWA, the other requirements in that Act could be relevant to their interpretation. In addition various regulations are amended from time to time and the Health and Safety Executive (HSE) publish new or revised guidance on them as an ongoing exercise. Therefore there is a constant need to keep up to date, as ignorance of the law would not be a very good excuse in court. Fortunately the HSE has a fairly good web site that provides up to date information, particularly on the large number of

*Applicable to England and Wales only.

publications and videos they produce. Some of these publications are described in the Further Reading section at the end of this chapter.

All the regulations reviewed apply at some point to construction work except that the Construction (Design and Management) Regulations do not fully apply where the site is *not* notifiable or where there are less than five persons at work at any one time (CDM 3(2)). One reason why it need not be notified could be because the construction phase will not last 31 days, or will involve less than 501 person days of construction work (CDM 2(4)). These regulations also do not apply where the local authority is the enforcing authority (CDM 3(4)).

19.2.1 The status of material other than the statutory requirements published by the HSE

The Health and Safety Commission (HSC) and Executive (HSE) publish large quantities of material related to safety each year. This varies from information on the number of accidents that have occurred, reports on major incidents. This shows the numbers of persons injured or suffering ill health in various industries and the major areas of risk. In addition research reports are available on work carried out or commissioned by HSE. Finally there is a large amount of guidance material published on how to deal with specific risks involved in different areas of work such as scaffolding or roofing. This published information can often be of great help when carrying out risk assessments as it not only describes how failures have happened or could happen but also how they can be avoided.

The guidance documents are useful in having 'off the shelf' solutions that will generally satisfy inspectors. For instance in the introduction to HS(G) 33 on Roofwork there is the following note: 'This guidance is issued by the Health and Safety Executive. Following the guidance is not compulsory and you are free to take other action. But if you do follow the guidance you will normally be doing enough to comply with the law. Health and safety inspectors seek to secure compliance with the law and may refer to this guidance as illustrating good practice.'

Another types of material published are Approved Codes of Practice (ACOP) and commentaries on the law. While this material does not strictly impose legal duties it does represent a legal interpretation of the law, as against guidance which advises how the law might be complied with. Approved Codes of Practice are permitted by Section 16 of HSWA. This says that the HSC may issue them if they think they are suitable for providing practical guidance on Sections 2–7 of the Act, the safety regulations or any of the existing statutory provisions, whether such codes have been prepared by the commission or not. The consent of the Secretary of State and other government departments is required before any such code is approved or withdrawn. Section 17 of HSWA states that a failure to observe any provision in an ACOP would not itself render a person liable to prosecution but could be used to show a contravention if the court thought that provision was relevant and the matter it referred to had not been complied with in another way.

Examples of approved codes of practice are those dealing with asbestos (L127), the Management of Health and Safety at Work (L21), the Construction (Design and Management) Regulations (L144) and the Confined Spaces Regulations (L101). None of these Approved Codes of Practice are reviewed in this chapter.

19.2.2 Outline of the broad main health and safety requirements

Although there is a large number of regulations, approved codes and guidance, the main essentials of the legislation can be generally grouped into three main areas. These are:

(1) duties placed on persons or companies for ensuring health and safety;
(2) the identification of the hazards and the assessment of the risks; and
(3) the measures required to control these risks.

Much of what then is within the law dealing with accident prevention and securing good health is therefore an expansion of these three areas. While Sections 9.4–9.6 give a précis of the main details of the law these three areas are described in a little more detail below.

19.2.3 Who carries what duties?

The Health and Safety at Work etc. Act places duties on employers, the self-employed, those in control of premises, manufacturers and suppliers, and employees. While in many cases the reference to employers or manufacturers etc. would generally be taken as the company, the Act also allows managers etc. within a company to be personally prosecuted if it can be shown that they made the decision or omission that led to a contravention. In addition where an offence by one person can be shown to be because of what someone else did or failed to do, then that other person could be liable for the offence. As already noted in effect there is no one who does not carry some duties for health and safety.

19.2.4 Hazard identification and risk assessments

The Management of Health and Safety at Work Regulations require employers and the self-employed to make suitable and sufficient assessments of the risks involved with the work (MHSWR 3). This is then implicit in all the regulations where duties are imposed concerning the avoidance of risks, e.g. the Construction (Design and Management) Regulations require designers to avoid foreseeable risks in their designs, (CDM 13), the Health and Safety at Work Act require those who design, manufacture, imports or supplies articles for work to ensure that these are safe and without risks to health (HSWA 6).

Risk assessments are intended to identify potential harm that could be caused by hazards. The risk will therefore depend on the nature of the hazard, including how aware the workers would be of it, the degree of exposure to that hazard and the possible consequences that could develop from it. Assessing any risk will involve making subjective evaluations of these variables, balancing them to arrive at some overall figure. While advice in the HSE guidance stresses that this is something that most people do quite often almost without thinking, clearly it is important for anyone preparing a risk assessment to have a good understanding of the factors involved. Where the risk is assessed as being negligible then it could be assumed to be acceptable and no further action is required. Otherwise action is required to remove the risk, control it or protect persons from it.

A philosophy promoted by HSE to carry out a risk assessment would be to follow a safety hierarchy. The first step in doing this would be to eliminate of the risk altogether. Hazardous substances could be replaced by none hazardous ones and work could be done at ground level rather than at height. Where this is not possible (or reasonably practicable), then the next step would be to guard the hazard. For example, perimeter fencing could be used to prevent falls or access to moving machinery. The advantage of providing guarding to the hazard means that the workers need not take any special measures and that the guarding will be available to all who might be affected by the hazard. If this is also not possible or reasonably practicable then the workers must be protected. Eye goggles and ear defenders will protect the worker from injury or possible hearing loss but their effectiveness would depend on the worker using them properly and such measures would generally require effective training for the workers and possibly higher levels of supervision. Finally if this is not possible then measures must be taken to mitigate the consequences should the risk being realised. The provision of suitable eye washes or fall arrest equipment could be an example of this.

It is important that the safety provisions adequately deal with all the factors involved in a risk, i.e. those arising from the situation, the job itself, the competence of those involved, the level of supervision, the access equipment used (its type, condition and reliability) and how environmental factors may influence the work. It is important to remember that generally while high risk situations can be controlled by suitably competent persons, even the safest systems can be undermined by incompetent ones.

19.2.5 The measures required to control risk

Having identified that their could be a significant risk it is then a legal requirement to either prevent, control or mitigate it. It is also a general legal requirement to do this by following the principles of prevention. These are set out in Schedule 1 of the Management of Health and Safety at Work Regulations and shall be followed. They are as follows:

(1) Avoid risks
(2) Evaluate those risks that cannot be avoided
(3) Combat risks at source
(4) Adapt the work to the individual
(5) Adapt to technical progress
(6) Replace the dangerous with the non-dangerous or less dangerous
(7) Develop a coherent overall prevention policy
(8) Give collective protective measures priority over individual protective measures
(9) Give appropriate instructions to employees.

In the Work at Height Regulations there is a similar set of requirements designed to avoid risk when working at height. In Regulation 6 it gives the following list of requirements:

i.   work is not to be done at height when it is reasonably practicable to do it other than at height.
ii. where work is done at height suitable and sufficient measures to be taken, so far as is reasonably practicable, to prevent anyone falling from a height likely to cause injury

iii. suitable measures shall include working from an existing place complying with schedule 1.

iv. otherwise to provide sufficient work equipment to prevent, so far as is reasonably practicable, a fall occurring;

v. the work equipment is to minimise the distance or consequence of a fall

vi. provide suitable and sufficient training and instruction.

This list is further expanded in Regulation 7 which requires:

vii. collective protection is to be given priority over personal protection.

Having a good understanding of these three areas should provide a good way of understanding most of the details within the law.

19.3 Some definitions and common legal terminology

When reading the various legal regulations it is important to understand that some of the terms and expressions regularly used have well defined meanings in law, either through the evolution of case law or as written in the regulations etc. themselves. The significance of some of the more important terms and phrases are explained below.

Where the word **'shall'** is used, then the requirement is absolute and must be complied with, without any exception unless the context of the word is linked to alternative methods etc.

The word **'practical'** means that all measures which are currently available, must be taken. Where there have been new developments in equipment, materials, substances etc, that will reduce the risk to health and safety than more traditional ones, then these new methods must be adopted no matter what they might cost. Hence there is an important need to be aware of current developments and to understand how they might have beneficial health and safety results.

Where the phrase **'so far as is reasonably practicable'** is used then this implies that a balance should be struck between the risk and the cost or effort required in dealing with it. Where the risk is low, then only that cost or effort necessary to ensure it remains within acceptable limits is legally required. For instance where the work being done on site is almost finished and is of a generally low risk, then the absence of the supervisor or managers for a short length of time, might not contravene the duty to ensure the work is properly supervised and controlled. On the other hand where the risk is high, then much greater effort and commitment is required to deal with it.

This is a phrase that occurs far more often in the legislation than 'shall' and 'practical' and allows the duty holder to make a balanced judgement so that where it is appropriate only those measures need be provided to reduce the risk to within reasonable limits, perhaps minimising the spending on effort and cost. In the following text this phrase is abbreviated to (sfairp). However, a word of warning: although it may have been deemed that it was not reasonably practical to provide additional controls over a risk while the work progresses without incident, this may not be the case if an accident occurs. The accident itself could be taken as proving that sufficient reasonable and practical measures had not been taken.

Clearly, therefore, the interpretation of what is 'reasonably practicable' can sometimes be open to question. As already discussed, to be able to justify minimum health and safety provisions would require a very good understanding of the risks, how these might come about and the most effective way of dealing with them. In addition in the event of an accident very good and clear details would be useful concerning how the decision came to be made in implementing the safety system and on how that was managed.

But as already described there is little protection in trying to cover a lack of understanding by following a blanket approach to safety. This will not only result in excessive resources being thrown at the health and safety problem than are necessary often to the detriment of other aspects of the work, yet still not ensure that there is effective control over the risks.

The question of what is **'reasonable'** would normally be based on what might be thought reasonable for the individual person making a decision or carrying out some work in a particular situation. Therefore what might be a reasonable course of action would vary depending on who was involved in doing what kind of work.

In the phrase **'suitably and sufficient'** suitable applies both to the purpose for which it is being used and for the person using it. If only one person is designated to use a piece of equipment it need only be suitable for that person but for no other person. Equipment may still be suitable even if it fails to give protection, provided it is well adapted both for the process and the person using it.

A **'client'** is defined for the purposes of the Construction (Design and Management) Regulations as a person who in the course of business seeks or accepts the services of another to carry out a project or carries out the project himself.

A definition of **'competent person'** is given in Regulation 8 (3) of the Management of Health and Safety at Work Regulations which could be the basis for a general definition of competence. In this sub paragraph it states that a competent person is one who can implement the duties in the regulation through having 'sufficient training and experience or knowledge and other qualities'.

A definition of **'construction work'** is given in Regulation 2 of the Construction (Design and Management) Regulations (CDM) as meaning carrying out of any building, civil engineering or engineering construction work. It includes:

- alteration, conversion, fitting out, commissioning, renovation, repair, upkeep, redecoration or other maintenance (including cleaning that involves the use of water or an abrasive at high pressure or the use of corrosive or toxic substances), de-commissioning, demolition, and dismantling of a structure;
- the preparation for any structure (but not site survey), and excavation, and the clearance or preparation of the site or structure for use or occupation at its conclusion; the assembly on site of prefabricated elements to form a structure or the disassembly on site of prefabricated elements which, immediately before the disassembly, formed a structure;
- the removal of a structure or of any product or waste resulting from demolition or dismantling of a structure or from disassembly of prefabricated elements which immediately before such disassembly formed a such a structure;
- the installation, commissioning, maintenance repair or removal of mechanical, electrical, gas, compressed air, hydraulic, telecommunications, computer or similar services which are normally fixed to a structure;

- but does not include the exploration for or extraction of mineral resources or activities preparatory thereto.

A definition of **'structure'** is given in Regulation 2 of the Construction (Design and Management) Regulations as including the following. Any building, timber, masonry, metal or reinforced concrete structure, railway line or siding, tramway line, dock, harbour, inland navigation, tunnel, shaft, bridge, viaduct, waterworks, reservoir, pipe or pipe-line, cable, aqueduct, sewer, sewage works, gasholder, road, airfield, sea defence works, river works, drainage works, earthworks, lagoon, dam, wall, caisson, mast, tower, pylon, underground tank, earth retaining structure or structure designed to preserve or alter any natural feature, fixed plant and any structure similar to these.

It also includes any formwork, falsework, scaffold or other structure designed or used to provide support or means of access during construction work.

A definition of **'design'** is given in Regulation 2 of the Construction (Design and Management) Regulations as includes drawings, design details, specification and bills of quantities (including specification of articles or substances) in relation to the structure, and calculations prepared for the purpose of a design.

A **'hazardous substance'** is defined in Regulation 2 of the Control of Substances Hazardous to Health Regulations as a substance that has the intrinsic potential to cause harm to the health of a person.

'Work at height' is defined in Regulation 2(1) of the Work at Height Regulations as work in any place, including a place at or below ground level; obtaining access to or egress from such a place except by a staircase in a permanent workplace, where if measures required by the regulations were not taken, a person could fall a distance liable to cause injury.

19.4 Summary of the duties imposed on various persons and companies

Section 19.1 of this chapter explains the principal philosophies of the health and safety law which generally apply to everyone involved with work, have been explained. This section looks in more detail at the responsibilities placed on different persons or corporate bodies which have been developed following these philosophies. In looking at these responsibilities, site engineers should remember that the agents for these corporate bodies or individuals are expected to help to manage the duties placed on their employers, i.e. that site engineers could be seen in some circumstances as acting for their employer. Therefore failure to deal with any contravention of the law or to notify their employers of that contravention could make the site engineer liable to prosecution.

This section looks at the more specific responsibilities placed upon the following broad groups of persons although in some cases the law cannot be always compartmentalised in this way. This means that the principle of a legal requirement could be applied to someone who might not fit properly into one of the groups listed below. In addition the responsibilities of persons or companies listed in this section are not limited to just these, but depending on the circumstances much of the requirements given in Sections 9.5–9.6 could well be seen as also being specific to an individual or company.

(1) Employers
(2) Clients etc.
(3) CDM coordinators
(4) Principal contractors
(5) Designers
(6) Contractors, construction managers and supervisors
(7) Sub contractors and self-employed workers
(8) Persons in control of premises
(9) Suppliers
(10) Individuals

Note: The abbreviation (sfairp) sometimes used in this section means 'so far as is reasonably practicable'. An explanation of this term is given in Section 9.3 of this chapter.

Generally the duties imposed by the Act and by the regulations are directed towards the 'employer', whether an individual person or a company. However there are certain duties directed towards other persons or companies as illustrated in this section. Readers should remember that an individual or a company could have duties under several of the following categories at the same time. For instance, a person could have duties as an individual, an employer and as a contractor. Therefore while this summary, and the regulations etc., themselves, appear to draw distinctions between persons with different jobs, in practice the situation would be likely to be much more fluid.

As so often the case there are instances where this may not be the case. For instance, in the case of the Construction Design and Management Regulations while many duties and responsibilities imposed by the regulations are targeted at particular persons or positions there are some preliminary requirements that apply more broadly and set the tone for quite a lot of the detail that follows. These general requirements are summarised below.

Every person who has a duty under the regulations shall seek the co-operation of others on the same or adjoining sites so far as it is necessary to enable him to carry out his duties under the regulations **CDM 5(1)**. They are also to co-ordinate their activities to ensure (sfairp) the health and safety of persons involved or affected by the work **CDM 6**. Every person having a duty in relation to the design, planning and preparation of the project, shall take into account the 'principles of prevention' when carrying out their duties **CDM 7(1)**. While those having duties in relation to the construction of the project shall ensure (sfairp) that the general 'principles of prevention' are followed **CDM 7(2)**.

From previously, therefore, while the summary of the duties in the following sub-sections appears to refer to different groups of persons, in many cases the requirements could be applied quite generally and that those duties imposed on the employers, for instance, could be relevant to most persons with any sort of authority on a construction project.

Employers

All employers have an overall and wide ranging duty to their employees imposed by the Health and Safety at Work etc. Act 1974 which can be summarised as follows:

- An employer is to ensure the health and safety of all his employees, including the provision and maintenance of plant and systems of work, that (sfairp) are safe and without risks.
- He is to make safe arrangements (sfairp) for the use, handling, storage and transport of articles and substances, and to provide (sfairp) necessary health and safety training, supervision, information and instruction for his employees.
- He is also to maintain places of work, and providing and maintaining access to these, where under the employer's control, in a condition that (sfairp) is safe, and to provide and maintain (sfairp) a working environment that is safe and healthy, having adequate welfare facilities **HSWA s2(1) and (2)**.
- Employers also have a responsibility to others who are not employed by him but who may be affected by his activities and an employer must therefore ensure (sfairp) that his undertaking does not expose these others to risks to their health and safety **HSWA s3(1)**.

Other general duties in effect stem from these requirements with the main duty to assess risks and then arrange for suitable measures to control those that are significant.

For instance, every employer is to make suitable and sufficient assessment of the risks to the health and safety of his employees and those who could be affected by his undertaking so as to be able to identify the preventative measures required. These preventative measures should be then put into practice and the employer is expected to control and monitor them to ensure that they remain effective as the risk assessments should be reviewed if there was any reason to suspect they were no longer valid or if there had been significant changes in the work procedures. Where the employer employs five or more persons he is to record the preventative measures **MHSW 3(1), (3) and 5 (1) and (2)**.

He shall provide comprehensive information for his employees on

- The risks to health and safety identified by the assessment.
- The preventive and protective measures.
- Procedures to deal with serious and imminent danger.
- The identity of the person(s) nominated to implement these procedures and who would deal with serious and imminent danger.
- Any risks arising from sharing a workplace with others.

Where necessary every employer shall make appropriate procedures to deal with serious and imminent danger to persons at work in his undertaking, nominating a sufficient number of competent persons to implement these procedures. These shall (sfairp) include informing those at risk of serious and imminent danger of the hazard and the precautions to be taken, enabling them to go to a place of safety and to prevent them from returning to work while the danger exists. Further the employer is to ensure none of his employees have access to areas restricted because of the risks to health and safety unless they have adequate instructions on how to safely deal with these **MHSW 7, 8 and 10**.

Every employer shall ensure that the employer of anyone from an outside undertaking working in his undertaking is provided with comprehensive information on the risks to their health and safety and the measures to be taken to deal with these and that anyone who is not his employee is provided with appropriate health and safety instructions.

Where two or more employers share a workplace each is to cooperate with the others to coordinate the measures to comply with the regulations and take all reasonable steps to inform the other employers of the risks to the health and safety to their employees arising out of his undertaking **MHSW 11 and 12**.

Every employer shall appoint one or more competent persons to assist him in complying with the regulations ensuring these have the necessary information and resources to carry out their function and are able to cooperate with each other. This requirement will not apply where a self-employed employer or individual employer(s) who have sufficient training, experience or knowledge and other qualities to carry out this duty themselves **MHSW 7**.

Every employer shall take into account the capabilities of his employees as regards health and safety in entrusting tasks to them. He shall ensure they are provided with adequate health and safety training when they are recruited and when they are exposed to new or increased risks (because they have been transferred or given new responsibilities, introduced to new work equipment or systems). This training shall be repeated at intervals where appropriate, be adopted to take account of the risks and take place in working hours **MHSW 13**.

In addition to these broad requirements the employer has other duties more specific to the activities he is controlling. Where work is to be carried out at height employers are to ensure it is properly planned, supervised and carried out safely (sfairp). This also includes the planning for emergencies and rescue **WAH 4**. They are to ensure that no one is involved with work at height or the equipment for work at height, including its planning, organisation, and supervision, unless they are competent to do so. If they are being trained for such work then they must be under the supervision of a competent person **WAH 5**. They are also to ensure that work at height is only carried out when the weather conditions will not affect the health and safety of those doing it **WAH 3**.

The employer has specific duties to control the risks to health from noise and hazardous substances generally having the same duties to any person as to his own employees (sfairp) **COSHH 3(1)**.

An employer must make a suitable and sufficient assessment of the risks to the health of his employees from any substance and the measures needed to protect them, before it is used **COSHH 6(3)**.

He is not to carry out work that could expose his employees (and others) to asbestos unless he has carried out a risk assessment identifying the type(s) of asbestos, determine the nature and degree of exposure in the course of the work and set out steps to prevent or reduce exposure to the lowest reasonably practicable level, ensuring that any control measures are properly used, kept clean, in good repair and working order. **CAR 6(3), 12(1) and 13(1)**.

An employer shall not carry out work that is liable to expose anyone to lead unless he has made a suitable and sufficient assessment of the risks, the steps required and implement these steps **CLW 5(1)**. He shall ensure that exposure to lead is either prevented or, where this is not reasonably practicable, adequately controlled **CLW 6(1)**. Where it is not reasonably practicable to prevent expose to lead the employer shall apply protective measure in order of priority by design of the work process, controlling the exposure at source or by providing suitable personal protective equipment **CLW 6(3)**. All control measures shall be maintained in a clean condition, efficient state, working order and good repair. Adequate steps shall be taken to ensure (sfairp) that his employees do not eat, drink or smoke in a place which is liable to be contaminated by lead. Any personal protective equipment that has become contaminated with lead when removed from the contaminated area shall be kept apart from any uncontaminated clothing or equipment **CLW 8(6)**. Where an employer carries out work that could expose persons to noise above a lower exposure action value he shall make an assessment of the risks and any control measures **CNW 5(1)**. If the noise level is at or above an upper exposure action value the employer is to reduce the noise level to as low as is reasonably practicable by using organisational and technical measure other than by personal protective equipment **NWR 6(2)**. If this cannot be done the employer shall provide personal hearing protection to those exposed **NMR 7(2)**. If an employee is likely to be exposed to noise at or above a lower exposure action value an employer shall make personal hearing protection available on the request of the employee **NMR 7(1)**. If an employee works in an area where he is likely to be exposed to noise at or above an upper exposure action value the employer shall designate that area a Hearing Protection Zone identifying this zone on the ground and restricting access to it **NWR 7(3)**.

Where employee's exposure to lead is liable to be significant, the employer is to ensure the concentration of lead in air is measured with a suitable procedure **CLW 9(1)**.

All these assessments shall be reviewed if it is suspected that they are no longer valid or if there has been any significant change in the work or if the results of monitoring show this to be necessary. **COSHH**. The assessments shall be reviewed where it is suspected the existing assessment is no longer valid or where there has been significant change in the work. The employer is to prevent his employees being exposed to asbestos (sfairp) and where this is not reasonably practicable then reduce this to the lowest reasonably practicable level by means other than respiratory protective equipment, ensuring that the number of persons exposed is as low as is reasonably practicable. He is to ensure his employees are given adequate information, instruction and training when they are liable to be exposed to asbestos **CAR 10(1)**.

Work with asbestos is not to be undertaken without a suitable plan of work **CAR 7(1)**.

He is the monitor the exposure to his employees to hazardous substances where the assessment indicates this is required **COSHH 10(1)**.

The measures taken to protect the employees shall primarily be to prevent their exposure to a hazardous substance, or if this is not reasonably practicable, to ensure the risks are adequately controlled **COSHH 7(1)**. An employer shall eliminate the risks from noise at source or, if this is not practicable, reduce it to as low a level as is reasonable practicable **CNW 6(1)**.

Every employer shall prevent the spread of asbestos or reduce the spread to the lowest reasonably practicable level if this is not possible **CAR 11(1)**.

The employer shall notify the enforcing authority in writing with the particulars in the schedule to the regulations at least 14 days before the work starts and prepare a plan of work for the removal of asbestos before starting work **CAR 9(1)**.

Every employer who provides control measures, PPE or other facilities to comply with the regulations, shall take reasonable steps to ensure these are properly used or applied. He will also ensure these are maintained in an efficient state, working order and in good repair. In the case of PPE, they shall also be kept in a clean condition properly stored, checked, repaired or replaced where necessary **COSHH 8(1) and 9(1)(5)**.

Every employer is to comply with the electricity regulations so far as it is within their control **EWR 3(1a)**.

Every employer shall avoid the need for his employees to undertake manual handling operations which involve a risk of injury (sfairp). Where this is not reasonably practicable a suitable and sufficient assessment of the manual handling operations are to be made considering the tasks, loads, working environment, individual capacity and any other factors, taking into account such things as the physical capability of the employee, the clothing and footwear he is wearing and his knowledge and training. If so, appropriate steps should be taken to reduce the risk of injury to the lowest practicable level, giving general indications, or where reasonably practicable, precise information on the weight of each load and the heaviest side where the centre of gravity is offset, on the basis of this assessment. These assessments shall be reviewed if they are no longer valid or if there have been significant changes in the manual handling operations and changes made where necessary **MHO 4**.

Where safety signs are used to warn employees of risks to their health and safety because collective protection measures cannot adequately reduce these risks, the employer shall provide and maintain appropriate safety signs and ensure (sfairp) that appropriate hand signals or verbal warnings meet with the requirements of the regulations.

Every employer shall ensure that each employee receives suitable and sufficient instruction and training in the meaning of safety signs and the measures to be taken in connection with them **HS(SSS)R 4 and 5**.

Employers shall provide each employee with suitable head protection, maintaining and replacing this as necessary, and (sfairp) and will ensure these are worn by giving directions unless there is no foreseeable risk of head injury other than falling **C(HP)R 3(1), 4(1) and 5(3)**. Every employer who has control over any other person at work shall ensure that they wear suitable head protection unless there is no foreseeable risk of head injury other than from falling. He may give directions requiring this so far as is necessary to comply with the regulations **C(HP)R 4(2) and 5(4)**.

Every employer shall ensure that suitable personal protective equipment appropriate to the risk and conditions, is provided to his employees exposed to a risk to their health or safety unless these risks have been adequately controlled by equally or more effective means. Where more than one type of PPE is required to be worn simultaneously for more than one type of risk, the employer is to ensure the equipment is compatible with each other and remains effective **PPE 4(1) and 5(1)**.

Before choosing any PPE an employer shall assess whether it will be suitable. This assessment shall be of any risks not avoided by other means and the characteristics required of the PPE to deal with those risks, comparing these to the characteristics of the PPE. If the employer thinks the assessment is no longer valid or that there has been a significant change in the situation, then the assessment shall be reviewed **PPE 6**.

Employers shall ensure that any PPE provided to his employees is maintained in good order, in an efficient state and working order, being replaced or cleaned as appropriate and is being properly used. Appropriate accommodation shall be provided for any PPE when it is not in use **PPE 7(1), 8 and 10(1)**.

Where PPE is provided employers shall ensure his employees are given adequate and appropriate training, comprehensible instructions and information on the following. The risks the PPE will avoid, the purpose and manner in which it is to be used and any actions that must be taken by the employee to keep it in an efficient state, working order and good repair **PPE 9**.

An employer shall ensure that adequate first aid equipment and facilities appropriate to the circumstances are provided for his employees should they be injured or become ill at work. This generally will include providing a suitable number of persons who have undergone first aid training qualifying to the minimum levels approved by HSE and such additional training as may be appropriate. The employer must inform his employees of the arrangements made for first aid provision **HS(FA) R 3 and 4**.

Where the employer is the responsible person as defined by the regulations, where a person has an accident and dies, suffers a major injury or suffers an injury that requires hospital treatment due to work activities, whether that person is an employee or not, then the employer should inform the relevant enforcing authority as quickly as possible, e.g. by telephone, and make a report on the incident within 10 days. Where a person dies within one year after an accident, the employer is to notify the enforcing authority as soon as he becomes aware of this whether or not the accident was reported when it originally occurred **RIDDOR 3–5**. Accidents that cause injuries which lead to a person being absent from work more than 3 consecutive days after the day of the accident, should be notified as soon as practicable but within 10 days.

Similarly the employer must report any dangerous occurrence listed in the schedule to the regulations (which includes the collapse of lifting machinery, scaffolding, building or structure and an explosion or fire).

Where a person suffers from any of the occupational diseases listed in the schedules to the regulations while involved in one of the specified activities and where this has been confirmed by a medical practitioner, the employer as the responsible person shall immediately send a report to the relevant enforcing authority.

The client

The client is the person who seeks or accepts the services of others for carrying out a project for him; or who carries out the project himself **CDM 2(1)**.

The client is to take reasonable steps to ensure that the arrangements for managing the project are suitable so that the construction work can be done without any risks to the health and safety of any person, that suitable welfare provisions are made and that any structure being used as a workplace take account of the Workplace (Health, Safety and Welfare) Regulations 1992. **CDM 9**. The client is to promptly provide pre-contract information that he can reasonably obtain for designers and contractors to assist them in their duties under the regulations **CDM 10**. This information shall also be given to the CDM coordinator including the minimum time that has been allowed for the principle contractor to plan and prepare for the work **CDM 15**, plus information that is likely to be required for inclusion in the Health and Safety file **CDM 17(1)**.

He shall appoint a CDM coordinator and principal contractor as soon as practical after initial design work or other preparation for the construction work has begun

CDM 14(1). After appointing a CDM coordinator and when enough is known about a project to be able to choose a suitable person, the client shall appoint the principal contractor **CDM 14(2)**. These appointments may be changed or renewed as necessary so that the positions remain occupied until the end of the construction phase. If these positions are not filled it will be assumed that the client has taken these duties and responsibilities himself **CDM 14(3) and (4)**.

No person shall appoint or engage a CDM coordinator, designer, principal contractor or contractor unless he has taken reasonable steps to ensure that they are competent—accept such appointment unless competent—or arrange for a worker to manage, design, construct or carry out work unless the worker is competent or is under the supervision of a competent worker **CDM 4(1)**.

The client shall ensure that the construction phase will not start until the principal contractor has prepared a construction phase plan that complies with Regulations 23(1) and 23(2) and that he is also satisfied that the requirements for welfare facilities will be met **CDM 16**. The client is to take reasonable steps to ensure that after the construction phase, information in the Health and Safety File is revised to include any new relevant information and is kept available for inspection by any person who might need that information. The client shall ensure that all the information relating to any particular project, site or structure can be easily identified where a single Health and Safety file relates to more than one project, site or structure **CDM 17(2)**. Where the client disposes of his interest in the structure, he could comply with his duties under Regulation 17(3) by giving the file to the new owner ensuring that the new owner is aware of the purposes of the file **CDM 17(3) and (4)**. Where there is more than one client on a project, one or more can elect to be the 'client' for the purposes of the regulations. The remaining clients will then have no duties except those concerning any information they may possess CDM 8.

CDM coordinator

The CDM coordinator shall ensure that as many of the details specified in Schedule 1 of the regulations, as are available at that time, are given to the Executive as soon as is practicable after his appointment **CDM 21(1)**.

The CDM coordinator is to take all reasonable steps to ensure that designers comply with their duties under the regulations **CDM 20(2)**, and that the designers cooperate with the principal contractor during the construction phase concerning any design or change in design **CDM 20(2)**.

The CDM coordinator shall give suitable and sufficient advice and assistance to the client to assist him comply with the regulations **CDM 20(1)**. He is also to ensure that suitable arrangements are made and implemented for the co-ordination of health and safety measures during the planning and preparation of the construction phase including the co-operation and co-ordination between the various persons involved and that the principles of prevention are followed **CDM 20(1)**.

The CDM coordinator shall liaise with the principal contractor concerning any information required by the principle contractor so he can prepare the construction phase plan and any design development that could affect the construction works **CDM 20(1)**.

The CDM coordinator is to take all reasonable steps to collect pre-construction information and provide any information from this in a convenient form, that is relevant to every designer or contractor working on the project **CDM 20(2)**.

The CDM coordinator is to prepare or review and update the health and safety file containing information required to ensure the health and safety of any person during any subsequent work, and pass this information to the client at the end of the project **CDM 20(2)**. In order to do this the CDM coordinator is to liaise with the principle contractor **CDM 20(1)**, and receive relevant information from the designers **CDM 18(2)**, and the client **CDM 17(1)**.

Principal contractors

The Principal Contractor is appointed by the client after he has appointed the CDM coordinator as soon as practical after the client knows enough about the project to be able to select a suitable person **CDM 14(2)**.

The Principal Contractor shall plan, manage and monitor the construction phase of the works to ensure (sfairp) it is carried out without risks to the health and safety of those involved or who might be affected by the works. The Principal Contractor is to liaise with the CDM coordinator to ensure that the welfare provisions comply with the requirements of Schedule 2 of the CDM regulations, drawing up health and safety rules for the site, giving reasonable directions to any contractor on site to enable the Principal Contractor comply with his duties. The Principal Contractor is to ensure that every contractor on site is informed of the minimum time that will be allowed for planning and preparation (where necessary consulting the contractor before finalising the construction phase plan), also ensuring that every contractor is given access to that part of the construction phase plan that involves him, together with any further information, in sufficient time to allow the contractor to properly prepare for his work. The Principal Contractor should also identify for each contractor the information that that contractor should provide for inclusion in the health and safety file. The Principal Contractor is to ensure that the details included in the sites notification to the authorities (as given in Schedule 1 of these regulations) are displayed so that they can be read by any worker and take steps to prevent access to the site by unauthorised persons **CDM 22(1)**.

The Principal Contractor is to take all reasonable steps to ensure that every worker is given a suitable site induction, the information and training required by the regulations and any further training or information necessary **CDM 22(2)**.

The Principal Contractor shall prepare the construction phase plan before the construction work starts, incorporating the information from the designer(s) and client required by these regulations, to ensure that the work is planned, managed and monitored so that the works will be without risks to anyone's health and safety (sfairp) **CDM 23(1)(a)**. The Principal Contractor shall arrange for the construction phase plan to be implemented in a way that will ensure (sfairp) the health and safety of all persons doing the work or who might be affected by it **CDM 23(1)(c)**. The Principal Contractor shall take reasonable measures to ensure the construction phase plan identifies the risks to health and safety and the measures necessary to control these, including the preparation of any site rules **CDM 23(2)**. The Principal Contractor shall review, revise and refine the construction phase plan to ensure it is sufficient for its purpose **CDM 23(1)(b)**.

The Principal Contractor shall make and maintain arrangements to enable him and the workers, to cooperate effectively in promoting and developing measures to ensure health and safety and to check the effectiveness of these measures **CDM 24(a)**. The

Principal Contractor is to consult with the workers or their representatives in good time on matters which may affect their health and safety to the extent that their employers have not consulted them **CDM 24(b)**. The Principal Contractor shall ensure that workers or their representatives can inspect and take copies of any information the Principal Contractor has relating to the planning and management of the project or which may affect their health, safety or welfare **CDM 24(c) Demolition**.

Designers

A Designer is any person (including a client, contractor or other person referred to in the regulations) who prepares or modifies a design or who arranges for or instructs any person under his control to do so, that relates to a structure, product, mechanical or electrical system intended for a particular structure **CDM 2**. The definition of a 'design' has been explained earlier.

The duties on designers preparing a design for construction work shall be carried out (sfairp) taking into account other relevant design considerations **CDM 11(2)**.

The designer is to take all reasonable steps to provide with the design sufficient information about the design, its construction and maintenance to adequately assist clients, other designers and contractors, to comply with their legal duties **CDM 11(6)**. Where the project is notifiable the designer is to provide sufficient information to assist the CDM coordinator with his duties **CDM 18(2)**. Where a design is prepared or modified outside the United Kingdom, the person who commissions it (if established in the United Kingdom or if not any client for the project) shall ensure that the duties of designers in Regulation 11 are compliant with **CDM 12**.

No design is to be prepared unless reasonable steps have been taken to ensure the client is aware of his duties **CDM 11(1)**.

The designer shall avoid foreseeable risks to the health and safety of any person who carries out the construction work, subsequent cleaning and maintenance work, who may be affected by the work or who will use the structure as a workplace. In doing this the designer shall eliminate hazards if they may cause risks and reduce the risks from any remaining hazards giving priority to those measures that will protect all persons rather than those that only protect individuals. In designing structures to be used as a workplace the designer will take account of the requirements of the Workplace (Health Safety and Welfare) Regulations 1992 relating to the design of and materials used in the structure (e.g. Re-stability and solidity—Reg 4a, maintenance—Reg 5, ventilation—Reg 6, indoor temperatures—Reg 7, lighting—Reg 8, cleanliness—Reg 9(2), condition of floors and traffic routes etc—Reg 12) **CDM 11(3), (4) and (5)**.

Where a project is notifiable, the designer shall not start work, other than preparing initial design drawings, unless a CDM coordinator has been appointed **CDM 18(1)**.

Contractors, construction managers and supervisors

No contractor is to start work on site unless reasonable steps have been taken to prevent unauthorised access **CDM 13(6)**, or until the client is aware of his own duties under the regulations **CDM 13(1)**.

Where the project is notifiable, no contractor shall start construction work unless he has been given the names of the CDM coordinator and principal contractor and he has been given access to that part of the construction phase plan relevant to him. Nor shall he start work until notice of the project has been given to the HSE or office of Rail Regulation as the case may be **CDM 19(1)**.

Contractors shall plan, manage and monitor their work to ensure it is without risks to health and safety (sfairp) **CDM 13(2)**. They shall also ensure that the requirements for welfare, as outlined in Schedule 2 of the regulations, with respect to anyone working under his control **CDM 13(7)**.

Contractors shall ensure that any contractors they employ are informed of the minimum amount of time they will be allowed to plan and prepare for their work **CDM 13(3)**.

Every contractor shall provide every worker under their control any information and training they may need to carry out their work safely and without risks to their health. This will include a suitable site induction (if not provided by others) information on the risks to his health and safety, the measures identified by the contractor to deal with these risks, any site rules, the procedures to be followed in the event of serious and imminent danger, and the identity of those who are to implement these procedures **CDM 13(4)**.

Every contractor shall comply with the requirements of Regulations 26–32 and 34–44 as far as they affect him or any person under his control or relate to matters under his control **CDM 25(1)**.

Every contractor is to promptly provide the principal contractor with information which might affect the health and safety of any person, might justify a review of the construction phase plan or has been identified for inclusion in the health and safety file, identify any contractor he engages comply with the directions given by the principal contractor and comply with the health and safety plan. Every contractor is to cooperate with the principal contractor, providing him with information on anything which might affect the health and safety of any person (sfairp), comply with the directions given by the principal contractor and the site rules. The contractor is to promptly supply the principal contractor with the information concerning any death, injury, condition or dangerous occurrence that the contractor is required to report under RIDDOR **CDM 19(2)**.

Every contractor is to take all reasonable steps to ensure that the construction work is carried out in accordance with the construction phase plan or if it is not possible to comply with the plan to ensure health and safety, and notify the principal contractor of any significant findings which requires the construction phase plan to be altered **CDM 19(3)**.

Where a contractor has control of any lifting equipment, a person who manages or supervises the use of lifting equipment, or the way lifting equipment is used then he is to comply with the Lifting Operations and Lifting Equipment Regulations **LOLER 3(3)b**.

Every person who is responsible for, in control of, or whose acts or omissions could affect the health or safety of persons involved in a diving project shall take reasonable measures to ensure the regulations are complied with **DWR 4**.

No person shall act as a diving contractor unless his details as given in the schedule have been supplied to HSE. No person shall dive unless there is only one diving contractor employing all the divers on the project or is a self-employed diver in a team of self-employed divers who has been jointed appointed by them all to act as the diving contractor. The diving contractor shall ensure the diving project is planned, managed and carried out so as to protect the health and safety of those taking part (sfairp). He shall ensure a diving project plan is prepared before the start of the diving project and is updated as necessary. He shall also appoint a supervisor, making a written record of this and confirming that appointment in writing, supplying the relevant parts of the diving project plan, before any dives. The diving contractor shall ensure that there are sufficient persons competent to carry out the project; there is available, suitable and sufficient plant which could be needed for the diving project, maintained in a safe working condition; including that which may be required for any foreseeable emergency; that any person taking part in the diving project complies with the diving project plan and the legal requirements for the work (sfairp) and that a record is kept for each diving operation, retaining this for at least two years after the last entry **DWR 5–7**.

Anyone appointed or acting as a diving supervisor shall be competent and suitably qualified where appropriate. Only one supervisor shall be appointed at any one time. The supervisor shall not dive while supervising a diving operation. The supervisor shall ensure a diving operation is carried out without risks to the health and safety of both the divers and those who may be affected by the work, in accordance with the legal requirements and diving plan; ensure every person involved in the diving operation is aware of how the diving project plan relates to that operation; write up the diving operation record during the course of the operation. He may give directions to any person involved in the diving operation or who could affect its safety **DWR 9, 10 and 11**.

Sub contractors and self-employed workers

Every person, other than a contractor, who controls the way any construction work is carried out, shall comply with Regulations 26–44 as far as they relate to matters within his control **CDM 25(2)**. Similarly concerning the use of electricity, every self-employed person is to comply with the regulations as far as it is in their control **EWR 3(1a)**. The requirements placed on the employer by the WAHR shall also apply to a self-employed person or other person for themselves and to anyone under their control **WAHR 3(3)**. A similar requirement applies concerning hazardous substances **COSHH 3(2)**, control of asbestos **CAR 3(4)**, lead **CLW 3(2)**, vibration **CVW 2(4)** and noise **NWR 2(2)**.

Every self-employed person shall ensure that the employer of anyone from an outside undertaking working in his undertaking is provided with comprehensive information on the risks to their health and safety and of the measures to be taken to deal with these **MHSW 10**.

Where the project is notifiable, no contractor shall start construction work unless he has been given the names of the CDM coordinator and principal contractor and he has been given access to that part of the construction phase plan relevant to him **CDM 19(1)**.

Every person who has duties under the regulations cooperates with others to enable himself to carry out those duties **CDM 5(1)**. All persons having duties under the regulations shall co-ordinate their activities with each other to ensure (sfairp) the health and safety of others **CDM 6**.

The self-employed are to conduct their undertaking (sfairp) so that he does not expose himself, or others not employed by him, to risks to their health and safety

HSWA s3(2). Every self-employed person shall make suitable and sufficient assessment of the risks to his own health and safety at work and to those who may be affected by his work. He is then to use this to identify the measures he needs to take in order to comply with his legal duties. This assessment shall be reviewed if there are reasons to suspect they are no longer valid or if there has been significant changes in the work procedures **MHSW 3(20 and 3(3)**.

Where he uses lifting equipment at work he is to comply with the Lifting Operations and Lifting Equipment Regulations **LOLER 3(3)a**.

Every self-employed person shall ensure he is provided with suitable PPE against risks to his health and safety while at work unless the risks are controlled by other equally or more effective means. Where there is more than one risk requiring the simultaneous use of several items of protective equipment, these must be compatible and continue to be effective and efficient against all the risks **PPE 4(2) and 5(2)**. Before choosing any PPE a self-employed person shall assess whether it will be suitable. This assessment shall be of any risks not avoided by other means and the characteristics required of the PPE to deal with these risks. If the self-employed person thinks the assessment is no longer valid or that there has been a significant change in the situation, then the assessment shall be reviewed **PPE 6**. A self-employed person shall ensure that any PPE provided to him is maintained in good order, in an efficient state and working order, being replaced or cleaned as appropriate **PPE 7(2)**. Where PPE is provided, appropriate accommodation is required for it when it is not in use **PPE 8**. A self-employed person is to make full and proper use of any PPE provided, returning it to the accommodation provided for its storage **PPE 8, 10(3) and 10(4)**.

Every self-employed person shall provide himself with suitable head protection, maintaining and replacing this as necessary. He shall wear this unless there is no foreseeable risk of head injury other than falling, or when required to do so **C(HP) 3(2), 6(2) and 6(3)**. Where he has control over another person at work he shall ensure that person wears suitable head protection unless there is no foreseeable risk of head injury other than falling. He may give directions to those at work under his control, requiring them to wear suitable head protection **C(HP) 4(2) and 5(4)**. Every self-employed person required to wear suitable head protection shall make full and proper use of it and take all reasonable steps to return it to the accommodation provided when it is not being used **C(HP) 6(4)**.

A self-employed person shall either provide adequate first aid equipment for himself or ensure it is provided appropriate to the circumstances while at work **HS(FA) 5**.

Persons in control of premises

Anyone who has control of or part control of premises, including any plant or substances in these, is to ensure (sfairp) that they do not present a risk to the health and safety of others, not their employees, who work there **HSWA s4(2)**.

The person in control of a site may make rules concerning the wearing of suitable head protection. These rules shall be in writing and brought to the notice of those who may be affected by them **C(HP) 5(1) and 5(2)**.

Where the person in control of premises is the responsible person as defined by the regulations, where a person has an accident and dies, suffers a major injury or suffers an injury that requires hospital treatment, then he should inform the relevant enforcing authority as quickly as possible and make a report on the incident within 10 days.

Similarly the person in control of the premises must report any dangerous occurrence listed in the schedule to the regulations (which includes the collapse of lifting machinery, scaffolding, building or structure and an explosion or fire).

He must also report the death of an employee where this occurs up to 12 months after an accident whether or not the accident was reported when it originally occurred.

The person in control of premises as the responsible person shall immediately send a report to the relevant enforcing authority where a person suffers from any of the occupational diseases listed in the schedules to the regulations while involved in one of the specified activities where this has been confirmed by a medical practitioner **RIDDOR 2–5**.

Suppliers of equipment and materials

Any person, or company, who designs, manufactures, imports or supplies any plant for use or operation by a person at work, or any component of that plant, to ensure (sfairp) that this will be safe at all times when being set, used, cleaned or maintained. In order to be able to do this he must carry out all testing and examination as may be necessary. Where the person designs or manufactures the plant then (sfairp) he is to carry out any necessary research to discover and eliminate or minimise any health and safety risks **HSWA s6(1) and 6(2)**. Articles for work in any premises are to be (sfairp) designed etc so that they can be erected and installed without risks to health and safety **HSWA s6(3)**. Any person, or company, who manufactures, imports or supplies any substance for use at work is to ensure (sfairp) that this will be safe at all times when being used. In order to be able to do this he must carry out all testing and examination as may be necessary. Where the person manufactures the substance then (sfairp) he is to carry out any necessary research to discover and eliminate or minimise any health and safety risks **HSWA s6(4) and 6(5)**.

Individuals

Every employee to take reasonable care for both his own health and safety and of others who may be affected by his acts or omissions. He is also to cooperate with his employer or any other person who have duties imposed on them as far as is necessary to enable that duty to be complied with **HSWA s7**. While at work every employee shall cooperate with his employer in complying with the Electricity at Work Regulations and comply with them himself so far as it is within his control **EWR 3(2)**.

Every person under the control of another person shall report to that person any defect that may be a risk to the health and safety of himself or any other person **CDM 25(3)**. Similarly, any shortcomings in the health and safety arrangements **MHSW 12(2)**, and when working at height **WAH 14**. No one is to intentionally or recklessly interfere with, or misuse any thing provided for health and safety purposes **HSWA s8**. Similarly when provided with suitable head protection shall take reasonable care of it and report to his employer if it is lost or damaged **C(HP) 7**, or with other PPE to report loss or obvious defect **PPE 11**, **COSHH 8(2)**, or ear protection **NWR 8(2)b**, or measures to control asbestos **CAR 12(2)b**.

Every person shall use any work equipment or safety device provided for work at height in accordance with any training and instructions **WAH 14(2)**, Every employee shall use any machinery, equipment, dangerous substance, transport equipment, means of production or safety device provided by his employer, in accordance with any training or instructions received **MHSW 14(1)**. Every employee who has control over another person at work shall ensure that person wears suitable head protection unless there is no foreseeable risk of head injury other than falling. He may give directions to those at work under his control, requiring them to wear suitable head protection **C(HP) 4(2) and 5(4)**. Every employee provided with suitable head protection shall wear it when required to do so making full and proper use of it and take all reasonable steps to return it to the accommodation provided when it is not being used **C(HP) 6(12) and 6(4)**. Every employee shall make full and proper use of any control measures, other thing or facility provided in accordance with COSHH. He shall take all reasonable steps to ensure relevant equipment is returned to any accommodation provided for it after use **COSHH 8(2)**. Every employee shall make full and proper use of any PPE provided in accordance with any training and instructions and is returned to the store provided for it when not in use **PPE 10(2), 10(3) and 10(4)**.

Employees shall not eat, drink or smoke in any place which they have reasons to believe may be contaminated with lead **CLW 7(2)**.

Every employee shall make full and proper use of any control measures provided to comply with the regulations, other thing or facility provided and take all reasonable steps to ensure it is returned to any accommodation provided for it after use **CAR 12(2)**.

Each employee shall make full and proper use of any system of work provided to reduce the risk of injury to the lowest reasonably practicable level from manual handling work **MHO 5**.

So far as is practicable employees shall fully and properly use personal hearing protectors and any other protective measures provided for them **NWR 8(2)a**

No one shall dive in a diving operation unless he is has an approved qualification and is competent to carry out any activity involved in the operation and has nothing which would make him unfit to dive. He should have a valid medical certificate for fitness to dive. He shall comply with any directions given by the supervisor and any applicable instructions in the diving project plan where these do not conflict with the supervisor's directions. He shall maintain a daily record of his diving, keeping this for at least two years after the last entry **DWR 12 and 13**.

19.5 A summary of the legal requirements when involved in various activities

The various legal requirements for various working situations and activities given below are set out in the following order:

(1) Planning the work (including emergency planning)
(2) Management and supervision (including inspection)
(3) Safe places of work
(4) Safety signs and signals
(5) Use of personal protective equipment
(6) Training and competence. Inspection
(7) Working with, and the control of, hazardous substances and situations (including noise, fire and explosives)

(8) Work with or near live services
(9) Manual handling
(10) Control of noise
(11) Confined spaces
(12) Excavations
(13) Work over or near water
(14) Work at height (including roof work)
(15) Stability of structures (including temporary structures)
(16) Demolition
(17) Work with plant and machinery
(18) Lifting operations
(19) Diving operations
(20) Control of traffic
(21) Welfare

Many of the duties placed upon individuals concern assessing risks to the health and safety of those concerned with or affected by work activities, and then planning or managing those risks that cannot be avoided. However there is also another area of health and safety law that stipulates specific actions or procedures for particular activities and these must be followed even if the risk to health and safety if they were not followed appears slight. While the responsibility for complying with these duties is generally aimed at the employer, depending on the circumstances virtually anyone concerned with the work could find that they also were expected to comply with these specific requirements and could be liable to prosecution if they had not done all they could to ensure that the requirements were followed.
Note: The abbreviation (sfairp) sometimes used in this section means 'so far as is reasonably practicable'. An explanation of this term is given in Section 9.2 of this chapter.

Planning the work

Every employer shall ensure that work at height is properly planned, appropriately supervised and carried out in a safe manner (sfairp) including planning for emergencies and rescue **WAH 4(1) and 4(2)**. Every lifting operation shall be properly planned by a competent person, be adequately supervised and carried out in a safe manner **LOLER 8**.

Where necessary suitable and sufficient arrangements shall be prepared and implemented for dealing with any foreseeable emergency including procedures for the part or total evacuation of the site. In doing this account shall be taken of the type of work, the size and location of that work, the equipment being used, the number of persons employed and the physical and chemical properties of any substances or materials being used. Those for whom such procedures are planned should be informed of them and the procedures shall be tested at suitable intervals **CDM 39**. A sufficient number of suitable emergency exit routes and exits are to be provided to enable any person to quickly reach a place of safety in the event of danger. These routes etc. shall lead as directly as possible to an identified safe area and be identified by suitable signs. They shall be kept clear of obstructions at all times and if necessary have emergency lighting. **CDM 40**.

Work at height is to be properly planned, including planning for emergencies and rescue, appropriately supervised and carried out (sfairp) in a safe manner. Such work shall only be done when the weather conditions allow it to be done safely **WAH 4**.

Management and supervision

Every lifting operation that uses lifting equipment is to be planned by a competent person, and be appropriately supervised and carried out in a safe manner **LOLER 8**.

Any fire-fighting equipment, fire detection or alarm systems are to be examined and tested at suitable intervals and be properly maintained **CDM 41(3)**.

Work shall not be carried out in an excavation that has supports or battering unless the excavation and any work equipment that affects its safety has been inspected by a competent person at the start of a shift, after any event that could affect the strength and stability of the excavation, and after unintentional dislodgement or fall of any material, and the inspector is satisfied that work can be done safely **CDM 31(4)**. If the inspector has informed the person in charge of the work of anything with which he is not satisfied, then no work is to be done in the excavation until that matter is rectified **CDM 31(5)**. A coffer dam or caisson and any work equipment that affects its safety shall only be used if it has been inspected by a competent person at the start of each shift and after any event that could affect its strength and stability **CDM 32(2)**. Where the inspector informs the person in charge of matters about which he is not satisfied, no work shall be carried out in the coffer dam or caisson until those matters are rectified **CDM 32(3)**.

Before lifting equipment is put into service for the first time it is to be thoroughly examined for any defect unless it has not been used before, has an EC declaration of conformity not more than 12 months old or has been obtained from someone else who has provided evidence of an inspection **LOLER 9(1)**. Where the safety of the lifting equipment depends on the installation conditions it is to be thoroughly examined after installation or after assembly and before being used **LOLER 9(2)**. Where lifting equipment is exposed to conditions that cause deterioration, it is to be thoroughly examined at least every 12 months, or 6 months if used to lift people, in accordance with an examination scheme, and after exceptional circumstances that could have affected the safety of the equipment. If appropriate it should be inspected by a competent person at suitable intervals between thorough examinations **LOLER 9(3)**.

Where the safety of work equipment depends on the installation conditions it shall be inspected after installation or assembly, and before being used. Work equipment that is exposed to conditions that could cause deterioration is to be inspected at suitable intervals and after exceptional circumstances that could have affected its safety. The results of these inspections are to be recorded and kept until the next inspection is recorded **PUWER 6(1), 6(2) and 6(3)**.

Safe places of work

There is to be suitable and sufficient safe access to and egress from every place of work or place provided for use by someone at work (sfairp), which shall be properly maintained. Every place of work shall (sfairp) be made and kept safe without risks to the health and safety for anyone who works there. The place of work shall provide sufficient working space and be suitable for the work being done. Suitable and sufficient steps shall (sfairp) be taken to prevent anyone using places of work or the access and egress to them if those places do not comply with the regulations **CDM 26(1), 26(2), 26(3) and 26(4)**. Every part of a construction site shall (SFAIRP) be kept in good order and every part used as a place of work shall be kept in a reasonable state of cleanliness **CDM 27(1)**. Where necessary, in the interests of health and safety, a site shall either have its perimeter clearly marked, or be fenced off, or both (sfairp) **CDM 27(2)**.

Any indoor working place is to be kept at a reasonable temperature (sfairp), suitable for the work undertaken there **CDM 43(1)**. Every outdoor workplace shall be arranged (sfairp) to give protection from adverse weather, having regard for the work carried out there and any protective clothing provided as necessary for the health and safety of any person at work **CDM 43(2)**. Every place of work, the access to it and any traffic route shall (sfairp) have suitable and sufficient lighting. Preferably this will be by natural lighting, but any artificial light used shall be of a colour that will not alter the perception of any signs etc. If there would be a risk to the health and safety of any person if the lighting failed there should be suitable and sufficient additional secondary lighting **CDM 44(1), 44(2) and 44(3)**. Suitable and sufficient steps shall be taken to ensure (sfairp) that every place of work and access to it, has sufficient fresh or purified air to ensure it is safe and without risks to health **CDM 42(1)**. Where necessary, any plant used to supply this air shall include an effective device to give a visible or audible warning of any failure to that plant **CDM 42(2)**.

No timber or other material with projecting nails or other sharp objects shall be used, or allowed to remain in place if they are a source of danger **CDM 27(3)**.

Persons who use work equipment including those who supervise its use, are to have adequate health and safety information on the use of the equipment **PUWER 8(1) and 8(2)**.

Safety signs and signals

Safety signs are to reinforce other collective safety measures where these cannot avoid or adequately reduce the risks to employees. These signs are to be in accordance with Parts 1–7 of the schedule to the regulations. This schedule sets out the requirements for colour, shape, positioning and when they should be used **HS(SSS) 4(1) and Schedule 1**.

Communicating warnings verbally is to reinforce other collective safety measures where these cannot avoid or adequately reduce the risks to employees. Such verbal communications are to be in accordance with Part 8 of the schedule to the regulations. This schedule sets out the requirements for the style of the command words and the ability of the hearers to understand them **HS(SSS) 4(1) and Schedule 1**.

Hand signals are to reinforce other collective safety measures where these cannot avoid or adequately reduce the risks to employees. These signals are to be in accordance with Part 9 of the schedule to the regulations. This schedule sets out the requirements for distinguishing signals and the signalman's duties. Reference is made to the standards, codes etc operating where cranes are used or in agriculture and the fire services **HS(SSS) 4(1) and Schedule 1**.

The position of any fire-fighting equipment shall be indicated by suitable signs **CDM 41(7)**.

All emergency routes and exits are to be indicted by suitable signs **CDM 40(5)**.

Use of personal protective equipment

Employees are to be provided with suitable head protection and the self-employed are to provide their own. Before choosing the head protection an assessment shall be made of its suitability. The assessment shall include a definition of what would be the suitable characteristics and a comparison of these with those that are available with the head protection selected. This assessment is to be reviewed where it is suspected it is no longer valid or when the nature of the work has changed **C(HP) 3(1)–3(6)**. Appropriate accommodation is to be provided for the head protection when this is not in use. Every employee and self-employed person is to take reasonable steps to return the head protection to this store when the head protection is not in use. PPE is to be kept in the accommodation provided when not in use **C(HP) 3(7) and 6(4)**.

Where it is not reasonably practicable to prevent exposure to a hazardous substance and adequate control cannot be achieved by the design and use of appropriate work processes or the control of exposure cannot be achieved at source, then suitable personal protective equipment shall be provided **COSHH 7(3)c**. Personal protective equipment shall be appropriate to the risks, the conditions where used and the health of the persons using it. It should be capable of fitting the person correctly, be ergonomically efficient, be effective in controlling the risks as far as practicable and comply with any legal requirement for that equipment **PPE 4(3)**. Before choosing any personal protective equipment it is to be assessed to ensure it is suitable for its purpose **PPE 6(1)**. PPE is to be maintained in good order, in an efficient state and working order being replaced or cleaned as appropriate **PPE 7(1) and 7(2)**. Except where disposable equipment is provided, respiratory protective equipment provided to comply with the regulations, shall be tested and examined at suitable intervals **COSHH 9(3), CAR 13(2)**. Personal protective equipment is to be used properly according to any training and instructions provided any where provided be returned to any accommodation when not in use **PPE 10**.

Suitable respiratory protective equipment is to be used to adequately control exposure to asbestos in the air inhaled by the employee to a concentration below the control limits only when other control measures cannot do this by themselves **CAR 11(3)**. Personal protective equipment shall be suitable for its purpose and comply with any applicable provision in the Personal Protective Equipment Regulations or if no provision applies, be of a type approved by HSE **CAR 11 (4)**. All control measures provided shall be fully used or applied properly, and returned to any accommodation provided when not in use **CAR 12**.

Suitable personal protective equipment is to be used in conjunction with other control measures to achieve adequate control of exposure to lead only when other control measures cannot do this by themselves **CLW 6(3)**. Any PPE provided shall comply with any relevant provision in the Personal Protective Equipment Regulations or if this is not relevant, to standards approved by the HSE **CLW 6(7)**. All control measures provided shall be fully used or applied properly, and returned to any accommodation provided when not in use **CLW 6(8)**.

Any personal hearing protectors provided to eliminate the risk to hearing or to reduce it to as low level as is reasonably practicable shall be selected after consultation with the employees or their representatives **CNW 7 (4)**. Employees are to make full and proper use of any personal hearing protectors provided **CNW 8(2)**.

Training and competence

Account shall be taken of an employee's capabilities regarding their health and safety, when giving them tasks **MHSW 13(1)**. Employees are to be provided with adequate health and safety training when recruited and when exposed to new or increased risks through being transferred or given new responsibilities, introduced to new work equipment, to a change in use of existing equipment or introduced of new or changed systems of work. **MHSW 13(2)**. Training is to be repeated periodically where appropriate, be adapted to changes to the risks to health and safety, and be carried out in working hours **MHSW 13(3)**. No person shall appoint, engage or accept an appointment as a CDM coordinator, designer, principal contractor or contractor unless they are competent **CDM 4(1) a and b**. No person shall arrange or instruct a worker to carry out or manage construction work unless the worker is competent or under the supervision of a competent person **CDM 4(1)c**.

No person is to be engaged in any activity related to work at height, including using equipment, planning or supervision, unless competent or being trained under the supervision of a competent person **WAH 5**. A cofferdam or caisson, and any work equipment and materials that affect its safety, shall only be used if it has been inspected by a competent person **CDM 32(2)a**. Construction work shall not be carried out in an excavation unless it and any work equipment and materials that affect its safety, has been inspected by a competent person **CDM 31(4)a**.

Scaffolds may only be assembled, dismantled or significantly altered if under the supervision of a competent person and by persons who have received appropriate and specific training in the operations envisaged **WAH Schedule 3, Part 2 (12)**.

Personal fall protection systems shall only be used if the user and a sufficient number of available persons have received adequate and specific training, including rescue procedures **WAH Schedule 5 Part 1(1)b**.

Every person on site (sfairp) shall be instructed in the correct use of any fire-fighting equipment he may need to use **CDM 41(5)**. Where a work activity may cause a risk of fire, a person may not do that work unless he has been suitably instructed **CDM 41(6)**.

Employees are to be trained in the meaning of the safety signs and other measures, together with the course of action they are to follow **HS(SSS) 5**.

The surface, parapet, permanent rail or other such fall prevention measures shall be checked on each occasion before that place is used for work at height **WAH 13**. Where the safety of work equipment depends on how it is assembled or installed it is not to be used until it has been inspected **WAH 12(2)**. Where work equipment is exposed to conditions that cause deterioration and this results in a dangerous situation, it shall be inspected at suitable intervals and after any exceptional circumstances which could have jeopardised the safety of the equipment so any deterioration can be detected in good time and the equipment maintained in a good condition **WAH 12(3)**. Any working platform from which a person could fall 2 m or more, shall not be used unless it has been inspected within the previous seven days **WAH 12(4)**. No work equipment other than lifting equipment to which Regulation 9(4) of LOLER applies shall leave an undertaking or if received from another undertaking, is used, unless there is physical evidence that the last inspection required by the regulations has been carried out **WAH 12(5)**. The results of any inspection are to be recorded and kept until the next inspection is carried out **WAH 12(6)**. A person who carries out an inspection of a working platform shall prepare a report on that inspection before the end of the working period during which the inspection took place and within 24 hours. The inspector shall provide the person on whose behalf the inspection was undertaken with a report or a copy of that report on his inspection **WAH 12(7)**. The person receiving this report shall keep it at the site where the inspection was carried out until the construction work is complete and then at his office for a further three months **WAH 12(8)**.

Those carrying out inspections of excavations, coffer dams and caissons shall before the end of the shift within which the inspection is completed, inform the person on whose behalf the inspection was carried out about any matters with which he was not satisfied, and prepare a report of his inspection in accordance with Schedule 3 of these regulations and submit this within 24 hours of the inspection. There is no requirement to prepare such a report more than once every seven days where the inspection is a routine inspection at the start of each shift. The person for whom the inspection was carried out shall keep the reports of the inspections on site and for three months after the works have been completed. Where the inspector is an employee or under the control of someone else, then his employer or person in control, shall ensure that the inspector carries out his duty **CDM 33**.

Unless any lifting equipment has not been used before or is accompanied by a certificate showing it has been examined by the previous owner/user in the proscribed period, it must be thoroughly examined for any defect. Where the safety of the lifting equipment depends on its installation then it is to be thoroughly examined after installation or assembly **LOLER 9(1) and 9(2)**. Where lifting equipment is exposed to conditions that could cause it to deteriorate and become unsafe it shall be thoroughly examined by a competent person at least every 6 months if used to carry persons or otherwise every 12 months. It should also be examined after any exceptional circumstances that could jeopardise its safety **LOLER 9(3)**. No lifting equipment is to be passed onto another user unless it is accompanied by physical evidence that the last thorough examination has been carried out **LOLER 9(4)**.

Where engineering controls are provided to meet the requirements of Regulation 7 COSHH, the employer shall ensure that the controls are thoroughly examined and tested at least every 14 months for local exhaust ventilation or in other cases at suitable intervals. Where respiratory protective equipment is used a thorough examination and, where necessary, testing shall be carried out at suitable intervals. A suitable record of these examinations and tests, together with any repairs carried out following those examinations or tests, shall be kept for at least five years **COSHH 9(2), 9(3) and 9(4)**.

All persons who use work equipment and their supervisors, are to receive adequate training for health and safety purposes including how the equipment is to be used, the risks involved and precautions to be taken **PUWER 9**.

Working with, and the control of, hazardous substances and situations

Explosives shall be stored, transported and used safely and securely (sfairp) **CDM 30(1)**. An explosive charge shall only be used or fired if suitable and sufficient steps have been taken to ensure that no one is exposed to injury from the blast or flying debris **CDM 30(2)**. Suitable and sufficient steps shall be taken to prevent (sfairp) the risk of injury from fire or explosion **CDM 38a**. Where necessary for the health and safety of any person at work on a construction site there shall be suitably located suitable and sufficient fire-fighting equipment and fire alarm and detection systems **CDM 41(1)**. Any fire-fighting equipment that does not come into use automatically shall be easily accessible **CDM 41(4)**.

Suitable and sufficient steps shall be taken to prevent (sfairp) the risk of injury from any substance liable to cause asphyxiation **CDM 38c**.

Work shall not start where it is likely to involve anyone being exposed to a hazardous substance without suitable and sufficient assessment of the risks and the steps necessary to comply with the regulations being made first. The assessment shall consider, amongst others, the hazardous properties, information on health effects, the level, type and duration of exposure the working circumstances and subsequent maintenance, and the results of monitoring. The assessment shall be regularly reviewed as well as immediately it is suspected the assessment is no longer valid or if there has been a substantial change in the work **COSHH 6**.

Employees are to be prevented from being exposed to hazardous substances. Where this is not practicable then the exposure is to be adequately controlled. When preventing exposure, substitution will take preference. Where it is not possible to prevent exposure the protection measures will give priority to the design of work processes, controlling the exposure at source, and where this is not possible, by the provision of suitable personal protective equipment **COSHH 7(1), 7(2) and 7(3)**.

The risk of damage to the hearing of employees or other persons at work of hearing damage from noise caused by work shall be eliminated or, where not reasonably practicable, reduced to the lowest reasonably practicable level **CNW 6(1)**. Exposure to noise at or above an upper exposure action value of 85 dB daily or weekly exposure and 137 dB peak sound pressure, should be controlled by a programme of measures, including the provision of personal hearing protectors, to as low a level as is reasonably practicable appropriate to the activity **CNW 6(2)**. Such measures should also include using quieter work equipment, the layout of the work place, correct use of equipment, limitation of the duration and intensity of noise levels and reduction of technical means **CNW 6(3)**. So far as is practicable, everything provided to reduce the risk of damage to hearing shall be fully and properly used and maintained in an efficient state, working order and in good repair **CNW 8(1) and 8(2)**.

Exposure to lead is to be prevented, or if this is not reasonably practicable, adequately controlled **CLW 6(1)**. In preventing exposure, substitution is to be preferred. Where this is not practicable protection measures will be applied appropriate to the activity and consistent with the risk. Where these are not likely to reduce exposure sufficiently then personal protective equipment shall be used **CLW 6(2), 6(3), 6(4), and 6(5)**.

Where an assessment shows that asbestos is liable to be present in any part of the work place the risk from the asbestos should be determined and a written plan prepared identifying the areas where the risk is present and the control measures that will be put in place. These measures are to include monitoring, maintaining or removing the asbestos and ensuring that information about the asbestos, its condition and location is given to anyone likely to disturb it **CAR 4(8) and 4(9)**.

Risks from vibration are to be eliminated or if this is not reasonably practicable, reduced to as low a level as is reasonably practicable. In doing this the principles of prevention shall be followed **CVW 6(3)**. Where an employee is exposed to vibration above an exposure limit the reason why this occurred is to be identified and the work procedures modified accordingly **CVW 6(4)**.

Work with or near to live services

All systems shall be of such construction as to prevent danger at all times, and be maintained to prevent such danger (sfairp) **EWR4(1)**. Any equipment required by the regulations to be provided to protect persons at work on or near electrical equipment shall be suitable for its use **EWR 4(4)**, be maintained **EWR 4(2)** and used properly **EWR 4(4)**. Electrical equipment shall not be used so that dangers will arise if its strengths and capabilities are exceeded **EWR 5**. Electrical equipment shall be of such construction or protected to prevent dangers where it is reasonably foreseeable that it could be exposed to mechanical damage, the effects of the weather, natural hazards or pressure, the effects of wet, dirty, dusty or corrosive conditions or any flammable or explosive substance, including dusts, vapours or gases **EWR 6**. All conductors that could be a source of danger shall be either suitably covered with insulating material to prevent (sfairp) that danger or placed or have other precautions so as to prevent that danger (sfairp) **EWR 7**. Where it is reasonably foreseeable that a conductor could become charged through use or through a fault causing a danger then it shall be earthed or made safe by other suitable means **EWR 8**. Breaks or high impedance shall not be introduced into a conductor connected to earth or other reference point that might be reasonably be expected to cause danger unless suitable precautions are taken **EWR 9**. Every joint and connection in a system shall be mechanically and electrically suitable for use **EWR 10**. Suitably located efficient means shall be provided to prevent excess current in any part of a system being a danger **EWR 11**. Suitable means shall be available to prevent danger for cutting off electrical supply and for isolating electrical equipment from every source of electrical energy in a secure manner. This does not apply to electrical generators themselves although in these cases precautions shall be taken to prevent danger **EWR 12**.

Work, including the operation, use and maintenance of a system or near a system, shall(sfairp) be done so as to avoid danger **EWR 4(3)**. Precautions shall be taken to prevent danger through electrical equipment becoming live while work is carried out on or near it **EWR 13**. No work shall be carried on or near an insufficiently insulated live conductor where this may cause danger unless, it is unreasonable for it to be dead, it is reasonable in all the circumstances for it to be live and suitable precautions, such as suitable protective equipment, are taken to prevent injury **EWR 14**. Adequate lighting, means of escape and working space shall be provided at all electrical equipment on which or near which work is being done, to prevent injury **EWR 15**. Only persons who have the necessary technical knowledge or experience to prevent danger or injury shall carry out work unless they are under an appropriate degree of supervision **EWR 16**.

Energy distribution installations shall be suitably located, checked and clearly indicated where necessary to prevent danger. Where there is a risk from electrical power cables these shall be directed away from the area of risk, or the power shall be isolated and where necessary earthed, or if these measures are not reasonably practicable, suitable warning notices are to be displayed and barriers used to exclude work equipment that is not needed, or where vehicles pass under the cables suspended protection or other measures shall be used that give an equivalent level of safety **CDM 34(1) and 34(2)**. No construction work will be carried out that is liable to create a risk to health and safety from an underground service or cause damage or disturbance to it unless (sfairp) suitable and sufficient steps have been taken to prevent the risk **CDM 34(3)**.

Manual handling

When assessing whether manual handling operations could cause a risk of injury, the factors that must be considered are the task being done, the magnitude of the loads, the working environment, the capability of the individual and any other relevant factors **MHO Schedule 1**.

Confined spaces

No person shall enter a confined space to work unless it is essential **CS 4(1)**. Any work in a confined space must be in accordance with a safe system of work **CS 4(2)**, this includes suitable and sufficient arrangements for rescue in an emergency **CS 5(1)**. The above arrangements must reduce (sfairp) the risks to the health and safety of those carrying out the rescue and where resuscitation is envisaged, suitable resuscitation equipment is kept available **CS 5(2)**.

Excavations

All practicable steps are to be taken to prevent, where necessary, danger to any person including the provision of supports or battering, to ensure that any excavation or part of an excavation does not collapse, or that any material from the sides, roof or adjacent to an excavation is dislodged or falls and that no person is buried or trapped in an excavation by falling material **CDM 31(1)**. Suitable and sufficient steps are to be taken to prevent any person, work equipment or accumulation of material from falling into an excavation **CDM 31(2)**, or of the ground around an excavation being overloaded by work equipment or materials **CDM 31(3)**.

Every cofferdam and caisson shall be of a suitable design and construction, be equipped so workers can shelter or escape if water or materials enter it, and be properly maintained **CDM 32(1)**. Suitable and sufficient measures are to be taken to prevent any vehicles from falling into any excavation, pit or water, or from over running any embankment or earthwork **CDM 37(6)**.

Work over or near water

Suitable and sufficient steps shall be taken to prevent (sfairp) the risk of injury from flooding **CDM 38b**. Where any person could fall and drown in any water or other liquid, suitable and sufficient steps shall be taken to prevent (sfairp), that person falling, minimise the risk of drowning in the event of a fall and to ensure suitable rescue equipment is provided, maintained and used to promptly rescue a person who has fallen **CDM 35(1)**. Suitable and sufficient steps shall be taken to ensure the safe transport of any person to or from their workplace. Any vessel used for this purpose shall be suitably constructed, properly maintained and not be overcrowded or overloaded **CDM 35(2) and 35(3)**.

Work at height

Suitable and sufficient steps are to be taken (sfairp), to prevent anyone from falling a distance liable to cause personal injury **WAH 6(3)**. Where a workplace contains an area where there is a risk of someone falling a distance or being struck by a falling object, it is to be equipped (sfairp) with devices preventing unauthorised entry **WAH 11**. Work at height is not to be undertaken where it is reasonably practicable to work other than at height **WAH 6(2)**. Suitable alternative working positions shall include working from an existing place of work or existing means of access **WAH 6(4)a**.

Suitable existing places of work shall be stable and of a suitable strength, area and size to allow safe passage to the work position. They shall also have suitable means of preventing a fall and have a surface with no gaps through which a person or materials could fall and have a surface which (sfairp) would prevent a slip or trip. It shall also be such that a person would not get trapped between it and another structure. If it has moving parts then it shall have means of preventing an inadvertent movement **WAH Schedule 1**. Where it is not reasonably practicable to work from an existing structure sufficient work equipment will be provided to prevent (sfairp) a fall occurring WAH 6 (3)b. Where the above measures would not eliminate the risk of falling then sufficient equipment shall be provided (sfairp) to minimise the distance and consequences of a fall providing such additional training and instruction, or take other additional suitable and sufficient measures to prevent (sfairp) a person falling a distance liable to cause injury **WAH 6(5)**.

In selecting any work equipment priority shall be given to that which provides collective protection over personal protective equipment. Account shall also be taken in selecting work equipment of the working conditions and the risks to the safety of persons where the equipment is to be used; where the equipment is to be used for access, the distance to be negotiated; the distance and consequence of a fall; the duration and frequency of use; the facility for evacuation and rescue and the additional risks of using, installing and removing the equipment **WAH 7(1)**. Work equipment is to be selected which is appropriate for the work and allows adequate passageway **WAH 7(2)**.

Where the means taken to prevent someone falling consist of a guard-rail, toe-board, barrier or other similar means they should be suitable, and have sufficient strength and dimensions and rigid enough for the purposes for which they are being used. Where these are attached to, or supported on any supporting structure, then this too must be suitable and of sufficient strength. These barriers should (sfairp) be placed and secured so they will not be accidentally displaced and so that (sfairp) they will prevent any person, material or object from falling.

The height of a barrier must be at least 950 mm above the edge from which a person could fall (or 910 mm where the height has been fixed prior to the regulations coming into force), the toe-boards are to be suitable and sufficient to prevent the fall of any person, object or material from the place of work and the vertical gap between any guard rail, intermediate rail and toe board etc, should not exceed 470 mm.

Barriers used to prevent falls are not to have a gap in them except where necessary at the point of access to a ladder or stairway. Barriers etc. may be temporarily removed as necessary for the movement of materials provided they are replaced as soon as practicable afterwards. When the barriers are so removed work shall not be carried out at that place unless compensating measures are in place **WAH Schedule 2**.

A collective safeguard to arrest falls, such as safety nets or air bags, should only be used after a risk assessment has shown that the work can be done safely (sfairp) and the use of safer equipment is not reasonably practicable and that there is a sufficient number of adequately trained persons available including in being able to carry out rescues. The safeguard shall be of sufficient strength and, if it distorts when impacted, have sufficient clearance. In the event of a fall (sfairp) the safeguard itself should not cause injury **WAH Schedule 4**.

Suitable and sufficient steps shall be taken (sfairp) to prevent the fall of any materials or objects where there is a risk these might injure any person. No materials or objects are to be thrown or tipped from a height where this would create a risk of injuring anyone. Where it is not reasonably practical to prevent falling objects, suitable and sufficient steps shall be taken to prevent any person being struck by falling materials or objects liable to cause injury **WAH 10**.

Suitable devices are to be used to prevent a person falling down a shaft or hoistway **LOLER 6(2)**.

A ladder shall only be used where a risk assessment has shown that more suitable equipment is not justified because of the low risk and the short duration of use or because there are existing features on the site which cannot be altered. Any ladder used shall rest against surfaces that are sufficiently strong, firm and stable so that the rungs remain horizontal. A portable ladder shall be prevented from slipping during use by securing the stiles near their upper or lower ends, by using an effective anti-slip device or any other equally effective means. A suspended ladder shall be attached in a secure manner so it cannot be displaced. An interlocking or extension ladder shall have its sections prevented from moving relative to each other. A mobile ladder shall be prevented from moving before it is stepped upon. Where used to give access to another level, a ladder shall extend a sufficient height above that level to give a safe handhold unless an alternative handhold is available. Where a ladder or a run of ladders rises more than 9 m vertically, where reasonably practicable, sufficient rest platforms or landings shall be provided at suitable intervals.

Every ladder shall be used so that a secure handhold and support are always available **WAH Schedule 6**.

No person shall work on, from or near a fragile surface or pass across or near it when it is reasonably practicable not to do so. Where it is not reasonably practicable then suitable and sufficient platforms, coverings, guard rails or similar shall be provided. If after providing such equipment there still remains a risk of falling then suitable and sufficient means shall also be taken to minimise both the distance and consequence of a fall. In addition, where a person may pass across or near a fragile surface, or work on, from or near it, then warning notices shall be fixed on the approach to the fragile surface or if this is not reasonably practicable, persons must be made aware by other means **WAH 9**.

Personal fall protection systems shall only be used if other, safer equipment is not reasonably practicable, the work can be done safely using personal fall protection equipment, and there are sufficient number of persons available who have been adequately training including training in rescue procedures. The system shall be suitable and of sufficient strength. Where necessary it will fit the user correctly, be designed to minimise injury and be adjustable to prevent the user slipping from it or being subject to an unplanned or uncontrolled movement **WAH Schedule 5, Part 1**. Where the equipment is to be used for work restraint then it shall prevent the user getting into a position where a fall could occur **WAH Schedule 5, Part 5**. Where used as part of a work positioning system it should include a suitable back up to prevent or arrest a fall or if this is not reasonably practicable, then all practical measures are to be taken to ensure the system does not fail **WAH Schedule 5, Part 2**. Where used as part of a rope access system it has at least two separately anchored lines. The working line is to have a safe means of ascent and descent and the safety line is to have a mobile fall protection system. A single rope may be used if the use of a second rope would increase risk and appropriate measures have been taken to ensure safety **WAH Schedule 5, Part 3**. Where used as part of a fall arrest system it shall incorporate suitable means of absorbing energy. It shall not be used where there is a risk that the line could be cut, where there is not a clear falling zone including any pendulum effect, and where its performance may be inhibited or unsafe **WAH Schedule 5, Part 4**.

Stability of structures

All practicable steps are to be taken to prevent any new or existing structure becoming a danger to any person through any part of it becoming unstable or being in a temporary state of weakness due to construction work **CDM 28(1)**. Any buttresses, temporary supports or structure must be of a design and installed and maintained to be able to carry any foreseeable loads. Such structures must only be used for the purpose for which it was designed and installed **CDM 28(2)**. No part of a structure shall be so loaded as to make it unsafe **CDM 28(3)**.

Where the means taken to prevent someone falling consist of a working platform it should be suitably rigid, have sufficient strength for the purposes for which it is being used. It should be stable, being securely attached to a supporting structure (i.e. a scaffold), where necessary to prevent an accidental displacement that might endanger anyone.

Any supporting structure must be suitable and itself rest upon a stable surface which was both a suitable composition and sufficient strength for its purpose. The supporting structure should be suitable and of sufficient strength and rigidity. In the case of a wheeled structure it is to have suitable devices to prevent it moving while persons are at height. Otherwise it should be prevented from slipping. When the supporting structure or the working platform are erected, used, dismantled, altered or modified they should always remain stable and the platform should not be accidentally displaced when being dismantled.

The working platform should be of sufficient dimensions to allow the free passage of persons and the safe use of any equipment or materials and still be big enough for the work to be done from it. The surface of the platform should not have any gaps through which someone or something could fall and injure anyone below it, and be in a condition that will prevent anyone being caught between it and any adjacent structure. The risk of anyone slipping or tripping is to be prevented and is not to be overloaded such that the working platform and supporting structure could collapse or deform so that it was no longer safe to use **WAH Schedule 3, Part 1**.

Strength and stability calculations should be prepared for scaffolds unless there is available, a note of the calculations or if the scaffold is assembled in conformity with a recognised configuration. An assembly plan should be prepared for complex scaffolds which could be a standard plan modified for a specific use. This plan is to be available for those who build, use, alter and dismantle the scaffold. While a scaffold is not available for use it is to be marked with warning signs and have physical means to prevent access to a danger zone **WAH Schedule 3, Part 2; 7–11**.

Demolition

The demolition or dismantling of a structure or part of a structure, is to be planned and carried out so as to prevent danger or where this is not practicable, to reduce it to as low a level as is reasonably practicable **CDM 29(1)**. The arrangements for demolition or dismantling shall be recorded in writing before the work is started **CDM 29(2)**.

Demolition is not to start where persons could be exposed to asbestos until a suitable and sufficient assessment has been made of whether asbestos is present, its type and its condition **CAR 5**.

Work with plant and machinery

Work equipment is to be constructed or adapted to be suitable for its purpose. Regard is to be paid to the working conditions and any additional risks to health and safety posed by using that work equipment, in selecting it **PUWER 4(1) and 4(2)**. Work equipment is to be maintained in an efficient state and working order, and in good repair. Any maintenance log is to be kept up to date **PUWER 5**. Work equipment is to be constructed or adapted so that (sfairp) maintenance can be carried out while the equipment is shut down or maintenance can be carried out without exposing those doing it to risks to their health and safety **PUWER 22**. Any appropriate health and safety markings on work equipment are to be clearly visible **PUWER 23**.

Lifting operations

Lifting equipment is to be of adequate strength and stability for each load that it carries particularly at its mounting or fixing point. Every load and any pert of it used in lifting is to have adequate strength. The lifting equipment is to be positioned in such a way as to reduce the risk (sfairp) of it or its load, striking any person, of the load drifting, falling freely or being unintentionally released **LOLER 4 and 6(1)**. Any one being carried or working from lifting equipment shall (sfairp), be free of risks from being crushed, trapped, struck or falling from the carrying position. Any person being carried should not be exposed to danger and be capable of being freed if trapped in the carrying position. Every lifting equipment for carrying persons should have suitable devices to prevent the carrying position falling in the event of a failure. If such devices cannot be used then the suspension equipment shall have an enhanced safety coefficient and should be inspected by a competent person every working day **LOLER 5**. All machinery and accessories used for lifting loads are to be clearly marked with their SWL. Where this depends on the machinery's configuration it should be clearly marked for each configuration or the necessary information showing this is to be kept with the machinery. Accessories used in lifting are to be marked to show the characteristics required for their safe use. All lifting equipment to be used for lifting persons is to be clearly marked to this effect, and any equipment not intended to lift persons but which could be mistakenly used to do so, must be appropriately and clearly marked to show this **LOLER 7**.

Diving operations

No person shall act as a diving contractor unless his details as given in the schedule have been supplied to HSE. No person shall dive unless there is only one diving contractor employing all the divers on the project or is a self-employed diver in a team of self-employed divers who has been jointed appointed by them all to act as the diving contractor **DWR 5**. The diving contractor shall ensure (sfairp) that the diving project is planned, managed and conducted which protects the health and safety of those involved in the project **DWR 6(1)**. A diving project plan is to be prepared and only one supervisor appointed before the project starts and any person taking part in the project shall complies with the regulations and works to this plan **DWR 6(2)**. The diving contractor shall ensure that there are sufficient people with suitable competence to carry out the diving project without risks to health and safety and deal with any reasonably foreseeable emergency. The diving contractor shall ensure that there is suitable and sufficient plant is available including that needed for any foreseeable emergency, and is maintained in a safe working condition. The diving contractor shall ensure that any person involved complies with the diving project plan, that a record is kept of each diving operation and this is kept in his possession for at least two years after the last entry **DWR 6(3)**.

Control of traffic

Traffic routes shall be suitable for the persons and vehicles using them, sufficient in number and to be in suitable positions and of a sufficient size **CDM 36(2)**. Every site shall be organised so that pedestrians and vehicles can move without risks to health and safety. Sufficient traffic routes of a size and position suitable for pedestrians and vehicles that they can use them without causing any health or safety risks to persons nearby. They should be suitably signed where necessary, regularly checked and properly maintained **CDM 36(4)**. No vehicle shall use a traffic route (sfairp) unless the route is free from obstructions permitting sufficient clearance **CDM 36(5)**.

Any door or gate for pedestrians that leads onto a traffic route should be sufficiently separated from that traffic route so anyone using it can see oncoming vehicles **CDM 36(3)b**. Where it is unsafe for pedestrians to use any gate intended for vehicles, separate pedestrian doors should be provided in the vicinity that are clearly marked and kept free of obstructions **CDM 36(3)e**.

Sufficient separation is to be provided between vehicles and pedestrians to ensure safety or, if this is not reasonably practicable, by other means that will protect pedestrians and provide effective arrangements for warning any person liable to be crushed by the approach of any vehicle **CDM 36(3)c**. Loading bays are to have at least one exclusive pedestrian exit point **CDM 36(3)d**.

Suitable and sufficient steps are to be taken to prevent or control unintentional movement of any vehicle **CDM 37(1)**. Where an unintentional movement by a vehicle could be a danger to a person, then suitable and sufficient steps shall be taken to warn that person by whoever is in effective control of the vehicle **CDM 37(2)**. Any vehicle shall be driven, operated or towed in a manner that is as safe as the circumstances permit, and be loaded so that it can be driven, operated or towed safely **CDM 37(3)**. Persons may only ride on vehicles in a safe place provided for that purpose **CDM 37(4)**. No person shall remain on a vehicle while it is being loaded and unloaded with loose material unless a safe place has been provided for that purpose **CDM 37(5)**.

Welfare

(The Welfare regulations sometimes mention smoking areas but these references will generally be superseded by the ban on smoking in public places.)

Suitable and sufficient sanitary conveniences are to be provided or made available in readily accessible places that (sfairp) are adequately lit, ventilated and kept clean and orderly. Separate facilities are to be provided for men and women except where the door to a room containing separate conveniences can be locked from the inside **CDM Schedule 2; 1–3**.

Suitable and sufficient washing facilities, including showers where necessary for health reasons or because of the nature of the work, are to be provided or made available in readily accessible places that (sfairp) are adequately lit, ventilated and kept clean. Washing facilities must be provided both in the immediate vicinity of every sanitary convenience and by rooms used to change into special clothing. A supply of clean hot and cold, or warm water (running—sfairp), soap or other suitable means of cleaning, and towels or other suitable means of drying, shall be provided. Separate facilities are to be provided for men and women or where these facilities are contained in a room for only one person at a time, and the door to this room can be locked from the inside, other than where provided for washing hands, forearms and face only **CDM Schedule 2; 4–9**.

An adequate supply of wholesome drinking water shall be provided at readily accessible and suitable places, marked by a sign where necessary. A sufficient number of suitable cups or drinking vessels are to be provided unless the water is available in the form of a jet. Where necessary for health and safety reasons every supply of drinking water shall be marked with a conspicuous notice (sfairp) **CDM Schedule 2; 11–13**.

Suitable and sufficient changing rooms are to be provided or made available at accessible places if workers have to wear special clothing and cannot be expected to change elsewhere. Separate rooms are to be provided for men and women and are to seating and where necessary, facilities for drying any special clothes, personal clothing and effects. Suitable and sufficient facilities shall be provided where necessary, to enable a person to lock away any special clothing, their own clothes not worn during working hours, and their personal effects **CDM Schedule 2; 14**.

Suitable and sufficient rest facilities are to be provided or made available in readily accessible places that (sfairp) have arrangements to protect none smokers from tobacco smoke, be equipped with an adequate number of tables and adequate seating with back rests, for the number of persons likely to use them at one time. They are to have suitable arrangements for preparing food, have means for boiling water and be maintained at an appropriate temperature. Where necessary suitable rest facilities are required for pregnant or nursing mothers to rest lying down **CDM Schedule 2; 15**.

Employees shall not eat, drink or smoke in any place which may be contaminated with lead although non-contaminated drinking facilities shall be provided where these are required for the welfare of the employees **CLW 7**. Employees shall not eat, drink or smoke in an area designated as an asbestos area or respirator zone and arrangements are to be made for such employees to eat and drink in some other place **CAR 18(5)**.

Employees and the self-employed shall be provided with adequate first aid equipment and facilities that are appropriate to the circumstances. Such first aid will be intended preserve to life and minimise the consequence of injury or ill health until a medical practitioner or nurse takes charge, and give treatment of minor injuries where no medical practitioner or nurse will attend. This will include either providing a person trained in first aid, or appointing someone to take charge of the first aid equipment if this would be adequate **HS(FA) 2–4**.

(5) Training

Levels of training required	Where training, technical knowledge or experience is required to reduce the risks of injury to any person, then as may be appropriate, either those doing the work shall have these skills or they should be under the necessary degree of supervision of a person who does have them.	C(HSW)R Reg 28
	Adequate and appropriate training, comprehensive instruction and information shall be given where PPE is provided. This should deal with the risks that the PPE will avoid, the purpose and manner in which it is to be used and on any actions required to keep it in an efficient state, working order and good repair.	PPE Reg 9
	Suitable and sufficient information, instructions and training are to be given to employees and anyone carrying out work in connection with an employer's duties on the risks to their health where they may be exposed to hazardous substances. This is to include the risks created by the exposure and precautions to be taken. The information provided shall also include any results of monitoring and health surveillance undertaken.	COSHH Reg 12

(21) Work with plant and machinery

Doors and gates	Any permanent or temporary door, gate or hatch, that does not form part of any mobile plant, shall where necessary, be fitted with suitable safety devices to prevent the risk of injury to any person.	C(HSW)R Reg 16
	To comply with this requirement any sliding door etc, shall be prevented from coming off its track. Any upward opening door etc, shall be prevented from falling back. Any powered door etc, shall have suitable and effective features to prevent it trapping anyone and either be manually operable or to automatically open if the power fails.	
Plant and equipment.	All plant and equipment shall (sfairp), not cause risks to health or safety, be of good construction, be of suitable and sound materials, be of sufficient strength and be suitable for their purpose. All plant and equipment shall be used and maintained (sfairp),that it will not cause risks to the health or safety of anyone at all times.	C(HSW)R Reg 27

19.6   The various documents required by law

This section gives information on the various documents that are required by law. These include:

(1) Health and Safety Policy
(2) Formalised Risk assessments
(3) Health and Safety arrangements (Method statements etc.)
(4) Construction phase plan
(5) Health and Safety file
(6) Notifications to HSE and other enforcing authorities
(7) Notification of injuries, diseases and dangerous occurrences
(8) Site rules
(9) Other reports, records and certifications etc.

Within the Act and regulations there are numerous requirements for various documents. The requirement for some of these documents is based only on one piece of legislation, such as that for a Health and Safety policy, while in other cases, such as for reports and records, the requirement for these occur in various regulations etc. However, as can be seen in this brief summary, there is a broad similarity in these cases between what these types of documents are for and what they are required to contain.

Health and Safety Policy

Employers are to prepare and revise when necessary a written statement of their policy on health and safety including the organisation and arrangements for bringing this policy into force. The statement and any revisions to it are to be brought to the attention of his employees **(HSWA Section 2(3))**.

Formalised risk assessments

Where an employer employs five or more persons he is to record the significant findings of a suitable and sufficient assessment of the health and safety risks to his employees and others who may be affected by his work **MHSW Reg 3(1 & 6) and similarly RR(FS)O Article 9, CLWR Reg 5(4), COSHH Reg 6(4) and NWR Reg 5**. In these latter regulations an assessment of the risk from noise at or above a lower exposure limit and the measures to control the risks will be based on the working practices, relevant information on the probability of noise and, if necessary, measurements of the noise level. In all these regulations such assessments are to be reviewed when it is suspected that they are no longer valid or where there have been significant changes in the work situation. The record shall also identify any group of workers especially at risk. In addition records of any noise assessment are to be kept until a new assessment is carried out **NWR Reg 5(4)**.

Health and Safety arrangements (Method statements etc.)

An employer employing five or more persons is to record his arrangements for the effective planning, organisation, control, monitoring and review of the prevention and protective measures identified by the risk assessment **(MHSW Reg 5(2))**,

COSHH Reg 6(4) etc. As above there is also a requirement where five or more persons are employed for recording the necessary steps to guard against the identified significant risks. This is set out in slightly different formats for instance the responsible person is to prepare appropriate fire safety arrangements recording theses or where a licence is in force etc. **RR(FS)O Article 11**. A diving contractor is to prepare a Diving Project Plan before diving starts. The Diving Project Plan is to be based on a risk assessment and contain information and instructions as necessary to deal with the risks. It should identify each diving operation and the emergency and support services. **DWR Reg 9**.

The client is to provide every designer and contractor all the information he possesses or can reasonably obtain to assist them in carrying out their duties under the regulations and to determine the resources they will require to manage a project. This could include information about the site, the construction work, the proposed working use of the proposed structure, the minimum time before construction work starts and any existing health and safety file relevant to the construction work or the eventual use of the structure **(CDM Reg 10)**.

A written plan of work is to be prepared before starting any work with asbestos and a copy of this plan is to be kept at the place where the work is to be undertaken. Included amongst others in this plan are to be details of the location of the work, methods for handling the asbestos, means of preventing or reducing the exposure to asbestos and the means of cleaning the place where the work is being done **(CAR Reg 7)**.

Construction phase plan

The principle contractor shall prepare a construction phase plan where a project is notifiable, to ensure that the construction phase of the work is planned, managed and monitored so that as far as is reasonably practicable the construction work can be started without risks to health and safety. The plan is to include information concerning health and safety matters identified by the designers and the CDM co-ordinator and is to be reviewed, revised and refined as appropriate throughout the life of the project **(CDM Reg 23)**.

Health and Safety file

The CDM coordinator is to prepare a record of any information that may be needed during any future construction work, to ensure the health and safety of any person. Where such a record already exists then this is to be reviewed and updated. The record is to contain any relevant information supplied by the client, the designers and the principle contractor under these regulations. At the end of the construction phase the record is to be given to the client **(CDM Regulation 20(2))**.

Site rules

The principle contractor shall facilitate the preparation of rules where necessary for health and safety that are appropriate to site and the activities that will be carried out there as part of his duties to plan, manage and monitor the construction phase of a project **(CDM Regulation 22(1d))**.

Notifications to HSE and other enforcing authorities

The CDM co-ordinator shall ensure that notice of the project is given to the Health and Safety Executive as soon as practicable after his appointment **(CDM Reg 21(1))**. The particulars required include the site address, the name of the relevant local authority, a brief description of the project, the contact details for the client, CDM co-ordinator and principle contractor, the planned start date for and duration of the construction phase, an estimated maximum number of persons who will be employed on the site and planned number of contractors **(CDM Schedule 1)**. Where any of these particulars are not known because the principle contractor has not been appointed, they will be given as soon as practicable after he has been appointed and before any construction work starts **(CDM 21(2))**.

Note: a project is notifiable if the construction phase is likely to require more than 30 days work or involve more than 500 person days working on the project **(CDM Reg 2(3))**.

An employer shall not start any work involving asbestos until he has given at least 14 days notice to the appropriate enforcing authority, or a shorter period agreed with the authority, of the particulars of the work. Where the particulars of the work change then the employer is to inform the authority of that change in writing **(CAR Reg 9)**. Included in the particulars contact details of the employer, a brief description of the site, type and quantity of asbestos, the number of workers involved and the measures to be taken to limit exposure to the asbestos **(CAR Schedule 1)**.

Where ionising radiation operations are planned the work is not to start generally before the enforcing authority has been given at least 28 days notice. **IRR Reg 6(2)**.

Notification of injuries, diseases and dangerous occurrences

The person in control of the work is to notify the relevant enforcing authority where someone dies, suffers a major injury as a result of work or where there is a dangerous occurrence, whether the persons involved are employed with that work or not, by the quickest practicable means. A report of the incident is then to be forwarded to the authority within 10 days **(RIDDOR Reg 3(1))**. Where a person is absent from work for three consecutive days because of an incident at work, not including the day of the incident but including nine working days, the person in control of the works is to send a report of the incident as soon as is practicable but also within 10 days **(RIDDOR Reg 3(2))**.

Where an employee suffers an injury at work and dies within one year caused by that injury, his employer shall inform the relevant enforcing authority in writing of the death as soon as he is aware of it whether the accident was originally reported or not **(RIDDOR Reg 4)**.

Where a person suffers from an occupational disease specified in the regulations and his work involves one of the activities also listed, the person in charge of the site shall immediately send a report to the relevant enforcing authority **(RIDDOR Reg 5)**.

Records are to be kept by the person in charge of a site of any event that required him to report an incident or disease under the regulations. This record is to be kept for at least three years after the incident at either the site or the registered offices of the company **(RIDDOR Reg 7)**.

Other reports, records or certifications etc.

Where a person who has carried out an inspection of an excavation or coffer dam and is not satisfied that work can be safely done there, should not only inform the person for whom the inspection was done of his concerns but also prepare a report including the details given in **Schedule 3** to the regulations and submit this within 24 hours of the inspection **(CDM Reg 33)**. The particulars required in **Schedule 3** include details of the person on whose behalf the inspection was carried out, details of the place of work, details of those matters which could cause a risk to anyone, details of action taken and matters outstanding and the date, time and name of the person carrying out the inspection.

There is a similar requirement where a person carries out a thorough examination of lifting equipment **LOLER Reg 10(1)** this time including the details listed in **Schedule 1** of the regulations, in his report. The report is to be sent to the employer and the person from whom the equipment has been hired where applicable. In addition he the inspector considers that there is an imminent risk of injury a further copy of the report is to be sent to the enforcing authority. The employer is to keep the report available for inspection until he ceases to use the equipment, or in the case of an accessory for lifting, for two years. **LOLER Reg 11(2)(a)**. The details required in **Schedule 1** include name and address of the employer, address where the thorough examination was carried out, sufficient details to be able to identify the equipment (including date of manufacture where known), date of the last thorough examination, the safe working load, whether the thorough examination was carried out 6 or 12 months following the previous one or after exceptional circumstances, identification of any part having a defect and particulars of any repairs required or if there is not a danger an estimate of the length of time before there could be danger, the latest date for the next thorough examination or testing, and the date, name, address and qualification of the person making the report. The employer is to keep available the reports on his lifting equipment until he ceases to use it. Reports an accessories for lifting are to be kept for at least 2 years as are reports on lifting equipment where an inspection has been carried out concerning the possible deterioration of a part. **LOLER Reg 11 (2a)**.

Where a person has carried out an inspection of a working platform from which a person could fall 2 m or more he shall prepare a report containing the particulars given **Schedule 7** to the regulations, before the end of the working period and provide the report or a copy of it within 24 hours to the person on whose behalf the inspection was carried out **(WAH Reg 12(7))**. The details required in **Schedule 7** include the name and address of the person on whose behalf the inspection was carried out, location of the work equipment, a description of the work equipment, date and time of the inspection, details of those matters which could cause a risk to anyone, details of action required to remedy these, details of any further action and the name and address of the person carrying out the inspection.

A record of the results of inspections of lifting equipment carried out in between the thorough examinations should be sent to the employer as soon as practical after the inspection noting any defect. **LOLER Reg 10(2)**. These records are to be kept until the next one is made. **LOLER Reg 11 (2b)**. Employers are to keep suitable records of the examinations and tests of any exhaust ventilation equipment or respiratory protective equipment **CAR Reg 13(3), CLWR Reg 8(4)** and **COSHH Reg 9(4)**.

These records are generally to be kept for 5 years although in the case of **COSHH** the control measures are to be kept for 5 years. Other records required include that of the lead in air concentrations, kept for 5 years **CLWR Reg 9(4)**, and employers are to keep health records for each employee for at least 40 years after the last entry **CLWR Reg 10(5)**. A diver is to keep a daily record of his dives **DWE Reg 14(2)**. Suitable records of the monitoring of the exposure to hazardous substances is to be kept available for 40 years where the record is representative of the personal exposure of identifiable employees, or for 5 years in any other case. **COSHH Reg 10(5)**. Where workers are involved in or could be affected by ionising radiation procedures health records for each relevant employee is to be kept until the employee is 75 years old and/or for 50 years after the last entry in the record. **IRR 24(3)**. Records of the examination of respiratory protective equipment, noting its condition, are to be kept for at least 2 years. **IRR 10(2)**. The employer is to ensure that a health record is kept of each employee who undergoes health surveillance as a result of possible vibration injury. This record is to be kept available in a suitable form, the employee allowed access to this and copies sent to the enforcing authorities as required. **CVWR Reg 7(3–4)**.

Medical certificates in accordance with Reg 15 of the regulations are required confirming a person's fitness before being allowed to dive **DWR Reg 13(1)**.

Where appropriate measures, methods or procedures cannot adequately reduce risks safety signs warning of the risks and measures to be taken are to be positioned **HS(SS)R Reg 4**. Warning notices are required where a roof may contain fragile surfaces **WAHR Reg 9(3a)**.

It is the duty of any person who designs, manufactures, imports or supplies any article for work, to provide adequate information about the use and conditions to ensure it will be safe. **HSWA Section 6**.

Every employer shall ensure that work equipment incorporates appropriate unambiguous warnings or warning devices **PUWER Reg 24**. Where appropriate every employer is to prepare written instructions on the safe use of work equipment for employees and others who may use that equipment PUWER 8 (1).

Further reading

The following list is some of the guidance notes published by the HSE explaining the legal requirements in more detail. These are both examples of the extensive range of guidance published and a reference to that which might be most useful. Information may also be obtained from the HSE website at www.open.gov.uk/hse/hsehome.htm.

Management of Health and Safety at Work—Management of Health and Safety at Work Regulations. Approved Code of Practice and Guidance: Series No. L21, 1992.
Note: Deals with such matters as risk assessments, principles of prevention and information that should be provided for employees.

Managing Health and Safety in Construction—Construction (Design and Management) Regulations (CDM). Approved Code of Practice and Guidance: Series No. L144, 2007.
Note: An essential guide to these fundamental construction regulations.

Safe Use of Work Equipment—Provision and Use of Work Equipment Regulations. Approved Code of Practice and Guidance, 3rd ed.: Series No. L22, 1998.

Note: This deals with, amongst other things, the suitability of the work equipment, its maintenance and inspection; training, protection against specific hazards, markings and warnings.

Safe Use of Lifting Equipment—Lifting Operations and Lifting Equipment Regulations. Approved Code of Practice and Guidance: Series No. L113, 1998.

Note: General guidance including that on ropes and pulleys, scissor lifts and loader cranes.

Safe Work in Confined Spaces—Confined Spaces Regulations. Approved Code of Practice and Guidance: Series No. L101, 1997.

Note: Deals with such matters as rescue and resuscitation, fire safety.

Construction (Head Protection) Regulations. Guidance on Regulations, 2nd ed.: Series No. L102, 1989.

Note: Deals with head injuries, safety helmets.

The Management of Asbestos in Non-Domestic Premises—Regulation 4 of the Control of Asbestos at Work Regulations. Approved Code of Practice and Guidance: Series No. L127, 2006.

Note: Advises on how to manage the risks and protect workers and others who may come across asbestos in their day-to-day activities.

A Guide to the Reporting of Injuries, Diseases and Dangerous Occurrences Regulations: Series No. L73, 1995.

Note: Summarises the main requirements of the regulations and gives detailed guidance notes.

Health and Safety in Construction, 3rd ed.: Series No. HSG 150, 2006.

Note: Updated with the Work at Height Regulations incorporating the latest advances and provides examples of good practice.

Health and Safety in Roof Work, 3rd ed.: Series No. HSG 33, 2008.

Note: Guidance for clients, designers etc. as well as for those carrying out all types of roof work.

Protecting the Public. Your next Move, 2nd ed.: Series No. 151, 1997.

Backs for the Future—Safe Manual Handling in Construction: Series No. HSG 149, 2000.

Note: Basic principles for dealing with manual handling risks and ideas for their solution.

Avoiding Danger from Underground Service: Series No. HSG 47, 2000.

Notes: Outlines the dangers from underground services and gives advice on how to reduce the risks.

Essentials of Health and Safety at Work, 4th ed.: 2006.

Notes: Explains what the law requires and how to put it into practice with chapters on slips and trips, work at height and contractors and agency workers.

Manual Handling—Solutions You Can Handle: Series No. HSG 115, 1994.

Notes: Guidance on how to avoid manual handling using lifting devices and trolleys, and how to reduce the risk of injury by considering the weight, reach and twisting of any manual lifting.

Fire Safety in Construction Work: Series No. HSG 168, 1997.

Notes: Gives advice for properties with substantial fire risks. Deals with rubbish disposal, protective coverings, LPG etc. storage, oxyfuel equipment, electricity, bonfires and arson. It also deals with escape routes, fire alarms and emergency signs.

Electrical Safety on Construction Sites: Series No. HSG 141, 1995.

Notes: Advice on precautions to reduce accidents.

A Step by Step Guide to COSHH Assessment: Series No. HSG 97, 2004.

Notes: General advice to employers on assessing their activities as required under the COSHH Regulations 2002.

Safety in the Use of Abrasive Wheels: Series No. HSG 17, 2000.

Notes: Deals with training, mounting and use of abrasive wheels and personal protective equipment.

Health and Safety in Excavations: Series No. HSG 185, 1999.
Notes: Guidance for everyone involved in excavation work including clients and designers. Deals with pipe and cable laying, foundations, retaining walls, underpinning and trench less technology.
Successful Health and Safety Management: Series No. HSG 65, 1997.
Notes: Guidance for managers, health and safety professionals and employee's representatives who want to improve health and safety.
Safe Use of Vehicles on Construction Sites: Series No. HSG 144, 1998.
Notes: Guidance on preventing accidents by providing vehicle and pedestrian routes and public protection. Gives advice on vehicle selection.
The following publications by the HSE are all free when obtained as single copies.
Construction Site Transport Safety—Safe Use of Site Dumpers: Series No. CIS 52, 2006.
Note: Gives advice for managers and drivers on precautions when using forward tipping dumpers etc.
The Work at Height Regulations 2005 (amended) —A Brief Guide: Series No. INDG 401(rev 1).
Notes: Summarises how to comply with the regulations. Written so it can be used alone in most cases without having the regulations themselves.
Five Steps to Risk Assessment: Series No. INDG 163(rev 2).
A Short Guide to the Personal Protective Equipment at Work Regulations: Series No. INDG174, 1992.
Safe Work in Confined Spaces: INDG 258(rev 7).
Notes: Simple and practical guidance for anyone involved with working in confined spaces.
Safe Use of Ladders and Stepladders—An Employer's Guide. Series No. INDG 402.
Notes: When to use, selection of, using, maintenance and safety precautions.
Simple Guide to the Lifting Operations and Lifting Equipment Regulations 1998: Series No. INDG 290.
Notes: Summarises the main requirements of the regulations and approved code of practice.
Managing Health and Safety—Five Steps to Success: Series No. INDG 75.
Notes: Planning for health and safety, accident and incident investigation, monitoring, learning from experience. A tried and tested approach to managing health and safety.
Tower Scaffolds: Series No. CIS 10 (rev 4).
Notes: For users of mobile access towers and those who select and specify this equipment.
The Selection and Management of Mobile Elevating Work Platforms: Series No. CIS 58, 2008.
Notes: Advice on how to select and manage the use of this equipment.
Safe Erection, Use and Dismantling of False Work: Series No. CIS 56, 2003.
Notes: Gives advice on the management, design, planning and construction of these structures.
Inspection and Reports: Series No. CIS 47 (rev 1).
Notes: Gives advice on the inspection of excavations, coffer dams and caissons, existing places of work, work platforms (i.e. scaffolds), collective fall arrest systems, personal fall protection systems, ladders and step ladders; dealing with timing and frequency of checking. Suggested layout of reports.

Other related texts

Successful health and safety management: HSE, Sudbury UK, HSE Books, 1997.
Health and Safety statistics: HSE, Sudbury UK, HSE Books, 2008. [Available as a free download from the HSE website.]
Commercial diving projects offshore—Approved Code of Practice: HSE, Sudbury UK, HSE Books, 1998.

The operation and maintenance of bridge access gantries and runways, 2nd ed.: IStructE, London, IStructE, 2007.

Guide to inspection of underwater structures: CIRIA, London, 2001.

Experiences of CDM: W.S. Atkins, London, CIRIA, 1997.

Site safety handbook, 4th ed.: S. C. Beilby, London, CIRIA, 2008.

Construction—Health and Safety: CIOB, Ascot, UK, CIOB, 2008.

Refurbishing old buildings—management of risk under the CDM regulations: B. Nutt *et al.*, London, Telford, 1998.

Construction Safety Handbook, 2nd ed.: V. J. Davies and K. Tomasin, London, Telford, 1996.

Health and Safety in construction—guidance for construction professionals: J. Barber, London, Telford, 2002.

The role of safety in total quality management: A. Deacon, Safety & Health Practitioner, IOSH 1994.

Ensuring the safety of scaffolding: R. Johnson, Safety & Health Practitioner, IOSH: Vol. 16, No. 1, 1998.

Guidance on the use on rope access methods for industrial purposes: IRATA, Aldershot, IRATA, 2000.

Guidance note on safe working on fragile roofs—for inspection, maintenance, repair refurbishment and planning: ACR (Advisory Committee for Roofwork), London, ACR, 2002.

Chapter 20   OCCUPATIONAL HEALTH IN CONSTRUCTION

Francesca Machen

Basic requirements for best practice

- Fit, healthy and well trained operatives
- Good quality and well maintained equipment

This chapter will

- Provide an introduction to occupational health in the construction industry
- Outline key legislation and key health risk problems affecting employees
- Describe the industry-led occupational health scheme and standards for fitness for work for employees
- Describe the principles of effective health risk management
- Outline the role of the occupational health specialist

20.1   Introduction

Occupational health is a specific branch of medicine concerned with the impact of work processes and hazards on employee's health, and the effect of the individual's health on their ability to do their job. In this context, occupational health is concerned with health risk management.

The extent of occupational ill health in the construction industry and the challenges faced by construction companies in attempting to protect the health of their employees are well documented. At the end of their working lives the health of construction workers is worse than most groups and much of this is due to work related health problems. Construction workers are among the least likely of any occupational group to visit a GP. There are, therefore, strong arguments for introducing measures which are of benefit to individuals, companies and the industry as a whole.

Current limited evidence would appear to show that one third of construction workers have some health problem likely to impact on their ability to do their job. This might be an undetected health problem such as high blood pressure or a work related industrial disease, for example, noise-induced hearing loss or hand-arm vibration syndrome. According to the regulatory body, the Health and Safety Executive (HSE), the most common work related problems are back pain, skin conditions, breathing problems, and health effects caused by exposure to noise and vibration. The annual cost to the industry was estimated in 2008 by the HSE to be £760 million.

The legislative framework refers to the management of health and safety and Chapter 19 outlines the legal aspects of this subject, so this need not be covered in any detail in this chapter. Safety processes in the main are well understood. However, the management of health issues as they relate to groups of workers and individuals remains largely reactive. The reason for this may be that effects of ill health relating to a specific hazard are usually detected over time, so the impact is not felt until some time later. Clearly this is not always the case and an individual employee may suffer health effects relating to a substance or chemical with very immediate effects as in the case of skin conditions or lung problems. Taking this evidence into account, overall it would seem that the construction industry will need to go beyond compliance with health and safety legislation if significant improvements in the health of the workforce are to be achieved. Currently, however, the industry is faced with confusing information about how best to manage health risks.

20.2 Constructing better health—the industry-led scheme

The last couple of years have, however, seen significant progress in the construction industry's approach to health risk management and the advice available to enable employers to set up appropriate arrangements to protect the health of their employees.

To address occupational health problems in the industry The Constructing Better Health scheme (CBH) was set up in 2004. Following a successful pilot project in 2006 it has grown from an initial proposal for a national occupational health strategy into an established industry-led scheme, involving major contractors, the unions and occupational health professionals. In 2007 CBH published the *Occupational Health Standards for the UK Construction Industry*, Part 1 *Fitness To Work Standards* and Part 2 *Standards for Occupational Health Service Providers within the Construction Industry*. So that in addition to health and fitness standards for workers, a key proposal has been the setting up of a register for occupational health providers or approved suppliers. Providers have to demonstrate their relevant competencies in terms of qualifications and experience. There are now specific recommendations from both CBH and the regulatory body, the HSE, about the qualifications of those delivering services to the construction sector and this information is available on both websites (see www.constructingbetterhealth.com and www.hse.gov.uk).

CBH's broad objective is to improve the health of 2.2 million construction workers in the UK. Occupational health specialists working in the industry support the standards which reflect a very safety-based approach to health management. From the OH specialist's viewpoint the standards are all about addressing health inequalities. Improving health inequalities in the workplace means using a range of methods designed to obtain information and plan services aimed at improving the health of employees by effectively targeting resources and working collaboratively with all involved to achieve change. This means considering the demographic of this working group, as well as social or economic factors as they relate to ill health. This might include individual behaviour, such as smoking, and assessing medical conditions that impact on work.

In the construction industry, health inequalities will predominately affect men's health, so attempts to improve the health of this group will need to consider the current knowledge and evidence base. For example, disproportionately, less NHS funding is spent on men's health care. Men simply do not get or ask for as much of the allocated share as women. Many OH practitioners feel that if the current work related health issues in construction affected more women, much more would have already been achieved in terms of setting up preventive clinics, and even research into effective treatment.

Occupational medicine and work related health management is of course an important element of public health, so CBH is also concerned with the development of health practice and services which are delivered appropriately to those who need them, in such a way that they are encouraged to use them.

20.3   Occupational health service providers

Some construction employers have in-house occupational health teams, which might include an on site service to deliver first aid and a number of health assessments to comply with health and safety legislation and hazard exposure, in addition to general health screening for workers. There is a distinct difference here, which will be explained further in this chapter. The in-house team might consist of an occupational health nurse or general nurses, technicians and an occupational physician. The industry scheme CBH and the regulatory body, the HSE, are concerned about the number of services both in house and from outsourced providers which are being delivered by unqualified staff. However, this does not appear to be a concern to some construction companies who are happy with providing their staff with health assessments which deliver advice on general health problems such as high blood pressure and cholesterol testing, but do not help the employer comply with health and safety legislation. Such services have their place, as has already been noted, in terms of improving the health of the workforce and referring employees to their GP when a health problem is detected. In reality, however, those services which add real value are those which assist the employer to comply with the regulatory framework and health risk management. Most occupational health services are in fact delivered by outsourced providers. In practical terms, services are usually delivered on site with the use of mobile equipment in a custom-built vehicle, or the construction client provides facilities on site for use by the OH provider. Employees are seen for health assessment for an agreed number of clinical tests. For the major key health risks under discussion here, this could include, a lung function test, a hearing test, skin check and assessment and tests for those working with vibration. Services are either delivered by an OH practitioner working on site for an agreed number of days per week or as part of a managed project, over an agreed period, until all employees have been seen.

Based on current industry opinion, providers need to develop closer working relationships with their clients to understand the range of work processes, and should be involved at the earliest stage possible to give advice on suitable health management strategies, where risks to health are identified. Construction companies aim to manage the safety of their employees and contractors in such a way that this integrates with work processes. The same integrated and strategic approach should apply to managing the risks associated with health hazards, and those OH providers with experience of the industry will be able to assist with this process.

20.4 The regulatory framework

The Construction Occupational Health Management Essentials (COHME) section of the HSE website (see www.hse.gov.uk/construction/healthrisks) sets out the current regulatory framework for managing health risks in construction. The duty to manage health risks, like safety plans and arrangements is covered by Construction (Design and Management) Regulations (CDM) 2007. The regulations clearly set out roles and responsibilities for identifying, assessing and reducing health risks. These should be considered at the design stage rather than attempting to deal with risk reduction at the construction stage of a project. However, it is unclear from the current evidence whether this actually happens, in the same way as designing out safety risks with better engineering controls or improving on work processes. More broadly, the Management of Health and Safety at Work Regulations, Section 6 defines the duty to undertake risk assessments and to consider the hierarchy of controls measures in order to eliminate and reduce health risks. Regulation 6 states: 'Every employer shall ensure that his employees are provided with such health surveillance as is appropriate having regard to the risks to their health and safety are identified by the assessment'. In addition, further legislation or statute law deals with the key health risk areas. It is not useful to understanding health risks to explain the detail of the regulations in terms of assessing exposure to the hazard and this is not the role of an OH specialist, but may involve their skill in terms of understanding the health effects and exposure, and appropriate statutory health assessments at the design stage.

The four key health risk problems, as stated, are:

- back pain
- skin conditions
- lung function or breathing problems
- health problems relating to exposure to noise and vibration.

More recently, stress has been added to the list of construction health problems. Health hazards are therefore categorised as:

- physical
- chemical
- psycho-social
- biological

20.5   Key legislation

Construction (Design and Management) Regulations 2007 CDM coordinators have a key role in advising and assisting clients in respect of health risk management matters. They should manage the flow of information between other duty holders and have an understanding of the duties of others involved in the project.

Clients can remove or reduce risks to health; designers can identify and eliminate hazards, and reduce remaining risks; principal contractors can plan and implement a strategy to manage occupational health risks; and contractors can manage any occupational health risks that their workers may be exposed to.

* Management of Health and Safety at Work Regulations 1999.
* The Control of Noise at Work Regulations 2005.
* The Control of Vibration at Work Regulations 2005.
* The Manual Handling Regulations 1992.
* The Control of Substances Hazardous to Health (COSHH) Regulations 2002.
* The Reporting of Injuries, Diseases and Dangerous Occurrences Regulations (RIDDOR) 1995.

Failure to comply with these legal standards could result not only in criminal prosecution, but civil action claims against employers where an employee's health has been affected by their work. Prescribed industrial diseases, such as noise-induced hearing loss, occupational asthma, skin conditions such as dermatitis and hand-arm vibration syndrome are all reportable to the enforcement agency the HSE, under RIDDOR (1995). Employees have a limited time from the diagnosis of the disease to claim for the injury. Equally, once employers have been advised of the condition or health effect, which relates to an identified hazard, they should take all reasonable action to act on any health advice available, or further damage to the employee's health will be entirely foreseeable.

20.6   Health Risk Management

The principles for occupational health risk management are the same regardless of the sector or industry. However, striking the right balance between legislative requirements and being a responsible employer consumes valuable management effort, particularly in the Human Resource (HR) and safety departments. It is therefore essential that the structures that support health related activities are robust and include a framework into which any health management resource can be effectively integrated.

Just as organisations develop safety plans taking into account each aspect of a project or work process and who will be affected, developing a health risk management plan may allow for proper consideration of the range of health risks at the design stage of a project. Some organisations may collect information, which would enable them to devise a matrix of job profiles and related hazards. Safety-based activities and risk assessments may have identified those employees who need to be included in any statutory health assessment programme. The important issue here is that someone needs to gather this information—this may be a shared activity between the HR department, the safety team or operational managers, and the

OH Specialist. The revised CDM (2007) approach may mean that such activities are undertaken at an earlier stage. Equally, consultation with employees provides a lot of information about precisely how their job is undertaken and the processes involved.

20.7 An occupational health policy

While there is no legal requirement for a policy, this sets out roles and responsibilities and is seen as a statement of intent by operational managers, employees and their representatives. The policy should outline the process for establishing which employees should undergo statutory health assessment, based on risk assessment and safety-based activities. Equally, it should set out how occupational health advice will be implemented and detail how control measures will be reassessed in the light of occupational health advice, improved upon and how risk will be reduced for individual employees.

It will also need to outline the process for management of employees for whom an abnormal health result is obtained. The issues arising from statutory health assessments can be complex, and there may be a need to consider redeployment or dismissal on grounds of ill health for a very few employees, where the advice is that they are unfit to continue working. These are all issues which should involve a human resource professional, but are made all the more difficult for organisations who do not have access to such advice, which is common in the construction industry.

The policy might also cover those involved in the management of employee health. This will be a director of the company who has responsibility to implement the policy and gain support for its objectives among other board members. The policy might define the role of the **'responsible person'**—this might be a safety representative or line manager, who may have some duties within any statutory health assessment programme, such as undertaking to hand out simple health questionnaires to employees, and reporting under RIDDOR.

The **qualified person** is the specialist occupational health practitioner, or other appropriately qualified health professional whose role it is to undertake clinical assessment and a range of objective tests. For the major key health risks under discussion here, this could include, a lung function test, a hearing test, skin check and assessment, and tests for those working with vibration.

20.8 Aims and objectives

Statutory health assessments are concerned with the effects of specific hazards and work processes on employee health collectively and individually, and examine their impact on fitness to work. This is about the management of retained risk and compliance with health and safety legislation, including any implied or residual risk. There is limited evidence that they are effective only if part of an overall strategy on health risk management, which includes engineering and other control measures designed to reduce risks to workers.

Clear health parameters are known for most statutory health assessments. These include clinical and physiological assessment, and the use of some objective tests. The purpose is to advise the employer of the ongoing or continued fitness for exposure of that employee to the specific hazard, for example noise or vibration. While it may be necessary to obtain more in-depth health information for some groups of workers, relating to their fitness to carry out a specific task there is a clear distinction made between statutory, i.e. compliance based health assessments, and voluntary health assessments. Voluntary health assessments cover a range of health interventions from general health assessments to lifestyle health screening or targeted health promotion activities. Such assessments are often offered by employers as a 'nice to have benefit', for example cholesterol testing, or lifestyle advice on weight and blood pressure checks. When planning any health assessment programme, it is important to be clear on the aims and objectives. For example, is the plan to collect baseline health information on all employees likely to be exposed to health hazards identified at the design stage of a new project? Alternatively, is the programme to recheck the fitness for working with hazards on an ongoing basis for employees who have already undergone health assessment?

20.9 Are occupational health specialists needed?

The HSE initiative Construction Occupational Health Management Essentials (COHME) website gives a definition of 'health surveillance'. This activity referred to in the regulations actually means undertaking statutory health assessments for identified workplace hazards as opposed to voluntary health screening. Health surveillance 'is about systematic, regular checks on workers to identify early signs of ill health, and then acting on the results.' The HSE guidance on statutory health assessments (health surveillance) sets out a tiered approach to gathering health information about employee health and exposure to identified hazards. The idea being that having assessed all the health risks to their employees and found that some form of statutory health assessment is required, the company begins to collect information about health effects and symptoms. This process could be carried out by an operational manager or a trained 'responsible person' in the format of a simple health questionnaire given to each employee. If symptoms or health effects are reported which are likely to relate to the identified hazards then these are referred to an appropriately qualified OH practitioner. The guidance suggests that this is a cost effective way to start a statutory health assessment programme, but that involving an occupational health provider who can help with the design of the programme is another option.

20.10 Minimum standards for health assessments

CBH has published 21 standards (A–U) covering statutory health assessments and how these should be undertaken, as well as a range of general health tests and health indicators. The scheme includes a matrix, which can be used to identify the health assessments required for an individual undertaking a particular job.

Health assessment matrix—21 standards

A Baseline health assessment to assess fitness before exposure/pre-employment
B Blood pressure cardiovascular health
C Routine urine test
D Visual acuity
E Visual acuity-intermediate –DSE
F Colour perception
G Respiratory Health
H Hearing
I Hand-arm vibration syndrome
J Skin assessments
K Biological monitoring
L Chest x-rays
M Musculoskeletal assessment
N Appointed physician/ACOPs
O Stress
P Drug and alcohol testing
Q Fitness for work assessments for Safety Critical Workers
R Information training and instruction
S Cholesterol testing
T Lifestyle health promotion
U Body Mass Index

Statutory health assessments—underpinned by risk assessment or safety-based activities which identify the need for such assessments to be undertaken. Employers have a legal duty to provide these assessments and employees have to attend. It is recommended that some assessments are carried out before the person is exposed to health hazard at pre-employment and on an ongoing basis as advised by a registered occupational health specialist.

The information gathered must then be reported to management about the fitness of an individual to carry out a task when starting the job or during ongoing exposure to the identified hazard. The standards under this heading are: A, E, G, H, I, J, K, L, R.

Strongly recommended health assessments—a legal duty may exist based on individual/collective risk assessment. For example, this might include duty to undertake musculoskeletal assessment due to levels of physical fitness required and known risk in undertaking aspects of a job. The following standards could come under this heading: A, B, C, D, F, H, I, J, M, O, P.

Recommended health assessments—intended to give information to the employee regarding general health status but where there is no legal requirement. Only anonymised information is fed back to management. The following standards could come under this heading: B, C, D, O, S, T, U.

Safety critical workers—include all mobile plant operators, high-speed road workers, rail trackside workers, and others as identified by risk assessment. The following standards apply, in addition to other required statutory assessments: A, B, D, F, G, H, M, P, R.

20.11   The occupational health specialist's perspective

In the sense that occupational health is about the relationship between the effects of work processes and their impact on health, it has already been noted that it is about management of retained risk. This cannot be achieved in any meaningful way without establishing a baseline statutory health assessment for each employee, and this section explains why this is important.

Obtaining baseline health information, includes establishing whether or not there are signs and symptoms which might relate to a range of identified hazards, but also whether the employee's occupational history or current health status, places them at higher risk, which means action on control measures may be required. An assessment of general health is required so that a profile of a work group can be established, which is then used to inform any future health assessment programme. The statutory health assessment process can easily incorporate review of any other health problem such as untreated high blood pressure or any concern that the employee may have which is then referred to the GP. Without having established first a baseline for exposure to noise, vibration, skin or respiratory irritants and other key health risks, the employer will have no defence if the employee is then exposed to hazards which further damage their health. Construction employers often do not undertake any form of pre-employment health assessment and so do not know the health status of those joining their company. Many work related health problems are in fact legacy problems, which construction employers "inherit" from past employers. It is therefore important that statutory health assessment results are evidence based and defendable, thus the role of the occupational health specialist, as this requires significant clinical skill and expertise.

As outlined above, the HSE proposes that employers assess risks and current controls to decide if they require the input of a specialist occupational health provider. The HSE's proposal that construction companies begin the statutory health assessment programme with simple health questionnaires about hazards and health effects, given out by the manager, is unlikely to reveal very much in terms of information on how to progress the programme. In other words, it is the experience of occupational health practitioners working with the construction industry that employees, in the main, do not reveal health information directly to their employers. In part, this is because they do not understand the health effects that relate to workplace hazards, and may simply not be aware of their impact. More importantly, they often do not trust the process which is not always supported by a robust HR function, and simply fear that disclosure may lead to loss of employment. When asked by an occupational health practitioner: 'Do you think you have any problem hearing?', most employees with a hearing deficit frequently say no. The objective test then reveals that this is not the case. The advice to the employer in this instance might simply be to improve on hearing protection in order to protect the employee's hearing from further damage. But this cannot be established in the absence of an objective test. Just as the detection of symptoms which relate to exposure to vibration require rigorous clinical assessment, which would not be revealed on a simple questionnaire, or in the absence of a full occupational history and knowledge of past and current exposure to vibration. It is then possible to train employees or managers within an organisation to continue part of the programme with self-administered questionnaires as outlined by the HSE. Statutory health assessments are only any good if part of an overall risk management process.

20.12 Case study

Overview

This case study illustrates the health risk management process from meeting with the client to discuss an appropriate programme of statutory health assessments to delivering the results, health advice and recommendations. The company—an international construction company—had taken proactive measures to remove the health risks to their employees through improved business process assessments. The company has multiple sites around the UK and employs around 1000 construction operatives, in a variety of jobs and projects. Core business activities involve providing highway services for local authorities, and developing new materials and techniques for highway surfacing that are safe and sustainable. Overall the company has developed a strong safety culture by identifying the roles and responsibilities for safety, the training for all employees, and robust incident investigation. This in turn was supported by HR policies so that a health risk management programme could easily be incorporated and supported by the existing company policy.

Information gathering to design the statutory health assessment programme used risk assessments but also direct observation of some work processes. The occupational health practitioners worked closely with safety manager, HR team, and the CDM co-ordinator to identify those operatives for inclusion in the programme, to include subcontractors and their employees. The main health risks and hazards identified were exposure to vibration, noise, dusts, fumes, particulates and potential for skin irritants.

Communication

The aims of the health risk management programme were to collect baseline health information at one company site, as a pilot project before developing the programme across all operations. The HR manager communicated the aims to the board, and management briefings were held on site to outline the programme but also to inform managers of the likely outcomes of the programme and the actions required on control measures. Information leaflets were given to all employees in advance of the statutory health assessments.

The on-site statutory health assessments were delivered and 157 civil engineering operatives were seen. The findings of the statutory health assessments indicated health problems among those seen which were occupational in origin, the most significant relating to exposure to noise and vibration. Many employees seen had a short employment history with the company but a long occupational history working within the construction industry, and reported never having had a statutory health assessment in the past. When compared to the current knowledge base and evidence of ill health among those working in civil engineering across the industry these results are very similar.

Control Measures

Health data was matched with existing control measures and evaluated in view of the findings of the statutory health assessments, both collectively and for individual employees. The safety team worked with operational managers to look at the hierarchy of controls and how these could be improved. As outlined, the company had already made significant investment in plant and equipment to reduce risks for the most hazardous work processes in civil engineering. This had led to far greater output, in that the time taken to complete work using new plant and equipment was reduced by 75%. Employees reported significant reduction in back and upper limb pain and reduction in health effects related to working with vibration tools.

Training

Through management briefings and training operational managers were provided with information about the objectives of the health risk management programme and these covered the legislative framework and possible benefits to managers and employees. All managers were provided with appropriate information regarding those employees identified as having abnormal health. The training provided an explanation of the manager's role in workplace health management and how to get the best from the occupational health advice available.

Health records and fitness-to-work statements should be stored on site, as required, and not centrally with HR. Training covered storage, the use and retrieval of health information, and the provisions of the Data Protection Act 1998.

Occupational health champions

Following the health assessment programme, two employees from among the civil engineering operatives were selected to take on the role of occupational health champion. Both had a good understanding of work processes and were trained by the OH practitioners to understand the heath effects relating to specific hazards. All employees were made aware that they could report health problems to the Champions, upon which they would be referred to and assessed by the OH practitioners on their next visit to site. The Champions are also involved in the recording of exposure levels for vibration, risk assessment and provision of PPE.

The benefits for the company and their employees has been:

- A review of the most hazardous work processes and risk reduction, including engineering controls, investment in plant and equipment to eliminate risk for all employees exposed to significant vibration.
- A review of risk assessments and reduction of exposure where possible for all employees exposed to vibration and the provision of additional PPE.
- Implementation of a hearing conservation programme, including noise monitoring and the review of the use and provision of improved PPE.
- A process for ongoing health assessments and monitoring of the efficacy of control measures for all employees exposed to health hazards.
- An agreed occupational health strategy integrating occupational health with HR, safety and management processes, including ongoing policy development incorporating the CBH standards and commitment to redeployment.
- A 'case management' process has been agreed ensuring appropriate responses to employees with an abnormal health result.

Conclusion

As previously outlined there is currently limited published evidence that supports statutory health assessments as being effective if undertaken as part of an overall process of risk management measures. In the above case study, health risks to employees were evaluated in view of the objective tests and health information obtained as a result of an integrated health risk management program. This could not have been obtained without the involvement of occupational health specialists, working as part of a multi disciplinary team.

20.13 Final observations

Significant progress has been made in attempts to raise awareness of occupational ill health within the construction industry, and the industry-led scheme CBH has set high standards for fitness to work and for occupational health practitioners. Construction employers, however, often remain unclear about how to go about managing health risks. It is therefore possible that further research is required which would evaluate the different approaches and strategies for health risk management currently advised by the regulatory and advisory bodies. Robust evidence as to which practices are the most effective will thus lead to the improvements in employee health to which the industry aspires.

Further reading

Standards

Construction (Design and Management) Regulations 2007 (CDM), HMSO.
The Management of Health and Safety at Work Regulations 1999, HMSO.
Control of Vibration at Work Regulations 2005, HMSO.
Control of Noise at Work Regulations 2005, HMSO.
Control of Substances Hazardous to Health Regulations (as amended 2004), HMSO.
The Manual Handling Operations Regulations 1992 (as amended 2002), HMSO.

Other related texts

Constructing Better Health 2008—Occupational Health Standards for the UK Construction Industry, Part 1 Fitness For Work Standards: CBH, Crawley, 2008.
Clinical Testing and Management of Individuals exposed to hand transmitted vibration—An Evidence Review: Faculty Of Occupational Medicine, London, 2004.

APPENDIX A

Management of Demolition Projects Guidance for Clients & Administrators

Institute of Demolition Engineers

Introduction

In many instances, redundant buildings and structures are given little or no consideration by their previous owners or occupiers, regarding follow on activities when they have decamped, locked up and walked away. This is of course not an unexpected situation although it can be a costly process, e.g. to remedy or rectify the results of years of production at a factory.

Waste products, machinery and materials have to be handled and disposed of, often as part of the demolition process. This puts the onus of responsibility to comply with the Duty of Care on those tasked with dealing with the remnants of others activities. Although waste management is an important factor it has its equal in terms of service disconnections. Whereas leaving insitu hazardous waste for others to remove presents its problems, the termination and making safe of utilities will determine the start dates and very often the finish dates of the majority of demolition projects.

This guidance document has therefore been produced with these and other considerations in mind and is focused on the key areas that demolition engineers regard as critical in planning for and executing most demolition activities.

Terry Quarmby
Vice President
Institute of Demolition Engineers

Determining Responsibilities, Duties and roles

Initial stages

1. Planning of project—client

- Carry out feasibility study
- Implement surveys, i.e.:
    asbestos
    structural/dilapidations etc.

    ground conditions
    services
    local environment—aspect/impact
    access and egress constraints
- Provision of adequate resources
- Notification/planning consents, for example 'F10' and 'section 80'

Note: These stages will invariably be repeated by the appointed Principal Contractor.

2.   Appointment of key stakeholders

CDM Coordinator

- Knowledge, training and industry experience
- Production of adequate and suitable information specific to the project etc.
- Identification of significant hazards
- Liaison with consulting/informative contacts
- Early appointment
- Initial notifications
- Production of Health and Safety file

Principal Contractor

- Provision of adequate resources
- Knowledge, training and industry experience
- Production of Health and Safety plan
- COSHH assessments
- Liaison with authorities
- Compilation of relevant information and documents
- Control of sub-contractors
- Notifications/applications
- Creation of project teams

Designer

- Product knowledge
- Knowledge training and industry experience

Project stages

1.   Management of the project

- Implementation of plan
- Site set up
- Progress meetings
- Project safety meetings

- Audits and inspections
- Corrective actions
- Waste management
- Recycling, reclamation
- Liaison with authorities
- Meet client expectations
- Meet contract conditions
- Hand over

2. Management of subcontractors

- Service diversions etc.
- Health, Safety and Environmental implications
- Personnel management
- Methodology
- Audit and inspection
- Scaffolding and plant

3. Liaison with others

- Local Authority
- Health & Safety Executive
- Environment agency
- Police
- The public
- Emergency services
- British waterways
- Rail network
- Service providers
- Waste management
- Consultative bodies

Project Planning & Implementation Stages

Throughout all stages of the demolition process there are actions which must be implemented and clearly defined.

The timing of such actions are often dictated by the pace of the works or the safety, environment and contract conditions that exist. Formulating a checklist of possible needs or requirements will help to plan such actions and eliminate or reduce reactive "fire fighting" processes that make the management of projects stressful and difficult.

Listed below are examples of such actions, and although they are not to be regarded as exhaustive they show interdependence on each other and good practice.

Access & egress-security-emergency procedures

The following points are to be considered:

- Clearly defined site boundaries—preferably by utilization of existing walls and fencing and or by secondary Use of solid or anti-climb 'heras' fencing etc.
- Out of normal working hours security.
- Lockable gateways—of sufficient width to allow ease of access and egress by large vehicles or plant.
- Alternative traffic and pedestrian routes to retained areas of the site—other than access from shared routes, wherever practical and possible.
- Re-routing of emergency access or escape routes—where traditional routes have crossed the demolition areas. (In extreme cases this can mean the erection of covered walkways, fixing of temporary stairwells, arrangements with adjacent property holders etc.)

Asbestos-contaminated materials-waste

The following points are to be considered:

- A full type 3 asbestos survey—to be undertaken in all demolition areas (this is a statutory requirement where demolition and major refurbishment projects are to be carried out). (Refer to MDHS100.)
  Note: No waiver against the 14-day notice will be granted where further asbestos is discovered during the demolition phase in the absence of a Type 3 survey.
- Asbestos registers and details of previous asbestos removal etc.—to be made available prior to present occupier decamping.
- Contact detail of works engineers—familiar with all aspects and locations of asbestos, contaminated materials, oils, fuels, waste areas etc.
- Registers, drawings and or details of all contaminated areas—vessels, pipes and containers etc.
- Details of types and hazards—arising from such contaminated areas, products or materials.
- Details of any previous or intended decommissioning—purging, cleaning, testing and certification of safety etc or part thereof.
- Details of heavy, light or partial contamination—of site areas below and above ground.
- Details of statutory or commercial implications concerning the removal, handling and disposal of any such contaminated products and or materials either stored for use or as a by product of the works.

Services-Utilities-Waterways

The following points are to be considered:

- Details to include locations of all services, data cables etc.—entering, leaving or crossing the site. Services to mean: electricity, gas, water, sewerage, telecommunication. (It will be of assistance to provide details of all works engineers etc. having concern or responsibility for services, maintenance and repair to the above elements and to implement internal or external re-directions or cut offs to the site particularly where there appears to be a sharing of supplies to adjacent areas, cost and time being of paramount concern.)

- Details of all services sub-stations within the site boundary—and whether they are to be retained and protected or decommissioned and demolished.
- Details of services directly feeding retained areas—from within or crossing the demolition site boundary (where such conditions exist it is strongly recommended that re-direction of such supplies is sought prior to demolition).
- Details of or intention of notice to gas, electricity, water—vendor/shipper/meter owner etc. of termination/disconnection of services at the boundary of the site. **Note:** electricity and gas can take 8 weeks from initial notification to cut off.
- Details of any wayleaves—or other arrangements to maintain and retain service supplies to adjacent or opposite areas.
- Locations of utility service runs below and above ground—depth of services and whether they are to be removed as part of the demolition site clearance works, re-directed and or re-laid elsewhere. Utility services to mean; mains water, foul water, storm water, electricity, gas, data services, heating systems, fuel lines, internal communications etc.
- Details of requirements or arrangements with the Environment Agency, British Waterways or other interested bodies—for works near to or abutting rivers or canals, i.e. discharge points and water take off points concerning the removal of any pipes, pumps or other structural impediments. Also public access.

Statutory and public notices—local environment

The following points are to be considered:

- Details and timing of statutory notices—e.g. asbestos removal ASB5 to be submitted a minimum of 14 days notice to the Health & Safety Executive prior to the commencement of removal works. Section 80 notice of demolition of a structure, to local authority building control department, (many councils invoke a 6 week rule). Registering of site with the environment agency for removal of hazardous materials. (It should be noted that many hazardous waste landfill sites require at least 72 hours notice prior to delivery of waste from site.) Utility services notice (beware of possible 6–8 weeks delay).
- Party wall agreements—if any, with retained sections of the structure adjacent to occupied areas. (Note: the finalization and agreement with party wall surveyors etc. can be a prolonged and frustrating exercise. It is strongly recommended that where such agreements are required that matters are initiated at the earliest opportunity.)
- Details of any current or expected objections—to the proposed development where delay to the commencement of the demolition works may affect the programme.
- The requirement for the demolition contractor to address public concerns—over the proposed demolition works. (Note: many demolition contractors have the ability to allay public fears regarding demolition activities, by such as a PowerPoint presentation and evidence of previous works to show that issues such as noise, dust and damage etc. can be addressed and reduced to the lowest possible expec-

tations through careful and deliberate planning prior to execution. In the past this has proved to be an invaluable tool for the developer.

• Many brownfield sites—particularly those that may have lain idle for some time, have been adopted for habitat by wildlife. Where such wildlife are protected, i.e., in the case of bats, newts and birds etc., projected start dates may need to be revised and all parties advised of any constraints.

Guidance

The subject matter contained within this guidance is varied with many of the topics seemingly intertwined or running concurrently with others. It is important to consider the relevant statutory regulations and guidance available for these subjects when planning and managing the demolition process. The following lists have been compiled to assist the researcher but should not be taken as exhaustive.

Health, Safety & Welfare

Health and Safety at Work etc. Act 1974
Construction (Design & Management) Regulations 2007
Management of Health & Safety at Work Regulations 1999
Provision and Use of Work Equipment Regulations 2002
Lifting Operations, Lifting Equipment Regulations 1998
Control of Substances Hazardous to Health Regulations 2002
Control of Asbestos Regulations 2006

The Environment

Environmental Protection Act 1990
Part 1 Pollution Prevention
Part 2 Waste Management
Part 3 Statutory Nuisance
Waste Management Regulations 1996
Waste Management (Licensing) Regulations 1994
Waste Management (England & Wales) Regulations 2006
Hazardous Waste Regulations 2005

Miscellaneous

BS 6187: 2000 Code of Practice for Demolition
MDHS100 Surveying, sampling & assessment of asbestos
HSG198/1 Controlled asbestos stripping techniques
HSG189/2 Working with asbestos cement
L143 Work with materials containing asbestos
HSG151 Protecting the public
Section 80, Building Act 1984—Notice of Intended Demolition

Produced for publication by the Institute of Demolition Engineers May 2007
Guidance Document CDM001/2007 Website: www.ide.org.uk • E-mail: info@ide.
org.uk • Telephone: +44 (0) 1634 294255

Special Note: This is a guidance document only: you must take your own independent advice in respect of any contract into which you propose to enter.

APPENDIX B

British Standards

List of organisations holding reference copies

- Basingstoke Reference Library
- Chelmsford Central Library
- Farnborough Reference Library
- Middlesex University
- Poole Reference Library
- Portsmouth Central Library
- Highbury (Portsmouth) College of Technology Library
- University of Portsmouth
- Southampton Central library
- Hampshire County Library
- Woking Library
- Leicester City Council Library
- Staffordshire County Library
- Blackburn Central Library
- Darlington Borough Council
- Preston District Central Library
- South Tyneside Metropolitan Library
- University of Swansea
- Armagh: Southern Education and Library Board
- Belfast Education and Library Board
- ICE and IStructE libraries

Note: BSI members are provided with a year book in hard copy and CD format. This lists all BS, European and ISO standards. Check with BSI website for latest information.

Acronyms, abbreviations and websites

ABE	Association of Building Engineers	*www.abe.org.uk*
ACI	American Concrete Institute	*www.concrete.org*
ASCE	American Society of Civil Engineers	*www.asce.org*
asr	alkali-silica reaction	
ASTM	American Society for Testing Materials	*www.astm.org*
BBA	British Board of Agrément	*www.bbacerts.co.uk*
BCSA	British Constructional Steelwork Association	*www.bcsa.org.uk*
BDA	Brick Development Association	*www.brick.org.uk*
BITA	British Industrial Truck Association	*www.bita.org.uk*
BRE	Building Research Establishment	*www.bre.co.uk*

BSI	British Standards Institution	*www.bsiglobal.com*
BSI	Building Standard Institute	*www.buildingstandard institute.org*
BSRIA	Building Services Research & Information Association	*www.bsria.co.uk*
CAD	computer aided design	
CARES	Certification Authority for Reinforcing Steels	*www.ukcares.co.uk*
CDA	Copper Development Association	*www.cda.org.uk* see also: *www.copperinfo.co.uk*
CDM	Construction (Design and Management) Regulations 1994	
CERAM	CERAM Building Technology	*www.ceram.com*
CI	cast iron	
CIBSE	Chartered Institute of Building Services Engineers	*www.cibse.org*
CIOB	Chartered Institute of Building	*www.ciob.org.uk*
CIRIA	Construction Industry Research & Information Association	*www.ciria.org*
CITB	Construction Industry Training Board	*www.citb.org.uk*
	Construction Skills	*www.constructionskills.net*
CONSTRUCT	Concrete Structures Group	*www.aecportico.co.uk*
CORUS	CORUS Construction Centre	*www.corusconstruction.com*
CPA	Construction Plant Association	*www.cpa.uk.net*
CS	Concrete Society	*www.concrete.org.uk*
CWCT	Centre for Windows & Cladding Technology	*www.cwct.co.uk*
	Department for Transport	*www.dft.gov.uk*
	Eurocodes	*www.eurocodes.co.uk*
ECI	European Construction Institute	*www.eci-online.org*
fib	Federation International du Breton	*www.fib-international.org*
FPS	Federation of Piling Specialists	*www.fps.org.uk*
ggbs	ground granulated blast-furnace slag	
GMT	Greenwich Mean Time	
GPS	global positioning system	
GRP	glass reinforced plastics	
HA	Highways Agency	*www.highways.gov.uk*
HAUC	Highways Authority & Utilities Committee	*www.hauc-uk.org.uk*
HMSO	Her Majesty's Stationery Office	*www.hmso.gov.uk*
HSE	Health and Safety Executive	*www.hse.gov.uk*
HVAC	Heating Ventilating and Air Conditioning	
IoR	Institute of Roofing	*www.instituteofroofing.org*
ICE	Institution of Civil Engineers	*www.ice.org.uk*

ICES	Institution of Civil Engineering Surveyors	*www.ices.org.uk*
IEE	Institution of Electrical and Electronics Engineers	*www.ieee.org.uk*
ISO	International Standards Organisation	*www.iso.ch*
IStructE	Institution of Structural Engineers	*www.istructe.org*
IT	information technology	
JCT	Joint Contracts Tribunal	*www.jctltd.co.uk*
LEEA	Lifting Equipment Engineers Association	*www.leea.co.uk*
LOLER	Lifting Operation & Lifting Equipment Regulations	
LUL	London Underground Ltd	*www.tfl.gov.uk*
MAA	Mastic Asphalt Council	*www.masticasphaltcouncil. co.uk*
MCRMA	Metal Cladding & Roofing Manufacturers Association	*www.mcrma.co.uk*
Met Office	Meteorological Office	*www.met-office.gov.uk*
MPA	Mineral Products Association	*www.mineralproducts.org*
NASC	National Access & Scaffolding Confederation	*www.nasc.org.uk*
NBS	National Building Specification	
NCRA	National Roofing Contractors Association	*www.ncra.net*
NDT	non-destructive testing	
NEC	New Engineering Contract	
NFRC	National Federation of Roofing Contractors	*www.nfrc.co.uk*
NHBC	National House Building Council	*www.nhbc.co.uk*
NJUG	National Joint Utilities Group	*www.njug.org.uk*
NSCS	National Structural Concrete Specification	
NSSS	National Structural Steelwork Specification for Building Construction	
PASMA	Prefabricated Aluminium Scaffolding Manufacturers Association	
pfa	pulverised fuel ash	
PVC	polychloroethene or polyvinylchloride	
PWD	permanent works designer	
QA	quality assurance	
QC	quality control	
QPA	Quarry Products Association	See **MPA**
RIBA	Royal Institution of British Architects	*www.riba.net* see also *www.architecture.com*
RICS	Royal Institute of Chartered Surveyors	*www.rics.org*

RoSPA	Royal Society for the Prevention of Accidents	www.rospa.com
RQSC	Register of Qualified Steelwork Contractors	www.steelconstruction.org
SCCS	Steel Construction Certification Scheme	www.steelconstruction.org
SCI	Steel Construction Institute	www.steel-sci.org
SCQAS	Steel Construction Quality Assurance Scheme	
SPRA	Single Ply Roofing Association	www.spra.co.uk
TRADA	Timber Research & Development Association	www.trada.co.uk
TRL	Transport Research Laboratory	www.trl.co.uk
TSE	The Structural Engineer	
TWC	temporary works coordinator	
TWD	temporary works designer	
UKAS	United Kingdom Accreditation Service	www.ukas.com
uPVC	Unplasticised polyvinylchloride	
WRc	Water Research Council	www.wrcplc.co.uk
	Water UK	www.water.org.uk

See also sites such as *www.ascinfo.co.uk*, and links from institutional websites.

Imperial/Metric Conversions

Conversion Chart (to 3 Significant Figures)

Measure	Imperial to SI units	SI to imperial units
Length	1 yd = 0.914 m 1 ft = 0.305 m 1 in = 25.4 mm	1 m = 1.09 yd 3.28 ft 1 cm = 0.394 in 1 mm = 0.0394 in
Area	1 yd^2 = 0.836 m^2 1 ft^2 = 0.09290 m^2 1 in^2 = 645 mm^2	1 m^2 = 1.20 yd^2 = 10.8 ft^2 1 cm^2 = 0.155 in^2 1 mm^2 = 0.00155 in^2
Volume	1 yd^3 = 0.765 m^3 1 ft^3 = 0.0283 m^3 1 in^3 = 16400 mm^3 1 gallon = 4.55 litres	1 m^3 = 1.31 yd^3 = 35.3 ft^3 1 cm^3 = 0.0610 in^3 1 litre = 0.220 gallons
Mass	1 ton = 1020 kg = 1.020 tonne 1 cwt = 50.8 kg 1 lb = 0.454 kg	1 tonne = 0.984 ton 1 kg = 2.20 lb
Density	1 lb/ft^3 = 16.0 kg/m^3	1 kg/m^3 = 0.0624 lb/ft^3
Force	1 tonf = 9.96 kN 1 lbf = 4.45 N	1 N = 0.225 lbf 1 kN = 225 lbf = 0.100 ton

Pressure	1 tonf/ft^2 = 107 kN/m^2	1 kN/m^2 = 0.00932 ton/ft^2
	1 tonf/in^2 = 15.4 N/mm^2	1 kN/m^2 = 20.9 lbf/ft^2
	1 lbf/in^2 = 0.00689 N/mm^2	1 N/mm^2 (1MPa) = 145 lbf/in

For more detailed information on conversion from Imperial to SI units and *vice versa* see BS 350: Part 1, 1974. Chart taken from 1 STRUCTE Report 'Appraisal of Existing Structures' 2nd Edition 1996.

Table of atomic symbols

Element	Symbol	Element	Symbol
Actinium	Ac	Hydrogen	H
Aluminum	Al	Indium	In
Americium	Am	Iodine	I
Antimony	Sb	Iridium	Ir
Argon	Ar	Iron	Fe
Arsenic	As	Krypton	Kr
Astatine	At	Lanthanum	La
Barium	Ba	Lead	Pb
Berkelium	Bk	Lithium	Li
Beryllium	Be	Lutetium	Lu
Bismuth	Bi	Magnesium	Mg
Boron	B	Manganese	Mn
Bromine	Br	Mendelevium	Md
Cadmium	Cd	Mercury	Hg
Calcium	Ca	Molybdenum	Mo
Californium	Cf	Neodymium	Nd
Carbon	C	Neon	Ne
Cerium	Ce	Neptunium	Np
Cesium	Cs	Nickel	Ni
Chlorine	Cl	Niobium	Nb
Chromium	Cr	Nitrogen	N
Cobalt	Co	Nobelium	No
Copper	Cu	Osmium	Os
Curium	Cm	Oxygen	O
Dysprosium	Dy	Palladium	Pd
Einsteinium	Es	Phosphorus	P
Erbium	Er	Platinum	Pt
Europium	Eu	Plutonium	Pu
Fermium	Fm	Polonium	Po
Fluorine	F	Potassium	K
Francium	Fr	Praseodymium	Pr
Gadolinium	Gd	Promethium	Pm
Gallium	Ga	Protactinium	Pa
Germanium	Ge	Radium	Ra
Gold	Au	Radon	Rn
Hafnium	Hf	Rhenium	Re
Helium	He	Rhodium	Rh
Holmium	Ho	Rubidium	Rb
Ruthenium	Ru	Thallium	Tl
Samarium	Sm	Thorium	Th

Scandium	Sc	Thulium	Tm
Selenium	Se	Tin	Sn
Silicon	Si	Titanium	Ti
Silver	Ag	Tungsten	W
Sodium	Na	Uranium	U
Strontium	Sr	Vanadium	V
Sulphur	S	Xenon	Xe
Tantalum	Ta	Ytterbium	Yb
Technetium	Te	Yttrium	Y
Tellurium	Te	Zinc	Zn
Terbium	Tb	Zirconium	Zr

The Greek alphabet

Capital	Lower-case	Name	English transliteration
A	α	alpha	a
B	β	beta	b
Γ	γ	gamma	g
Δ	δ	delta	d
E	ε	epsilon	e
Z	ζ	zeta	z
H	η	eta	$\bar{e}$
Θ	θ	theta	th
I	ι	iota	i
K	κ	kappa	k
Λ	λ	lambda	l
M	μ	mu	m
N	ν	nu	n
Ξ	ξ	xi	x
O	o	omicron	o
Π	π	pi	p
P	ρ	rho	r
Σ	σ (ς at end of word)	sigma	s
T	τ	tau	t
Y	υ	upsilon	u
Φ	ϕ	phi	ph
X	χ	chi	kh
Ψ	ψ	psi	ps
Ω	ω	omega	$\bar{o}$

SUBJECT INDEX

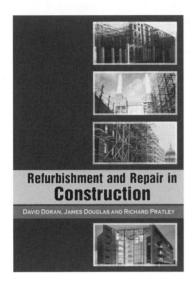

Refurbishment and Repair in Construction

DAVID DORAN, JAMES DOUGLAS AND RICHARD PRATLEY

Refurbishment and Repair in Construction
provides a companion volume to
Site Engineers Manual

Refurbishment and repair accounts for approximately 50% of annual construction turnover. The nature of refurbishment and repair is markedly different from new-build work since it is necessary to work within the restraints of a pre-determined situation. It may have been built to standards hardly recognisable when compared to those of today. It is also apparent that existing buildings may not conform to 21st century standards of structural analysis or stability – and yet have stood without distress for many years.

Refurbishment and Repair in Construction is a practical handbook and aide-mémoire for practitioners and students alike that fills a gap in construction literature. Failure to investigate the history of existing developments may add considerably to the cost of construction and, in the extreme, to structural failure and collapse involving injury or loss of life.

First and foremost it is essential for those involved in this type of work to gain an intimate knowledge of the structure under consideration, which can require a thorough forensic-style investigation. Guidance is provided to deal with how to assess the residual life of a refurbished or repaired building. The book proceeds logically through the necessary considerations and offers advice on risks; testing and monitoring in the discovery process; types of contract; materials; learning from the past and legal restraints. Best practice is illustrated by case studies and extensive references have been provided to assist those with the need for further research.

It is gradually being appreciated that refurbishment of existing constructions … is a more sustainable and preferable approach than demolition and … [It] is essential that the existing construction is fully understood through its material and structural properties … professionals can now gain the understanding they need from this book and its extensive reference sources.

—From the Foreword by Lawrance Hurst, consultant to Hurst Peirce + Malcolm

The authors: **David Doran,** Consultant, Civil/Structural Engineer, formerly Chief Structural Engineer, Wimpey plc, UK; with **James Douglas**, School of the Built Environment, Heriot-Watt University, Edinburgh; and **Richard Pratley**, Architect, London.

ISBN 978-1904445-55-5   234 x 156 mm   336pp   illustrated softback   £65

For further information or to order please contact:
Whittles Publishing • Dunbeath Mains Cottages • Dunbeath • Caithness • KW6 6EY
Tel: +44(0)1593 731 333; Fax: +44(0)1593 731 400
email: info@whittlespublishing.com; www.whittlespublishing.com